METAPHYSICS, EPISTEMOLOGY, AND TECHNOLOGY

RESEARCH IN PHILOSOPHY AND TECHNOLOGY

Series Editor: Carl Mitcham

Recent volumes:

Volume 9: Ethics and Technology

Volume 10: Technology and Religion

Volume 11: Technology and Politics

Volume 12: Technology and the Environment

Volume 13: Technology and Feminism

Volume 14: Technology and Everyday Life

Volume 15: Social and Philosophical Constructions of Technology

Volume 16: Technology and Social Action

Volume 17: Technology, Ethics, and Culture

Volume 18: Philosophies of the Environment and Technology

Volume 19: Metaphysics, Epistemology, and Technology

RESEARCH IN PHILOSOPHY AND TECHNOLOGY, VOLUME 19

METAPHYSICS, EPISTEMOLOGY, AND TECHNOLOGY

EDITED BY

CARL MITCHAM

Colorado School of Mines, Colorado, USA

Emerald

United Kingdom – North America – Japan
India – Malaysia – China

Emerald Group Publishing Limited
Howard House, Wagon Lane, Bingley BD16 1WA, UK

Printed and bound by CPI Group (UK) Ltd, Croydon, CR0 4YY

British Library Cataloguing in Publication Data
A catalogue record for this book is available from the British Library

ISBN: 978-0-7623-0681-7
ISSN: 0161-7249 (Series)

CONTENTS

EDITORS/EDITORIAL BOARD

Research in Philosophy and Technology is a refereed publication. Editorial correspondence should be addressed to Carl Mitcham, General Editor, Stratton Hall 301, Liberal Arts and International Studies, Colorado School of Mines, Golden, CO, 80401.

PREFACE

The last several issues of *Research in Philosophy and Technology* have emphasized the ethics and politics of technology, although with some reference to the metaphysical and epistemological dimensions of these issues. The present volume, by contrast, affirms the centrality of metaphysics and epistemology – although not without some reference to ethics and politics.

The theme section, Part I, contains fourteen articles all advancing important metaphysical and/or epistemological arguments. In the contemporary analytic and the phenomenological traditions, both of which are represented here, there is no bright line distinguishing metaphysics and epistemology. Thus it would be artificial to use the metaphysics/epistemology distinction to categorize this set of papers. On other grounds, as well, the richness of perspectives advanced is difficult to summarize. For introductory purposes, however, one may identify at least four overlapping subthemes to indicate in a preliminary manner the range of approaches to the metaphysics and epistemology of technology represented here: (a) proposals for a general reform of the philosophy of technology, (b) studies emphasizing the phenomenological tradition, (c) papers taking off from the metaphysical-epistemological work of some major philosopher, and (d) analyses of some specific metaphysical-epistemological issue.

With regard to general theories for advancement of the philosophy of technology, two contributions stand out. Andrew Light and David Roberts, in 'Toward New Foundations in Philosophy of Technology', argue that the biggest problem in philosophy of technology is a rush to social-normative assessments. Although Light and Roberts find something to approve in Carl Mitcham's recent attempt at what they term descriptivism, they criticize his tendency either to pile up descriptions to no philosophical effect or ultimately to subordinate descriptions to social-normative ends. By contrast, they argue that Ludwig Wittgenstein's "method of description is a more effective way of exposing and clarifying '[basic] problems' in the philosophy of technology." In Wittgenstein's motto, "Don't think, but look!" (*Philosophical Investigations*, sec. 66), they find inspiration for a reform and advance, in which the goal ought to be a "shared descriptive foundation of how we actually interact with technology."

Yannis Niadas' 'On the Cognitive Demarcation between Technology and Science' provides a serious epistemological reassessment of technological knowledge that points in a similar direction. With significantly more nuance than previous analyses, Niadas' conceives engineering as a knowing-how that incorporates and transforms scientific knowledge, which it deploys for the efficient achievement of socially set applications. Niadas is especially careful to try to separate his analysis from subservience to any social or normative assessment of technology. As he writes, his argument "attempts to subscribe to a philosophy of technology per se, in that the argumentation and the line of thinking followed does not depend on any particular metascientific approach or philosophical elaboration in order to assess technology and engineering." The result is an epistemology that may be interpreted in, say, either realist or pragmatic ways.

To some extent Peter-Paul Verbeek's 'The Thing about Technology' also argues for a general reform in the philosophy of technology, although his focus is more on the phenomenological tradition, and especially the influence of Martin Heidegger. Heidegger, according to Verbeek, despite an occasional verbal acknowledgement of the importance of things, fails truly to think technological artifacts. Reinterpreting Heidegger's teacher, Edmund Husserl, whose motto was to return to 'Sachen selbst', things themselves (*Logische Untersuchungen*, p. 6), Verbeek argues that "a philosophy of technological culture must take the material culture of technology into account." There needs to be a shift from concern about the thing in consciousness to real things.

Verbeek's contribution not only makes an argument for general reform in one particular approach to the philosophy of technology, it provides a bridge to the second major type of paper in the theme section: contributions seeking to advance the phenomenological tradition, especially in its metaphysical and epistemological aspects. Here Agustin Araya's 'Changed Encounters with Things and Ontological Transformations', Rafael Capurro's 'Hermeneutics and the Phenomenon of Information', Bill Hook's 'The Fate of Skills in the Information Age', and Craig Murray's 'Toward a Phenomenology of the Body in Virtual Reality' all make important contributions. It may be noted, as well, that these four papers highlight the significance of computers and information technologies as calling for special philosophical attention.

Araya's case study of the theory and practice of ubiquitous computing, as advanced during the 1980s and 1990s by the late Mark Weiser and associates at Xerox PARC, adapts Heidegger's ontological analysis of human experience to suggest that human-computer interactions are leading to the emergence of a new way of being-in-the-world. Basing himself on a more descriptive phenomenology of technological transformations in photography, Hook draws distinctions between service skill, device skill, and focal skill that extend Albert Borgmann's well known analysis of differences in engagement between thing and device. Further references to skill changes in music and in the law office complement Hook's argument, which parallel's Araya's, that something crucial has occurred in our worldly engagements when human activities become digitalized. Capurro's deft presentation of his thesis that the widespread presence of digital information calls forth the interpretative activity, provides something of a challenging counterpoint to both Araya and Hook.

Murray, building on a knowledgeable review of recent studies in the phenomenology of the body, questions some of the claims being made for virtual reality (VR). Body phenomenology, for instance, challenges precisely the kind of privileging of the visual that is often found in VR work, and phenomenological studies of persons with missing limbs, prostheses, paralysis, and plastic surgery have implications for VR as well. His argument is a cautionary one: "the extent to which such devices may be used to transform the phenomenal properties of the body, to (re)define experiential human morphology, remains an open question."

Two other papers that may be placed in this second category are Philip Brey's closely related papers on 'Technology and Embodiment in Ihde and Merleau-Ponty' and 'Technology as Extension of Human Faculties in Kapp, McLuhan, and Rothenberg'. Certainly Don Ihde and Maurice Merleau-Ponty are key figures in the phenomenological tradition. Brey's careful criticism of Ihde's use and misuse of Merleau-Ponty, and then his extension of Ihde's own arguments through a refinement of distinctions and broadening of concerns, provides a model of exegesis and interpretation. The same may be said of his subsequent analysis of relations between Ernst Kapp, Marshall McLuhan, and David Rothenberg. In both papers Brey also adroitly steers a course between subservience to either the phenomenological or the analytic philosophical traditions, effectively incorporating both to the metaphysical and epistemological benefit of each.

A third set of contributions further exemplifies this kind of reference to the existing work of major contributors to the philosophy of technology. Mary Bloodsworth's 'The Implications of Consistency', for instance, uses Heidegger's 'The Question Concerning Technology' to challenge standard interpretations of Plato's *Protagoras*. Bloodsworth thus uses Heidegger to advance studies in the interpretation of Greek philosophy, including metaphysics and epistemology.

Karina Stokes' article on 'Bakhtin's Concept of Individualist Ideolects as a Propitious Model for Communication among Human Beings and between Humans and Computers' takes the opposite tact. It brings the work of a philosopher who has not previously been thought to have much to say about technology, to bear upon technology. Stokes begins by pointing out an otherwise overlooked relation between human beings in the workplace and between humans and computers. Given the extent of human-to-human miscommunication in the workplace, and its generally recognized difficulty, why should the problem of human-computer communication be thought any less tractable? But with regard to human-to-human communication, the dialogic communication theory of Russian linguist Mikhail Bakhtin offers an alternative perspective that may, Stokes proposes, have implications for human-computer communication as well.

In this third group of papers it is also possible to place Ton van der Valk's comprehensive assessment of 'The Negative Teleological Metaphysics of Hans Jonas'. Van der Valk's provocative interpretation is that Jonas' thought exhibits both applied philosophical (meliorist) and cultural criticism (technological determinist) tendencies, and that only the applied philosophical aspects of his work remain viable. Three guidelines to be derived from Jonas — his rejection of progress as perfection in favor of progress as melioration, his apology for and acceptance of natural evils, and his basic egalitarianism — may be usefully applied to twenty-first century problems.

Finally, as a fourth classification in the theme section, two papers attempt to break new ground on particular issues: efficiency and time. José Manuel de Cózar's 'Toward a Philosophical Analysis of Efficiency' brings new analytic rigor to bear on what is a cliche of technological culture, the pursuit of efficiency. Theodore John Rivers' 'The Conception of Time and Its Relationship to Technology' calls our attention to time as a key phenomenon with intimate implications for technology. Both efficiency and time are otherwise neglected topics with metaphysical and epistemological implications for the philosophy of technology.

Part II, a set of five extended review essays, is especially important to *RPT* 19. In the first essay, Brian Rosmaita provides an extended critical analysis of Albert Borgmann's important new book, *Holding On to Reality* (1999). Borgmann's book is itself a model of epistemological and metaphysical reflection on technology, and Rosmaita relates it to the political claims of Dan Schiller's *Digital Capitalism* (1999) in a way that Borgmann himself would appreciate. Roderick Nicholls's essay review of *The Digital Phoenix* (1998), Terrell Ward Bynum and James H. Moor's edited collection of works, suggests that computer technology is transforming philosophy especially at theoretical – that is, metaphysical and epistemological – levels. In 'Philosophy of Mind and Technology?' Richard Liebendorfer tackles two recent books that use computer technology to reflect on the philosophy of mind. 'Living with Machines' is Sergio Sismondo's critical take on two books that advance our understanding of the techno-lifeworld, including its metaphysical and epistemological dimensions. Finally, Stuart Dalton's analysis of Mark Taylor's *Hiding* (1997) presents a metaphysics of the surface, one aspect of that pervasive reality characteristic of what may be described as the post-technological world.

Following these five review essays, Part III collects shorter reviews that continue the main theme of metaphysical and epistemological reflection. Six important new books are briefly assessed by Sismondo, Stephen Crowley, Raphael Sassower, John Collier, Drew Christie, and Albert Borgmann in ways that highlight precisely the concerns that emerge in the theme papers: metaphysics, epistemology and the proper approach to a general philosophy of technology.

The concluding Part IV includes nine short reviews that complement metaphysical and epistemological themes by continuing to explore the multiple dimensions of philosophy and technology. Taken together, this volume of RPT dedicated to the theme of 'Metaphysics, Epistemology, and Technology' thus makes a strong brief for the continuing relevance of fundamental issues.

ACKNOWLEDGMENTS

This volume of *RPT* witnesses the coming of a new book review editor, Mechthild Nagel. I want to thank Andrew Light for his good service in this capacity over the last few years, and to welcome Nagel as a new addition to the editorial staff. Let me also thank both Richard Devon, Director of the Science, Technology, and Society Program at Pennsylvania State University, and Arthur Sacks, director of the Division of Liberal Arts and International Studies at the Colorado School of Mines – of my former and my new institutional affiliations, respectively – for their support of the scholarly efforts that editing RPT entails. Last but by no means least, it must be acknowledged that without the hard work of my research assistants James Frazier and Abby Hoats this volume would not have appeared.

C.M.
Colorado School of Mines

PART I

THEME SECTION:
METAPHYSICS AND
EPISTEMOLOGY OF TECHNOLOGY

CHANGED ENCOUNTERS WITH THINGS AND ONTOLOGICAL TRANSFORMATIONS: THE CASE OF UBIQUITOUS COMPUTING

Agustin A. Araya

Our incessant involvement in everyday life with increasingly powerful technological devices, systems, and environments, amounting to what might be called a 'dwelling in techno-worlds', gives rise to the following question: Is it possible that in dealing with new kinds of useful things, requiring new kinds of engagements with them, we encounter the 'world' in novel ways which could ultimately lead to *ontological* transformations of us, that is, to *transformations of the way we are*? Thus stated, the question can elicit two very different kinds of responses. For some, even if they recognize the significant effects that new technological devices have on the activities in which they are introduced, the possibility that our involvement with them could lead to ontological transformations of us may seem too remote. For others, a growing awareness of the transformative powers of modern technology may lead them to answer this question immediately in the affirmative, with the consequence that a detailed inquiry into it may be considered unnecessary.

We take this situation to be an indication that the question itself is in need of clarification, possibly harboring in itself issues of great complexity and equal importance. In this attempt at exploring the meaning of this question and at taking some steps toward answering it, we start with a close examination of what is meant by 'ontological transformations'. Having established an initial

Metaphysics, Epistemology, and Technology, Volume, 19, pages 3–31.

ISBN: 0–7623–0681–5

context for our work, we continue with an analysis of what takes place in our encounters with things and technological devices. Next, we examine possible changes in our encounters with things by taking as an exemplary case a currently emerging technology, namely, Ubiquitous Computing, which constitutes a long-term strategy for the penetration of everyday life as a whole with computing. We conclude with general remarks on potential ontological transformations brought about by technology.

Readers who are not very familiar with ontological issues may want to examine first the section on Ubiquitous Computing given in the second half of this work, especially the beginning of the section, which expounds the basic tenets of that technology. An initial understanding of the principles of Ubiquitous Computing and their potentially far-reaching effects should help to motivate the discussion of ontological issues.

WHAT IS MEANT BY ONTOLOGICAL TRANSFORMATIONS

Ontological transformation is a notion that can be most properly situated in relation to what in the work of Martin Heidegger came to be known with the surprising title of the 'history of being'. In attempting to 'destructure' Western thinking as a whole, Heidegger traced significant transformations on fundamental 'notions' that characterize historical epochs, including such notions as truth, beings, essentia, existentia, beings as a whole, supreme being, and human beings.[1] In these various determinations of being, there is an interplay between what each of them unconceals and brings into the open, and what does not come forth with them and remains concealed. At the most encompassing level, as Heidegger suggested, in all of these transformations what most significantly remains concealed, even while making unconcealment 'possible', is 'being' itself; and Heidegger was most struck by the experience of such concealment. But, we need to add, this fundamental characteristic shared in common by these notions does not preclude the existence of significant differences among the transformed notions themselves. Ontological transformations, then, are transformations of what is unconcealed and what remains concealed, such that these various determinations of being open diverse dimensions of 'what is', giving place to various historical epochs.

Heidegger traced these ontological transformations by examining the central philosophical positions elaborated in the context of Western philosophy, from the early Greek thinkers to the late modern 'metaphysicians', and in doing so came to a certain proximity with Hegel's conception of history. One of the central differences among their respective notions of history of 'spirit' and history of being is that in Hegel, the movement of history exhibits an inner logic or rationality that accounts for the transformations of spirit leading to a moment of absolute self-knowing in 'absolute spirit'.[2] Such inner logic is totally absent from Heidegger's history of being. Rather, the various ontological transformations are seen as taking place in a 'discontinuous', emergent manner, as a result of 'sendings' or 'destinings' of being, which come to be revealed in terms of fundamental metaphysical positions.[3] Through philosophical thinking, although not exclusively through it, being comes to language, whereby new vistas of being are unconcealed, thus inaugurating and opening new historical epochs.[4]

But, we need now ask, are these transformations of fundamental notions the only kinds of ontological transformations that take place in the context of history? Significant transformations occur in all realms of beings, from the most immediate realm of useful things and environments with which we become involved in the course of daily activities, to the more encompassing realm of 'communities' and 'societies' to which we belong. Technological change which, partly because of synergistic effects among different technologies, is constantly accelerating, modifies existing things, systems, and environments and, in many cases, creates entirely new kinds of beings. To mention a single example, in the transformation that took place from slide rules to electronic computers we pass from one kind of being capable of computing a very small set of mathematical functions to another kind of being, a general purpose programmable computer, theoretically capable of computing any computable function.[5]

Are we justified in referring to these transformations as 'ontological'? Here the old Heideggerian distinction between 'existentials', i.e. fundamental ontological determinations of *Dasein*, and categories, i.e. ontological determinations of beings other than Dasein, can be brought to bear (e.g. Heidegger, 1996a, p. 42). While transformations taking place in the history of being would correspond to 'fundamental ontological' transformations, transformations in the ways of being as brought about by technological change would correspond to categorical transformations, that is, to 'ontological' transformations.

But the question remains: Are we entitled to call the transformations unfolded by technology, ontological, that is, categorical? This question leads to its converse: Can the traditional categories continue to characterize in essential ways the new kinds of beings that modern technology creates? To continue with examples drawn from computing technologies, let us consider the case of robotics, an area of technology emerging from the synergy between mechanics, electronics, and computing. Can robots be regarded as 'mere' things, or are they new kinds of beings exhibiting emerging forms of autonomy? Even if they are not living beings, the *possibility* exists that robots exhibit new forms of autonomy and agency which, for instance, in the context of an analysis of alterity relations of human users with technologies, have been characterized by Don Ihde as 'quasi-autonomy' (Ihde, 1990, p. 101). Or, to consider yet another emerging computing technology, the kinds of 'simulated worlds' developed in the context of Virtual Reality – which exhibit certain degrees of autonomy and allow non-trivial levels of 'experiential immersion' in them – can they be assimilated to what we usually call 'environments' or are they new kinds of beings?

As thought provoking as such transformations may be, they do not exhaust the kinds of transformations that take place because of the emergence of new technologies. All technological changes that transform things and create new kinds of beings *also transform our encounters with them*, that is, the specific ways in which we become involved with such things, including how we use, search for, and repair them. *It is these kinds of transformations that will be the primary focus of the present work*. Again, the important issue to consider is whether or not these transformations in our encounters with things have a fundamental ontological character.

We have now reached a point at which the question we raised at the beginning of this work can be restated in the following terms: Is it possible that fundamental transformations in our encounters with things, which accrue due to technological change, could contribute toward generating fundamental ontological transformations, that is, those kinds of transformations we referred to in the context of the history of being, in other words, transformations in 'what is'?

GENESIS OF ONTOLOGICAL TRANSFORMATIONS

To begin to approach this question, let us return to the notion of fundamental ontological transformation and pose the – admittedly – unusual question as to *how* these transformations could take place. Although never assigning to it the status of a leading question, in *Being and Time* Heidegger considered the 'ontological genesis' of several fundamental comportments and understandings. In particular, Heidegger focused on the "ontological genesis of the theoretical mode of comportment,"[6] by considering "the way in which circumspective taking care of things at hand changes over into the investigation of things objectively present found in the world" (Heidegger, 1996a, p. 327) as it happens, say, in the modern science of physics.[7] It should be noted that in pursuing this analysis, Heidegger was bringing to consideration genetic issues which, to a large extent, were regarded as antithetical to the phenomenological approach, with its emphasis on phenomena rather than on their 'causes' or genesis.

We now examine this analysis focusing primarily on the mode in which Heidegger asks the question, rather than on the specific results obtained regarding the particular, i.e. theoretical, comportment under consideration. After rejecting as inadequate the notion that theoretical discovery occurs by a 'disappearance' of praxis or circumspective concern, Heidegger's analysis focuses on a possibility that is inherent to praxis, namely, that of *saying* something about the thing in use while using it, that is, in the context of 'circumspect deliberation'. Although the notion of "talking to or conversing to oneself while doing something" is introduced here without further comment, that notion, in relation to the comportment of 'producing', is explicitly considered during a lecture course on Aristotle given a few years after the publication of *Being and Time*. Heidegger indicates there that "this is that *talking to oneself* which for the most part goes on silently or as a commentary which gets lost in the work and is often seen only from outside as a bunch of disconnected words."[8]

Although saying something like "the hammer is heavy" in the context of an activity still belongs to circumspective concern, it could become a saying of a different kind, namely, one in which the hammer no longer appears as 'ready to hand' but has been 'objectified', becoming an object with certain properties, including weight. This different kind of saying shows the hammer under a new 'light' because we are looking at it in a different way; our understanding of being has been transformed (Heidegger, 1996a, p. 330). Heidegger then goes on to suggest that this new way of saying does not yet constitute a scientific attitude and that, in the case of mathematical physics, what is crucial is an a priori projection of what natural beings are. In this projection, physical beings are 'removed' from the context in which they appear as ready to hand, so that now the 'place' which a thing occupies has become a 'point' in space.

How, then, does Heidegger understand the genesis of these transformations? As the example of the changing over from 'concerned circumspection' to the 'theoretical attitude' shows, Heidegger identifies 'genetic connections' between two kinds of comportments which, at first sight, may appear to be quite different from each other. By ontological genesis Heidegger means that a possibility rooted in the ontological constitution of Dasein – e.g. talking to oneself while using something – may become prominent and develop in such a way as to give emergence to a new kind of comportment. It is not that the transformation takes place because of changes in what is encountered, rather, the emphasis is put on the prominence acquired by 'internal' possibilities, residing in Dasein's ontological constitution.

It was in his subsequent work, while pursuing the 'destructuring' of Western ontology, that Heidegger came across what he termed "the history of being," and with it, reached a different understanding of what ontological transformations are. The latter are no longer understood primarily as possibilities rooted in Dasein's constitution, rather, they are ways in which 'being' comes to truth and is revealed. Consequently, the notion of ontology itself loses importance and, with the new focus on the truth of being, the notion of 'sending' or 'giving' of being becomes prominent. Admittedly a difficult notion, 'giving of being' attempts to do justice to the emergent character of how being 'gives' or obtains. By concentrating on such emergence Heidegger tries to enter into a closer proximity to 'being'.

But, it is fair to say, the notion of 'giving of being' is highly problematic – in the sense of being questionable – first of all to Heidegger himself. As stated in *The Principle of Reason* (Heidegger, 1996b, p. 62), the " 'giving of being' is not an answer but a question, among others the question of the essence of history, insofar as we think history as being and essence in terms of being." It is precisely this questionability of the 'giving of being' that makes it necessary to explore other paths, not necessarily to 'refute' such notion but to gain a further understanding about what ontological transformations are.

ORIGIN OF FUNDAMENTAL ONTOLOGICAL DISTINCTIONS

In fact, Heidegger himself had already explored other possibilities in addition to that mentioned earlier regarding the genesis of the theoretical comportment. On many occasions throughout his work, Heidegger refers to the importance of 'productive comportment' for Greek ontology.[9] During an extensive discussion of the *origin* of the distinction between the notions of 'essentia' and 'existentia' in *The Basic Problems of Phenomenology*, Heidegger suggests regarding "the productive comportment of Dasein toward beings as implicit horizon of understanding for essentia and existentia" (Heidegger, 1988, p. 106). The reference to production as providing a horizon for such fundamental ontological distinction suggests the possibility of a connection between Heidegger's question and the leading question of this present work, namely, the possibility that *using* things in changed ways could contribute to originate fundamental ontological transformations. For this reason, we will follow closely Heidegger's analysis to gain an insight into how Heidegger tries to answer this question, de-emphasizing references to the specific ontological notions he is concerned with, namely, essentia and existentia.

In the analysis, two indications are given at first which suggest of themselves that there is a relation between these ontological notions and 'productive comportment'. One of the usual conceptions of existentia is as 'actuality' and 'actualization', which contain a reference to an agent and its acting, from which that which is 'actual' follows (Heidegger, 1988, p. 101). But in such conception the nature of the agent's act remains unspecified, so that it could refer not only to production but to other kinds of acts or comportments, such as perception. A second indication is obtained by considering the way in which the scholastics understand actuality, which can be characterized in terms that actuality is 'that which is at hand as created by the creator' (Heidegger, 1988, p. 104). Here the notion of creator contains a more direct reference to production, although in this case the agent by whose act the actual is at hand is other than human.

As stated, these observations are to be regarded only as indications which, although making plausible a connection between such ontological notions and productive comportment, do not establish a more necessary link between them. To this effect, Heidegger goes on to consider the relation between 'eidos' and 'morphe', which appears to be reversed in Kantian epistemology and Greek ontology. Eidos, in an ontological sense, refers to "the look of a thing," where we see what a particular thing is. Morphe, refers to "the whole characteristic form impressed on a being from which we read off what it is" (Heidegger, 1988, p. 106). Succinctly stated, in Kant the 'form' of something determines its 'looks', while in Greek ontology the look determines its form. Heidegger suggests that this is such because in Kant the relation is, implicitly, seen from the comportment of perception, in which indeed the form of something determines its 'looks'. In Greek ontology, by contrast, the relation is seen from the comportment of production in which a prior 'model' of the thing being produced determines what the thing will be.

An analysis of other ontological notions, such as *ousia* and *einai*, corresponding to essentia and existentia, respectively, shows that it is possible to gain an added insight on them when seen from the perspective of productive comportment. In the final step, Heidegger suggests that a phenomenological analysis of the productive comportment should provide an ontological foundation for the two notions under consideration, that is, essentia and existentia. Productive comportment aims at letting the produced be a being by itself, released from the producer, available for use as at-hand in the surrounding world. Such a comportment offers an immediate understanding of the being in itself of beings. Even that which is not produced by a productive comportment, such as 'natural beings', can also be understood from that comportment, insofar as in producing 'materials' are used, where material is "that which is already there," that is, natural beings. Heidegger can then say that

> Productive comportment is not limited just to the producible and produced but harbors within itself a remarkable breadth of possibility for understanding the being of beings, which is at the same time the basis for the universal significance assignable to the fundamental concepts of ancient ontology (Heidegger, 1988, p. 116).

Heidegger concludes by stating a question which he does not go on to explore, on grounds that it would require a deep look into Dasein's constitution of being: "*why* ancient ontology interprets beings from exactly this direction," that is, from productive comportment?

On the one hand, in considering the ontological genesis of the theoretical mode of comportment, Heidegger suggested that a possibility rooted in the ontological constitution of Dasein could become prominent, developing into a comportment in its own right, that is, the theoretical comportment. On the other, the origin of notions such as essentia and existentia was understood as rooted in a particular understanding of being that obtains in productive comportment. While the first train of thought acknowledges the possibility of the emergence of new fundamental comportments in Dasein, the second emphasizes how such comportments harbor an understanding of being which is at the origin of fundamental ontological notions.

But what does it mean to say that "ontological notions are rooted in an understanding of being"? It could mean that in the case of Greek ontology, say, in Plato and Aristotle, fundamental ontological notions were elaborated taking, even if implicitly, production as a comportment of choice, such that these notions were unfolded from this comportment. But Heidegger means to suggest more than this. In productive comportment, which is a fundamental and pervasive comportment, an understanding of being emerges which is so primordial that it gives Dasein an immediate understanding of the beings which it encounters. Being such immediate understanding, it determines how 'what is' appears to us, while at the same time not becoming itself open for reflection, given that it itself contributes to make such reflection possible.

In spite of this important insight, to our knowledge Heidegger did not explore this notion any further and in his subsequent work he focused upon the "giving or emergence of being," as noted above. Would it be possible to say that while certain fundamental ontological notions have their 'roots' in an understanding of being that obtains in a pervasive comportment such as productive comportment, the particular way that such notions take as they emerge is not 'determined' by having such roots, but rather that in such 'giving of being' an ultimate freedom reigns? If this were the case, these two ways of conceiving the possible origin of fundamental ontological notions, namely, as rooted in an understanding of being and as emerging in a giving of being, are perhaps reconcilable. This issue will reappear, in a transformed way, in the next section, and we will have the opportunity to deal with it again there. In the remaining part of this work we will concentrate on possible ontological transformations in the context of our encounters with things. It is through such encounters that transformations of things brought about by technology may lead to changed understandings of being, hence, leading to fundamental ontological transformations.

ENCOUNTERS WITH USEFUL THINGS

We are always already encountering things, we are forever surrounded by them. In such encounters, we come up against some-thing, a something which may offer us resistance or may quietly acquiesce. Encounters are possible at all because we are open to things in such a way that they can approach us and offer themselves. In any encounter, the thing brings something to it; first of all, it brings itself, making itself noted. In approaching the thing we also bring something to the encounter, namely, a particular comportment, be it using a thing, repairing it, or searching for it. Things themselves can be powerful enough that we can encounter them not only when they are 'co-present' with us but, also and precisely, when they are 'absent', in which case we may miss them.

If we pay attention to what encounters are, rather than to the things encountered in them, we are witness to a phenomenon of singular richness. Although the primary notion of encounter that we will explore here refers to 'direct' encounters with things, such encounters take place in the context of totalities of things and encompassing activities, which themselves may be part of ulterior activities. Encounters with things also take place in the midst of particular surroundings which, in turn, belong in more encompassing surroundings or regions. Ultimately, an encounter takes place within a 'mode of revealing', that is, within a particular 'giving of being' that characterizes a historical epoch. In this work, our main focus will be on 'direct' encounters with things, but within certain limits we shall also take into account these other contexts in which encounters occur.

Because in the notion of encounter it is immediately apparent that it takes place when two kinds of beings, humans and things, get involved with each other, it is easy inadvertently to focus on either of them and their characteristics, rather than on encounters themselves. An encounter is a phenomenon that emerges from the concurrent, mutually affecting 'presence' of the parts to the encounter. What an encounter is cannot be found in any of the participants, but it is in itself nothing without them. What the participants bring to the encounter contributes to determine it, but this does not exclude the possibility that, as something that emerges from the mutually affecting 'presence' of the participants, encounters may possess unique characteristics, perhaps untraceable to the parts. As any phenomenon, encounters have certain characteristics, what might be called 'an ontological constitution', where the notion of constitution is not to be understood as some kind of 'fixed essence', but rather as something that emerges from the inter-play or counter-play of the participants.

As human beings, what we bring to the encounter is an 'openness', understood in the Heideggerian sense of a capability to dwell in an open region, where we encounter things and other kinds of beings. As Heidegger suggested in the essay *On the Essence of Truth*:

> All working and achieving, all action and calculation, keep within an open region within which beings, with regard to what they are and how they are, can properly take their stand and become capable of being said (Heidegger, 1977, p. 124).

We stand in this openness displaying a particular comportment, such as using a thing or searching for it. Encounters, then, emerge when in the context of an activity we become involved or engaged with a thing, as when in the course of reading we reach for our eyeglasses and put them on, keeping them securely in place, and then continue to read with their concourse.

DIMENSION OF OTHERNESS

If we now consider what an encounter is, that is, if we look at it from an ontological perspective, there is a particular trait that emerges of itself which, because of its multi-faceted character, we will refer to as a 'dimension', namely, *the ontological dimension of otherness*. Being an ontological dimension of encounters, and given that encounters emerge from the confluence of human beings and things, otherness is an emergent dimension that flourishes in the course of encounters and governs them. Otherness is rooted in and originates from the dual source of encounters, that is, things and human beings, and many traces of it can be found in everyday language. There is 'us' and the 'other' counter to us, that which we are not. There is 'that' thing and the 'other' thing. Whatever thing we encounter we understand it as being 'separate' from us, as something that is on its own. It also strikes us that the thing has a 'corporeity', just as we have ours, and that somehow there is a relationship between this corporeity and its being separate from us. Not being 'part' of us, and not being a figment of our imagination either, the thing affirms itself as something that has its own ways and its own whereabouts. Thus, if we are not careful enough, the thing may eventually become lost to us, and thus inaccessible, precisely when we most need it.

As Heidegger has shown in the analysis of 'useful things' (e.g. Heidegger 1996a, p. 65), while in use a thing somehow withdraws from being merely 'present' to us and we become concentrated in the activity being performed, such that, so to speak, the thing becomes one with us. But this should not be taken to mean that the character of being an 'other' of the thing has also vanished. In fact, such character remains in full swing and may hit us at any moment, as when suddenly the thing 'malfunctions', and one of the lenses abruptly dislodges itself from the frame of our eyeglasses and falls upon the table with a noise.

Yet there is another sense of 'otherness' which we need to uncover. In order to obtain a richer notion of what an encounter is, it is necessary to more amply 'admit' the thing in the encounter and what it has to offer by itself. Although our encounters with things take place primarily and usually in terms of using, finding, and repairing them, there are other possible ways of encountering them, ways which are well within our reach.

In a variety of circumstances, including those in which it malfunctions, the thing may take us by surprise and 'catch' our attention. In the case of our eyeglasses, we may suddenly become impressed by the simplicity and elegance of its various parts and by the precision with which they have been put together. Soon, we may realize that the precise disposition of the parts is fully determined by the shape, size, and characteristics of our own heads, and we may be surprised to discover that although the position of the eyes is the guiding factor in the overall architecture of the eyeglasses, the nose and ears play important though unusual roles as supporting pieces that keep the eyeglasses in place. Thus, the 'senses', the most significant visible elements of our heads are forcefully imprinted in the shape of the eyeglasses, which have themselves something of a mask, except that, instead of hiding behind them we use them to dismantle the fog all around us. It may strike us how the lenses, themselves made of solid pieces of material, can be so effortlessly traversed by light, and how the refractions that they induce, deliberately designed into them, compensate for our eyes' weaknesses. Thus, we may be led to realize how a thing as humble as a pair of eyeglasses can help to establish and make fit a relationship between us and the Sun, that distant purveyor of light.

All of this is meant to suggest that each and every thing harbors in itself a striking relationship with salient elements of our world and, ultimately, with the "world as a whole." Thus, our encounters with things span a broad space of possibilities, from being a thing in use almost 'invisible' to us, to being something pregnant with relationships to almost everything else. It will be instructive to consider how the thing fared in Heidegger's work. In *Being and Time* the analysis of useful things is performed with a primary focus on Dasein's "being in the world." Although in the course of the analysis Heidegger repeatedly refers to "Dasein encountering beings" and how this is possible because of Dasein's openness, the things themselves enter into the analysis in a very limited way, never occupying center stage. This is clearly reflected in the notion of 'world' that emerges from the analysis, a world which is rooted in human potentialities and complexes of references, ultimately leading back again to such potentialities, that is, a world that cannot let the thing in its full 'thingness' to come in.

Subsequently, the notion of world, which constituted "all that could be" for Dasein, finds a limit, an other to it, in the notion of *earth*. In a meditation on the work of art, world is understood as that which is open or comes to be open by the work of art. "By the opening up of a world, all *things* gain their lingering and hastening, their remoteness and nearness, their scope and limits" (Heidegger, 1971a, p. 45, italics mine). But if the world is self-disclosing openness, "the earth is the spontaneous forthcoming of that which is continually self-secluding and to that extent sheltering and concealing. World and earth are essentially different from one another and yet are never separated" (Heidegger, 1971a, pp. 48, 49).[10] Finally, in a meditation on *The Thing* (Heidegger, 1971b), which attempts to gain a more just notion of what a thing is beyond what a thing appears to be in our usual encounters with them, Heidegger focuses on the 'gathering' character of a thing. In a thing, and in a way which depends on its being a specific kind of thing rather than another, come to be gathered or reflected that which comes into play in the use of the thing. Thus, the thing becomes a gathering point of various 'regions' of beings including mortals, earth, and sky in such a way that in this gathering a *world* is disclosed.

Things have now come a long way. While in *Being and Time* the world was almost entirely human and the thing was like a non-thing, a *nothing*, in *The Thing*, the world has become that which emerges from the play of many 'participants', including the mortals, and the thing is encountered as that which is able to gather them and in which world comes to be. It would be shortsighted to take these extreme ways of encountering things as being the result of Heidegger's peculiar ways of thinking; rather, they constitute historical and genuine ways in which things can be encountered. We take these multiple possibilities for encountering things as constituting a dimension, in fact, as properly constituting what we called above the dimension of otherness. Again, we call this 'otherness' because what is at stake in this dimension is the extent to which in our encounters with things we admit them and let them be by themselves, that is, something that is other than us. It would be mistaken to believe that we are free to choose among these various ways of encountering things, and it is likely that technological change, in particular the penetration of everyday things with computing technology may bias our encounters with them toward a narrow range of possibilities. We will examine this issue later.

TRANSFORMABILITY OF OTHERNESS

In any encounter with a thing certain possibilities are cultivated and then flourish, while others are disregarded or never present themselves. What flourishes and what perishes in an encounter is contingent upon what the thing brings to the encounter, what our comportment is that leads to the encounter, and what a thing is for us. Appropriating a fundamental notion in Heidegger's thinking, while at the same time transforming it, we say that in any encounter 'something' is unconcealed and comes into the open, while 'something else' remains concealed, never emerging. While for Heidegger what remains concealed in this play of lights and shadows, even while making the unconcealed possible, 'is' being, for us what remains concealed are 'aspects' or 'possibilities' in the dimension of otherness. What, in an ontological sense, flourishes and perishes in any encounter are, then, possibilities in that dimension. That which takes place in any encounter, and as such comes to be 'cultivated', is otherness, understood in this case in relation to the thing being encountered. What does not emerge and remains concealed is forsaken.

Being an ontological dimension of encounters, otherness is a 'space' of possibilities that emerges in our encounters with things. As such emergence, otherness is nothing but what gives in such encounters, that is, in the concurrent, mutually affecting 'presence' of things and us. What emerges in encounters, what obtains of the other as *other*, how the thing is admitted, how we are admitted in the encounter, all of this comes to pass in secret ways, which are unknown to us. There is an *essential indeterminability of that which emerges*. But this is not to be understood as being only an indeterminability for us. Rather, what emerges of the other as other in an encounter emerges freely. In any encounter with a thing there is always the possibility that the thing may appear differently from what it has already been to us, that we might take it to be, in some sense, other than what it was. This essential indeterminability of what comes to pass in encounters suggests that otherness is essentially 'transformable', that is, that it has the capability of becoming different from itself. What this is pointing to might be called the 'transformability' of otherness.

At this point the following question emerges. Although it is a common 'experience' that in our repeated encounters with things they become different from what they were to us, could these changed encounters have an *ontological import*, or could they just be spurious and evanescent transformations? How could simple encounters with humble things be able to transform what things are for us?

Encountering a thing in the course of any kind of activity puts us in a concurrent 'co-presence' with the thing. Even in the case in which, while in use, "the thing withdraws and we may become one with it," the thing is for us. In encountering a thing, it stands counter to us, first by itself and for itself – in the sense that in its standing by itself the thing stands forth toward us, and 'strikes' us. But a thing, also, stands for 'what is'. Any encounter with a thing is an encounter with 'what is' as a whole, such that in the encounter, to a certain extent and in ways which are peculiar to it, the encountered stands as 'what is' for us. And so, when we grasp the eyeglasses with our hands, quickly inspecting them in passing, deftly placing them on our heads, and keeping them in place while reading and writing, the eyeglasses are to a certain extent 'what is' for us. We encounter them, thus encountering what is. They become for us what things are, standing not only for themselves but for what a thing is. Needless to say, we are always already encountering other things and, after a while, we will take the eyeglasses off and return them to the table.

Thus, the possible ontological import of our encounters with things resides in that in such encounters they stand for things as a whole and, insofar as our encounters with them are transformed, what things are for us may be consequently transformed. In the next section we pursue this issue further.

But there is something else that takes place while encountering a thing, which indicates that an 'understanding' of what a thing is emerges in such encounters, and that there is the possibility of a transformation of such understanding. In the previous section, while considering 'production' as a comportment, we referred to the notion of "conversing with oneself while doing something." Heidegger suggested that "*producing is intrinsically a talking to oneself and letting oneself talk*" (Heidegger, 1995, p. 124). When we grasp the eyeglasses to put them on, if the eyeglasses catch our attention in *unusual* ways, that is, in ways in which their use is 'suspended', we may engage in a conversation with ourselves about the eyeglasses. What transpires in that conversation, particularly when indeed an unusual encounter with the thing has taken place, can be understood as a *reformulation* of what this particular thing is for us. We briefly now examine this notion of reformulation.

In a work on translation which takes its initial directives from Heidegger's suggestions on translation and language, Parvis Emad has elaborated the notion of reformulation as a kind of 'innerlingual translation' (Emad, 1993). As Emad indicates, "reformulation occurs when the matter that appears in the initial formulation . . . reappears differently . . ." (Emad, 1993, p. 330). As innerlingual translation, reformulation occurs in the course of dialogues and monologues, when we say some matter in various and related ways. Thus, "the path that thinking takes in reformulation points out an altered appearing and a translating into this appearing" (Emad, 1993, p. 330). The explicit reference to 'thinking' should be taken as an indication that reformulation 'proper' occurs when there is an 'essentially' altered appearing, which does not occur necessarily in just any dialogue and monologue. Finally, then, and echoing Heidegger's notions of language, Emad suggests that "when language unfolds essentially, it allows things to show themselves and be manifest" (Emad, 1993, p. 329).

Returning now to the notion of encounter, we can 'see' that within the space of possibilities of encounters with things it is possible to encounter them in unusual ways, in which previously hidden aspects of things emerge. And it is in language that this altered appearance shows itself. Thus, we can say that encounters harbor in them the possibility of reformulating, in language, what things are, in particular, what things are in reference to the dimension of otherness. *The otherness of things can be reformulated.* Finally, then, reformulation points again toward an essential transformability of our encounters with things. And it is through reformulation that an 'understanding' of things is gained.

In this section we have identified an ontological dimension of encounters, namely, otherness, which is a space of possible ways of encountering things as being other than us. Such a play of possibilities gives or makes possible the 'transformability' of the way in which things appear in their being other. We then suggested that the transformed ways in which things appear in encounters with us had an ontological import in the following senses. First, in such encounters a particular thing not only stands for itself but also stands for 'what is' as a whole, in particular, for 'things'. And second, through 'reformulation', as it takes place in language, a transformed understanding of the thing as an other comes to pass. Although in our daily encounters with things the way in which they appear as being other tends to be remarkable uniform and unsurprising, the essential transformability of otherness harbors in itself the possibility that such encounters may change, in such a way that things in their otherness may become different.

Finally, let us consider how the transformability of otherness as elucidated in this section stands with respect to Heidegger's indications regarding the genesis of the theoretical comportment, presented in the previous section. Heidegger suggested that such comportment originates when a possibility inherent in Dasein's ontological constitution, namely, that of talking to oneself while doing something, becomes prominent. By contrast, in the case of encounters the space of possibilities constituting the dimension of otherness not only hinges on Dasein's ontological constitution, but also on the things themselves as they obtain in encounters. And it is through reformulation, as it gives in language, that new understandings emerge.

CHANGED ENCOUNTERS WITH THINGS

We suggested above that new understandings emerge irruptively in our encounters with things, in ways that are unknown to us. But what the thing itself brings to the encounter and what we bring with us, bias the encounter, such that in it certain aspects of otherness are unconcealed while other aspects remain hidden. What things and us bring to the encounter favors the cultivation of certain aspects of otherness while disfavoring others.

Encounters with things may change because the things themselves may have changed, due to technological 'innovation'. Given that otherness is an emergent dimension of encounters, the possibility exists that changes in the things encountered may bring about transformations in otherness itself. It is even possible that in some cases technological innovations are significant precisely because they transform otherness, although this may be obscure to the innovators themselves. At the same time, not all changes in things may have ontological import on otherness; for such an import to be possible changes in things should significantly alter the way we encounter them. In addition, as indicated in the previous section, encounters have ontological import because the things encountered in them stand forth as 'what is'. How pervasive and ubiquitous a technological innovation becomes, that is, the extent to which it penetrates everyday life, is significant because it bears upon the extent to which it contributes to determine 'what is'.

To gain a concrete understanding of the matter under discussion, namely, whether and how changes in things could ultimately lead to ontological transformations – in particular, regarding otherness – we will examine some proposals for a technology or set of synergistic technologies which, by their characteristics and purported scope, have the potential to effect such transformations. For this purpose, we will focus on the technology of Ubiquitous Computing. It should be noted that in what follows we do not intend to provide a comprehensive analysis of this proposed technology; rather, we will consider it as an exemplary case that can help us to understand the issues under discussion. (For a complementary analysis of Ubiquitous

Computing see Araya, 1998. For a related analysis of other computing technologies see Araya, 1997a and Araya, 1997b).

UBIQUITOUS COMPUTING

Ubiquitous Computing has been proposed since the early 1990s by members of a major research laboratory in the United States, as a broad-ranging vision for the future development of a variety of technologies centered around computing and communications, aiming at the penetration of everyday life as a whole with such technologies (Weiser, 1991, 1993; Weiser & Brown, 1997).[11] A central idea underlying the proposals is that the full potential of computing – that is, the potential to truly revolutionize everyday life activities – cannot be reached as long as computing technology remains encapsulated in computer machines, standing as a new kind of thing or device with which we interact in the performance of activities. As it currently exists, computing technology is such that to be able to use it in most cases we need to interrupt the course of an activity and approach a computer to interact with it. Such interruption or break in the activity is regarded as a serious impediment to the ability to penetrate everyday life with these technologies.

To overcome this difficulty, Ubiquitous Computing attempts to "bring technology to the things themselves." Specifically, by *embedding basic components of computing and communications technologies in things and environments*, a tight and smooth integration of computing in activities could be achieved, thus eliminating the interruption currently required by its use and its attendant disruptive effects. Taken to its logical limit, this smooth integration can be achieved fully only if *every* thing and environment is penetrated with such technology; otherwise, things remaining outside this penetration would act as barriers to the smooth flow of computation. Hence the importance of the ubiquitous deployment of the technology throughout things and environments. Considering the scope of the proposed technology, the authors refer to it as the "third wave of computation," beyond the first two waves characterized, respectively, by the computer and the Internet.

> The third wave of computing is that of ubiquitous computing, whose crossover point with personal computing will be around 2005–2020. The 'UC' era will have lots of computers sharing each of us. Some of these computers will be the hundreds we may access in the course of a few minutes of Internet browsing. Others will be embedded in walls, chairs, clothing, light switches, cars – in *everything*. UC is fundamentally characterized by the connection of things in the world with computation. This will take place at many scales, including the microscopic (Weiser & Brown, 1997, p. 77; emphasis added).

What distinguishes Ubiquitous Computing from Internet computing is that the latter, while achieving a high degree of interconnectivity, does not even begin to approach ubiquity at the envisioned level. To understand what the proposals mean by ubiquity, which is the central notion of the proposed technology, we need to attend to the 'literal' meaning of this word – *omni-presence, simultaneous existence everywhere*. The 'revolutionary' character of the proposed technology resides precisely in that it aims at achieving what we will call ultra-ubiquity, in which not only every thing and every environment is "connected with computation," but is connected in multiple ways, "including at the microscopic level," as Weiser & Brown suggest. As envisioned, the ubiquity of computing would far surpass the ubiquity of writing, which the proposals regard as a 'technology' that has achieved a very high level of ubiquity.

How could ultra-ubiquity be achieved? As indicated before, this can be obtained by embedding the fundamental components of computers, namely, micro-processors, memories, input devices, and 'displays' (monitors, screens) into things and environments, in a way that is appropriate to their specific characteristics. Connections among these primary components can be achieved by means of network technology, specially by wire-less networks, which are currently in use in cellular telephones. Finally, a number of other kinds of devices can be coupled with this basic infrastructure, including signal emitters, sensors, and 'actuators', the latter being primarily electro-mechanical devices that can perform actions and which are most conspicuously found in robotics technology.

Should Ubiquitous Computing be regarded as science fiction or, rather, as a technology that is beginning to emerge and whose first signs are already visible? The answer to this question must be mixed. On the one hand, we find signs everywhere of the increasing ubiquity of computing and communications technologies, of which the Internet is today the latest indication. In highly automated manufacturing centers not only the basic computing and communications infrastructure has been put into place, but actuators of all kinds, in particular automated or semi-automated robotic arms, are in routine use and human workers have already been 'wired' into the manufacturing complex by a variety of communications technologies. On the other hand, ultra-ubiquity is not only totally absent from everyday life but has not been achieved even in those highly automated centers.

Although the dissemination of computing technology is growing very strongly there is no assurance that this technology will evolve in the direction envisioned by the proposals for Ubiquitous Computing. Many obstacles remain on the road to ultra-ubiquity, including limitations of wireless networks, difficulties in the miniaturization of some of the embedded components, and complexity of the software required to integrate smoothly and manage astronomically large numbers of mobile components.[12] Moreover, it is not at all clear that human communities will willingly allow themselves to be 'wired' in such an indiscriminate fashion as envisioned by the proposals.

In the remainder of this chapter, we will consider Ubiquitous Computing as a proposed technology, some of whose elements are slowly emerging but whose full blooming is in no way assured. In spite of this uncertainty, Ubiquitous Computing is exemplary because it envisions an extremely pervasive penetration of everyday life things and activities with technology, giving it the potential for effecting ontological transformations of the kind we are concerned with. Our ontological focus should not be taken to imply that other perspectives such as ethical, political, and sociological are not relevant for the analysis of these proposals. Nor are we suggesting that Ubiquitous Computing is the only technology with potential for ontological transformations.

UBIQUITOUS COMPUTING SCENARIOS

Given the versatility of computing and communications technologies, these constitute a technological infrastructure capable of being embedded in things, environments, and activities, and which can be used for a variety of purposes. To illustrate such possibilities, the proposals include a variety of 'scenarios' depicting specific situations in which these technologies are deployed along the lines of Ubiquitous Computing. Let us now examine the following basic scenario, which we have composed from some of the scenarios included in the proposals:

> *All* things such as eyeglasses, cups and clothes, living beings such as trees and dogs, and human beings, have embedded or attached signal emitter devices which are permanently sending signals carrying the identity of their bearers. Insofar as these various kinds of beings are mobile, some of them autonomously, they move from environment to environment. *All* such environments, including office buildings, homes, cars, airplanes, highways, and parks, contain networks of sensors for detecting the signals emitted by these entities as they change location. In turn, the networks of sensors are interconnected, via wired and wireless networks, with computers distributed everywhere. The end result is that the identities and locations of all entities penetrated with this technology are constantly available in the web of computing, making it possible to locate and track anything. Computer programs, ubiquitously deployed, can then process this information for any purpose.[13]

This basic scenario can be extended in multiple ways. By embedding sensing devices in things and living beings, devices which are capable of detecting and transmitting relevant changes in 'external' and 'internal' states of these entities, information becomes instantaneously and constantly available in the web of computing. By embedding actuators, it becomes possible to perform actions remotely on these entities, actions which are computed using information about the states of these entities and transmitted via networks as commands to the actuators. And, as indicated above, what is crucial in these scenarios is the assumption of ultra-ubiquity, which implies that the sketched technological penetration has an all-encompassing character in its embrace of things, environments, activities, and people.

Having outlined the main tenets of Ubiquitous Computing and illustrated them in terms of a basic scenario, let us now consider potential transformations that could be elicited by this technology. To gain a better understanding of the basic scenario and its extensions, we will first perform a preliminary analysis of it and then will reexamine it from an ontological perspective. In the given scenario, all things and living beings have been penetrated with technological devices that constantly broadcast their identities; all environments are penetrated with networks of sensors capable of detecting the signals carrying these identities; such networks of sensors, in turn, are connected to other networks feeding processing nodes. In the extended scenario, not only identities but states of things are disseminated allowing, in turn, for the closing of the loop with commands that exercise the actuators to achieve the desired modification of these states.

What this scenario reveals is an extensive, ultra-ubiquitous apparatus, invisible to a large degree, aimed at making constantly available in the web of computing identities, locations, and states of things – purportedly, of every thing. Such a supra-entity, which we will call the 'ubiquitous web', transcends particular environments – constituting, in a limited sense, a unified environment into which each thing has been 'mapped', so as to leave a constant trace. Although if we attend to each thing in particular it would appear as if little has happened to it, the pervasive reach of this apparatus opens the possibility that the relationship of a thing and its surroundings may change essentially, thus possibly changing the thing itself, as will be explored in the next subsection.

With regard to our encounters with things, one of the most obvious consequences of this scenario is that whenever a thing becomes lost we can access the ubiquitous web in such a way that, after furnishing a description of the lost thing, we promptly receive an indication of its whereabouts. Not finding our eyeglasses, we turn around, avail ourselves of one of the ubiquitous input devices, and enter an appropriate command to search for them. Once submitted, the command is distributed over the web where software 'agents' search through networks for information about the location of the eyeglasses. Promptly, and contrary to our expectations that we had left them in the glove compartment of our car, we are informed that they are currently located in one of the overhead compartments of the airplane we took on our way home, which is currently flying half a continent away . . . .

More generally, it becomes possible that before we use a thing or, even better, before we initiate an activity, we 'check' through the ubiquitous web to determine whether all the things we need to perform the activity are available. If not, we ask for their locations. Our encounters with things, then, have become mediated by the web. Although the actual use of a thing may be little affected by this technology, because the way in which computing is embedded in things and environments is carefully designed so as not to disrupt their use in activities, other kinds of encounters, such as finding things and preparing for initiating activities, are mediated by the web.[14] Such mediation becomes particularly clear in the extended scenario, in which it becomes possible to examine and manipulate *everything* from *everywhere*. We can then engage in tele-encounters with things, which although admittedly clumsier than 'direct' encounters, may be appropriate in certain situations.[15]

We can then say that the existence of the ubiquitous web gives rise to new ways of encountering things. And even when we are using a thing 'directly', rather than in a tele-encounter, there is implicitly or explicitly another 'participant' in it, which is always there, behind the scene, ready to mediate our encounters. These have now become *web-mediated encounters*.

ANALYSIS OF UBIQUITOUS COMPUTING SCENARIOS

Let us now examine this scenario and its extensions, with their possible transformations of things, encounters, and understandings, from an ontological perspective, focusing in particular on the dimension of otherness. Our leading questions will be: Do web-mediated encounters introduce biases in our encounters with things? What do these encounters conceal and unconceal with respect to otherness? Considering that the purpose of this analysis is to illustrate possible ontological transformations rather than to give an ontological account of Ubiquitous Computing "as a whole," we will consider only certain aspects of otherness, which should suffice for this purpose.

A thing is 'other' than us. We can understand this characterization in multiple ways – in particular, in terms of a 'distance' 'between' the thing and us. In its most obvious but limited sense, distance refers to a 'physical' distance, that which can be measured with a measuring tape or device of some sort. There is also a distance that determines whether the thing is or not 'at hand' in the context of a particular activity. In this second sense, distance refers to the accessibility or inaccessibility of a thing for its use in the activity. It is this distance that Ubiquitous Computing attempts to 'overcome', as the basic and extended scenarios described above show. Through its pervasive penetration of things and environments the ubiquitous web acts like an invisible 'hand', aiming at making every thing at hand.

By attempting to bridge this distance, the ubiquitous web intervenes in the dimension of otherness, plays in that dimension, and thrives in it, specifically, by *concealing distance*. Bridging such distance *permanently* amounts to annulling it, as is particularly clear in the case of tele-encounters, in which the thing is actually 'handled' remotely, at a proximal 'distance'. At the same time, another aspect of the otherness of a thing, namely, its 'corporeity', is to some extent unconcealed, becoming a matter for 'concern'. Even if we can use the thing remotely, its 'material body' remains 'distant'. It is because of its corporeity that only traces of the thing travel in the ubiquitous web. Were it not for this corporeity, the web would bring to us not just the trace but the thing itself!

Another sense of distance that needs to be considered is that by being other than us, a thing is relatively 'unknown' to us. Even if we have a thing in our hands most of what it is remains hidden from us, but this is exacerbated when the thing is 'physically' distant. One way of bridging this distance is by trying to 'understand' the thing and, in a more limited sense, by attempting to 'know' about it. To a certain extent, Ubiquitous Computing also bears upon this sense of distance in that it allows us to know constantly the whereabouts of a thing. Admittedly, this is a very restricted kind of knowledge, but it has one significant characteristic, namely, that it is 'constant' knowledge, that is, we know the location of a thing *always*. In the extended scenario, where things are penetrated with sensors which capture information about the states of things which is then disseminated over the web, we are able to know constantly and instantaneously not just the location, but important properties or states of things, thus constituting a significant extension of our 'knowledge' of them.

Thus, in the scenarios under consideration the technology of Ubiquitous Computing bears upon the distance between things and us. It does so by the deployment of a ubiquitous web which – because it not only takes "the measure of all things" by capturing identity, location, and state information, but also allows for the return of ordering commands – has the character of an 'enframing' apparatus. This two-way movement of 'capturing' and 'ordering' made possible by the web enframes the thing in its embrace, 'pulverizes' distance, and *transforms the thing into an enframed thing*.

Turning now briefly to environments, we can say that knowing where a thing is becomes independent of the particular environment in which it finds itself. In a sense, *environments become 'transparent'* to us, because the web penetrates them by means of the networks of sensors deployed throughout them. Thus, the 'opacity' of environments, an aspect of their otherness, has been transformed into a certain transparency.

What about web-mediated encounters? Although the ubiquitous web attempts to make every thing immediately encounterable, because encounters, in all their diversity and in different ways, become web-mediated, what we are encountering foremost, constantly, and ubiquitously, is the web itself! This situation reveals a kind of *displacement*. Although in the course of activities we continue to encounter things, there is always in the background, as part of the context in which things are encountered, an invisible web, ready to come to the fore when needed. As previously indicated, in many cases before accessing the thing we go through the web, so that what we first encounter is the web itself. This many-sided phenomena of displacement is, then, a transformation that takes place in the dimension of otherness itself: we encounter a thing by way of another, supra-thing, the web. With the growing power of technology the supra-thing should become increasingly powerful, dwarfing the thing, thus intensifying the displacement.

So far, through the exploration of the basic and extended scenarios we have indicated possible transformations that could take place regarding things, environments, and encounters with things. But what authorizes us to say that they would indeed constitute 'ontological' transformations? In the discussion of the 'transformability' of otherness presented in the previous section, we suggested that encounters have an ontological import insofar as in encounters with things these stand not only for themselves but also for things as a whole, such that transformations in encounters make possible transformations of what things *are* for us. Taking into account that one of the fundamental characteristics of Ubiquitous Computing is precisely its ubiquity, that is, its penetration of everyday life as a whole – thus *pervasively permeating our encounters with things* – we can say that the transformations invited by such technology could indeed have such ontological import, transforming what things are for us. In addition, Heidegger's suggestion that 'productive' comportment opens "the possibility for understanding the being of beings," discussed in the second section, also holds here, in the following sense. Not only the productive comportment but other comportments in which we encounter things, such as using and searching, also bring with them an understanding of things. Thus, transformations on our encounters with them, such as those envisioned in the proposals for Ubiquitous Computing, open the possibility for transformed understandings of what things are for us.

Finally, we need to consider the question as to *what* those transformed understandings would be. In answering this question we need to keep in mind the following two points. First, the transformed understandings are related to what the changed encounters with things conceal and unconceal regarding otherness. But, second, what those understandings indeed turn out to be cannot be known in advance because of the essential indeterminability of their emergence, as was indicated in the previous section. Thus, we can say the following regarding the possible understandings that could emerge with Ubiquitous Computing technology: Independently of the presumed usefulness of ubiquitous computing in everyday life, the ontological relevance of this technology resides in that it intervenes in the dimension of otherness. From this perspective, Ubiquitous Computing is an attempt at substantially concealing certain aspects of the 'distance' between things and us, in such a way that through the mediation of the ubiquitous web things become 'enframed things'. Whatever understandings of things emerge from encounters permeated with this technology they are most likely related to this kind of transformation.

In the concealing of distance in the senses indicated above, that is, in relation to the accessibility of things and to our constant 'knowledge' of their states, it may appear as if things would now be 'closer' to us, as if we would have now bridged the distance that separated us from them. But even if in some respects this would be the case, still the concealing of distance amounts to an *attempt* at annihilating the otherness of the thing, that is, at annihilating the thing as an other, and ultimately, as a thing itself. This peculiar way of bringing things closer to us at the same time reveals itself as an attempt at annihilating them. By a peculiar displacement, the place of things is now occupied by a 'supra-thing', the ubiquitous web. It then makes sense to say that our understanding of things as *other* than us, should – in some way or another – be eroded. We would still continue to encounter things, but they would be to a notable extent what we put in them and what we get out of them, and no-thing else. We may even be able to say that, in the 'theoretical limit', there would be no more things, only us, humans. Thus, the possibility of significant transformations in our understanding of things is latent in this technology.

POTENTIAL ONTOLOGICAL TRANSFORMATIONS

As suggested by the analysis of Ubiquitous Computing, a technology may be oriented by design to alter our encounters with things in a deep way. Ubiquitous Computing intervenes directly in the dimension of otherness, itself a crucial dimension of encounters, by concealing and unconcealing aspects of it. Because of the transformability of otherness, that is, of the possibility that aspects of this ontological dimension may be transformed, the alteration of our encounters with things heralded by this technology may lead to ontological transformations. A crucial characteristic of this technology, namely, its ubiquity, points precisely in this direction. Ubiquity, that is, *simultaneous existence everywhere*, contributes to make 'what is' appear as what is encountered by the intermediation of this technology.

At the same time, it must be emphasized that a technology, while inviting certain kinds of ontological transformations, cannot by itself determine what these transformations will be. Such transformations emerge irruptively. But, if what our analysis suggests is justified, then Heidegger's notion of the 'sending' or 'giving' of being must be complemented precisely with the notion that what emerges as 'what is' does so within a space of possibilities opened by, among other 'contributors', what technologies conceal and unconceal regarding ontological dimensions. Givings of being, while preserving a freedom of emergence, do not take place in a vacuum.

It becomes important, then, for a better appreciation of the history of being to develop an understanding, from an ontological perspective, of the techniques and technologies that were predominant at various historical ages.[16] To achieve such understanding requires elucidating which ontological dimensions and which aspects of these dimensions are affected by these techniques and technologies. Albert Borgmann's notion of the 'device paradigm', although not explicitly developed from an ontological perspective, illuminates an important aspect of the current stage of the history of technology characterized by an increasing disassociation or separation between the 'machinery' of a device and the function that it provides (Borgmann, 1984, pp. 40ff.). Don Ihde's work explores, from phenomenological and hermeneutic perspectives, possible effects of technology, including effects on human self-conception (e.g. Ihde, 1983, pp. 9–23 and pp. 65–77). Recently, Tim Casey has explored the relationships between technology and metaphysics, focusing on the conceptions underlying the techniques associated with various kinds of technologies, specifically examining the 'mechanization of technique' itself (Casey, 1997). As relevant as these works are, they have the limitation that the perspectives from which they emanate are not primarily ontological, and thus are not necessarily able to illuminate ontological dimensions affected by modern technologies and techniques.

In the previous analyses, because of our focus on the dimension of otherness, useful things turn out to acquire what might be called 'a life of their own'. A related, although not necessarily similar, way of looking at things has emerged in studies of science and technology. Bruno Latour's work on "science in the making" (e.g. Callon & Latour, 1992) attempts to 'redistribute' the roles assigned to humans and non-humans – the latter referring to things, including 'natural things' and machines – in that process. Non-humans are regarded as 'actants' that have certain 'powers', which it is necessary to understand in order to appreciate the role they play in science and technology. Similarly, Andrew Pickering's *The Mangle of Practice* (1995), advances the notion of machine agency and suggests that human and material agency "are intimately connected with one another, reciprocally and emergently defining and sustaining each other" (Pickering, 1995, p. 17). But again, these works do not explicitly deal with fundamental ontological dimensions.

An important issue that arose in the analysis of Ubiquitous Computing can be succinctly stated as follows: Ubiquitous Computing attempts to bring things closer to us but, in so doing, it erodes their otherness. This leads to the following observation. Although the proposals for Ubiquitous Computing abound in remarks and examples of how this technology could bring things closer to us, they are almost entirely mute regarding considerations of otherness.[17] We interpret this overall situation of 'unconcern' regarding otherness not as being the result of negligence or unwillingness to consider certain matters but, rather, as being a consequence of the fact that at least certain aspects of the dimension of otherness are highly invisible, in any case, not visible to the naked eye. To see this situation more clearly, let us take an extreme case and assume that indeed the widespread penetration of everyday life with Ubiquitous Computing technology would lead to an understanding of things such that they would be taken to be ever closer to us, when rather, what would be actually taking place is an attempt at annulling their otherness, in a sense, at annulling the things themselves. We ask, then, what is the ontological significance of the concealment of the annulment of the thing?

As a means to considering this issue, let us return for a moment to our eyeglasses and ponder what the future history of things has in store for them. In their latest reincarnation, eyeglasses have become contact lenses, that is, the frames have now disappeared and we directly insert the lenses in our eyes to achieve the desired effect. Independently of the many justifications that can be given for the 'advantages' of contact lenses over eyeglasses, we regard as very likely the notion that this transformation constitutes a continuation of the erosion of the otherness of things, to the point that, in the particular case of eyeglasses, they have now become part of our bodies. In the next step, which goes totally beyond the scope of Ubiquitous Computing and which is already being made possible by the latest surgical techniques, the eyeglasses will simply vanish, having been made redundant as a result of transformations of the eyes themselves produced by eye surgery.

What then about the increasing concealment of the otherness of things that would be made possible by Ubiquitous Computing? Clearly, we already inhabit a world penetrated by a variety of technologies which in many ways contribute to the concealment of multiple aspects of otherness. Ubiquitous Computing takes as its point of departure a situation already characterized by such a concealment, and it furthers it, although to an unheard-of degree. In addition, because the dimension of otherness is so close to what we are, it tends by itself to be concealed from our understanding. We have then a situation in which an entire dimension of our encounters with things verges at the point of 'extinction'.

To conclude, then, our analysis suggests that by the protracted ways in which technology changes our encounters with things, entire ontological dimensions may vanish. It is unlikely that otherness, being such a fundamental dimension of encounters and of ourselves, may become concealed to such a point. Nonetheless, such an occurrence is within the realm of the possible.

ACKNOWLEDGMENTS

The author would like to thank two anonymous reviewers for their generous suggestions.

NOTES

1. Heidegger first presented the notion of 'history of being' in his work on Nietzsche. See, for example, 'Metaphysics as History of Being', in Heidegger, 1973, pp. 1–54, and 'The Essence of a Fundamental Metaphysical Position', Heidegger, 1991, vol. 2, pp. 184ff.

2. For a detailed comparison of Hegel and Heidegger on this issue see Michel Haar, 1993, pp. 67–77.

3. 'Sendings' and 'destinings' are translations of Geschick, a word notoriously difficult to translate. We will also render it as 'giving of being'. This notion is treated at some length in Heidegger, 1996b, and we will have more to say about it later.

4. It should be noted that, partly as a consequence of having reached an understanding of the history of being, Heidegger's own philosophical language was transformed and in his later work he abandoned talk of 'ontology' to refer to the domain of fundamental understandings of 'what is'. But we will retain this term here for its heuristic value and will continue to talk about ontological transformations.

5. As is well known, this property of general purpose computers was demonstrated by Alan Turing through the mathematical analysis of the so-called 'universal Turing machines'.

6. See Heidegger, 1996a, p. 327; translation slightly modified. In the German original Heidegger writes: "der *ontologischen Genesis der theoretischen Verhaltung* . . . ."

7. A related analysis in *Being and Time* 'ventures' an examination of the "ontological genesis of historiology as a science," which we will not pursue here. See Heidegger, 1996a, Section 76, pp. 358ff.

8. See Heidegger, 1995, Section 14, pp. 111ff.: "The extraordinary relationship of force and conversance . . . ." The quotation is from p. 124.

9. One of the most sustained discussions of this issue appears in *The Basic Problems of Phenomenology* (Heidegger, 1988, pp. 100–117). My subsequent presentation is based on this work. For several related considerations, see Robert Bernasconi 1996.

10. For a sustained discussion of the notion of earth in Heidegger, and its relations to physics and nature, see Haar, 1993, pp. 47ff.

11. Initially proposed by Mark Weiser of the Xerox Palo Alto Research Center in 1991, there appears to be an increasing number of researchers in many laboratories in the United States and elsewhere working in this new approach to computing. See note 12, next.

12. For recent views on these and related issues see Chandy et al., 1998.

13. Although scenarios are presented in Weiser, 1991, Weiser, 1993 and Weiser & Brown, 1997, they are most prominent in the first of these references.

14. We are not suggesting, by implication, that there are 'unmediated' encounters with things; we are simply pointing out a particular kind of mediation. Many authors have dismissed the notion of unmediated experience, most recently Petran Kockelkoren, in the analysis of seeing a landscape from a train in movement (Kockelkoren, 1998, pp. 50ff.).

15. Elements of this scenario have been in place for some time, as in the case of tele-controlled robots and other specialized equipment used in the exploration of dangerous or distant places, such as the bottom of the ocean or the surfaces of planets. What is distinctive about Ubiquitous Computing is, again, ultra-ubiquity, the notion that everything can be manipulated at a distance, from anywhere.

16. In the context of a work attempting to situate Heidegger's notion of 'equipment' or useful things, Hubert Dreyfus develops a sketch of the "history of the being of equipment" (Dreyfus, 1992). This work provides a high level view of broad stages in that history, without necessarily focusing on any technology in particular.

17. The closest that the proposals come to a discussion of otherness is in their consideration of privacy issues that may arise by the extensive application of this technology. These issues are immediately regarded as technical 'problems' requiring technical solutions, such as those provided by encryption technology.

REFERENCES

Araya, A. (1998). Encountering the World With Ubiquitous Computing. In: *Proceedings of the 4th International Conference on Ethics and Computing, Rotterdam, The Netherlands, March 25–27, 1998* (Rotterdam, The Netherlands: Center for Philosophy and Communication Technology), 33–43.

Araya, A. (1997a). Experiencing the World Through Interactive Learning Environments. *Electronic Journal of the Society for Philosophy and Technology, 3*(2), 1–22.

Araya, A. (1997b). Technological Change and Ontological Transformations: The Case of Virtual Reality. *Ludus Vitalis, Journal of Philosophy of the Life Sciences*, 2 (Special Issue), 221–239.

Bernasconi, R. (1993). The Fate of the Distinction Between Praxis and Techne. In: R. Bernasconi (Ed.), *Heidegger in Question: The Art of Existing*. (pp. 2–24). Atlantic Highlands, NJ: Humanities Press.

Borgmann, A. (1984). *Technology and the Character of Contemporary Life*. Chicago: University of Chicago Press.

Callon, M., & Latour, B. (1992). Don't Throw the Baby Out with the Bath School! A Reply to Collins & Yearley. In: A. Pickering (Ed.), *Science as Practice and Culture*. Chicago: University of Chicago Press, 343–368.

Casey, T. (1997). Technology and the Metaphysics of Technique. *Research in Philosophy and Technology, 16*, 73–86.

Chandy, M. et al. (1998). Virtual Roundtable: Ubiquitous Computing. *IEEE Internet Computing*, *2*(2), p. 96. See also the extended online discussion in IC Online, at URL: http:/ /computer.org/internet/v2n2/round.htm.

Dreyfus, H. (1992). Heidegger's History of the Being of Equipment. In: H. Dreyfus & H. Hall (Eds), *Heidegger: A Critical Reader.* Cambridge, MA: Blackwell, 173–185.

Emad, P. (1993). Thinking More Deeply into the Question of Translation: Essential Translation and the Unfolding of Language. In: J. Sallis (Ed.), *Reading Heidegger: Commemorations*. (pp. 323–340). Bloomington, IN: Indiana University Press.

Haar, M. (1993). *The Song of the Earth: Heidegger and the Grounds of the History of Being*. Trans. Reginald Lilly. Bloomington, IN: Indiana University Press.

Heidegger, M. (1971a). The Origin of the Work of Art. In: M. Heidegger (Ed.) and trans. A. Hofstadter, *Poetry, Language, Thought*. (pp. 15–87). New York: Harper and Row.

Heidegger, M. (1971b). The Thing. In: M. Heidegger (Ed.) and trans Albert Hofstadter, *Poetry, Language, Thought*. (pp. 163–186). New York: Harper and Row.

Heidegger, M. (1973). *The End of Philosophy*. Trans. Joan Stambaugh. New York: Harper and Row.

Heidegger, M. (1977). On the Essence of Truth. In: M. Heidegger & D F. Krell (Eds) and trans. John Sallis, *Basic Writings from Being and Time to The Task of Thinking*. (pp. 113–141). (New York: Harper and Row).

Heidegger, M. (1988). *The Basic Problems of Phenomenology*, rev. ed. Trans. Albert Hofstadter. Bloomington, IN: Indiana University Press.

Heidegger, M. (1991). *Nietzsche*. Vols. I and II. Trans. David Farrell Krell. New York: Harper and Row.

Heidegger, M. (1995). *Aristotle's Metaphysics Theta 1–3: On the Essence and Actuality of Force*. Trans. Walter Brogan and Peter Warnek. Bloomington, IN: Indiana University Press.

Heidegger, M. (1996a). *Being and Time*. Trans. Joan Stambaugh. Albany, NY: State University of New York Press.

Heidegger, M. (1996b). *The Principle of Reason*. Trans. Reginald Lilly. Bloomington, IN: Indiana University Press.

Ihde, D. (1983). *Existential Technics*. Albany, NY: State University of New York Press.

Ihde, D. (1990). *Technology and the Lifeworld: From Garden to Earth*. Bloomington, IN: Indiana University Press.

Kockelkoren, P. (1998). Toward a Technological Intimacy with Things. *Research in Philosophy and Technology, 17*, 45–57.

Pickering, A (1995). *The Mangle of Practice: Time, Agency, and Science*. Chicago: University of Chicago Press.

Weiser, M. (1991). The Computer for the Twenty-First Century. *Scientific American, 265*(3), 94–104.

Weiser, M. (1993). Some Computer Science Issues in Ubiquitous Computing. *Communications of the Association for Computing Machinery, 36*(7), 75–84.

Weiser, M., & Brown, J. S. (1997). The Coming Age of Calm Technology. In: P. Denning & R. Metcalfe (Eds), *Beyond Calculation: The Next Fifty Years of Computing*. (pp. 75–85). New York: Springer.

THE IMPLICATIONS OF CONSISTENCY: PLATO'S *PROTAGORAS* AND HEIDEGGER'S 'THE QUESTION CONCERNING TECHNOLOGY'

Mary Bloodsworth

INTRODUCTION

> The stages of [human] material progress were celebrated by all three great tragedians, as well as by philosophers like Anaxagoras and Democritus and the Sophist Protagoras. They might be associated with the name of Prometheus, patron saint of technology . . . if so, his first gift to [human beings] is sagacity of ingenuity, and the rest follows from that (Guthrie 1981, p. 18).
>
> Technology is not equivalent to the essence of technology. When we are seeking the essence of 'tree', we have to become aware that what pervades every tree, as tree, is not itself a tree that can be encountered among all other trees. Likewise, the essence of technology is by no means anything technological (Heidegger, 1977, p. 4).

The view dominating scholarly interpretation of Socrates's activity in Plato's early dialogues holds that Socrates's aim is to point out, or to generate, logical inconsistencies within the belief-sets of his interlocutors. Disputes among scholars have centered on how this generation of inconsistency takes place, while failing to challenge *whether* this is indeed Socrates's main objective. Socrates's means of generating inconsistencies has been termed the 'Socratic elenchus', although scholars agree that Socrates himself had no name for his method. The 'inconsistency view' is exemplified by Brickhouse and Smith, who maintain that "Throughout Plato's early dialogues, Socrates employs his

Metaphysics, Epistemology, and Technology, Volume, 19, pages 33–43.
2000 by Elsevier Science Inc.
ISBN: 0–7623–0681–5

elenctic method in such a way as to show his interlocutors that they cannot maintain the moral beliefs they have expressed consistently with certain other of their beliefs" (1984, p. 185). The elenchus is not only thought to be Socrates's method for generating inconsistencies within the belief-sets of others; some scholars consider it to be Socrates's means of scrutinizing his *own* belief-set. Vlastos maintains, for example, that Socrates has been "constantly exposing the consistency of his beliefs to elenctic challenge, ready to root out any belief which if allowed to stand would disturb the coherence of the system as a whole" (1984, p. 54).

Whether conceived as a method for uncovering inconsistencies in others's belief-sets or within his own system of belief, common to both accounts of the elenchus above is the idea that Socrates's activity is motivated by a desire to unearth *inconsistency*. To recognize that one subscribes to an inconsistent set of beliefs is considered, in this view, to be the worst situation for one to face. Formulated this way, the dominant activity in Plato's early dialogues is rendered on a coherentist model. That is, the elenchus is thought to expose inconsistencies within belief-sets, and these inconsistencies are thought to result when individually-held beliefs do not *cohere* to form a unified system. A unified system of belief is considered the ideal, while an (exposed) inconsistent set of beliefs is thought to present danger to that ideal. Exposure to inconsistency – or inconsistency itself – is presumed to pose the greatest threat to one's set of beliefs, and more significantly, to one's self (or in Socrates's terms, to one's soul).

In contrast to the prevailing view, I claim that it is not so much inconsistency as it is *consistency* that Socrates often regards as the greatest danger. Of fundamental importance to Socrates is not that beliefs cohere *with one another*, but that one's total system of belief not be at odds with a more basic, underlying approach or 'attitude' (for Socrates, an attitude of humility, rather than one of *hubris*). 'Danger', for Socrates, does not purely arise when beliefs are challenged by inconsistency; instead, the ultimate danger – to use the words of Heidegger – precisely is a challenging. Challenging [*Herausfordern*], Heidegger claims, is "the revealing that rules in modern technology;" it marks a worldview dominated by arrogance and defiance (1977, p. 14). The consistency of this challenging attitude means that the world only reveals itself in limited ways; for example, as merely present for [human] use or as available to fulfill [human] desires. Thus, Heidegger maintains that "Technology is no mere means. Technology is a way of revealing" (1977, p. 12). Relatedly, I propose that Socrates, in Plato's early dialogues, is not concerned with producing or exposing logical inconsistencies within interlocutors's belief-sets (or in his own belief-set) as much as with revealing the danger inherent in some very particular *consistent* views. The link between the implications of consistency and the question concerning technology may be found, I suggest, in Plato's dialogue, the *Protagoras*.

SECTION I: THE FIGURE OF PROTAGORAS

The figure of Protagoras – unlike many other Socratic interlocutors – is considered to have been an actual historical person, survived by specific assertions taking the form of fragments. Protagoras's two famous assertions are: "The human is the measure of all things," and "I know not whether gods exist . . ." (Wheelright, 1966, pp. 239–240). In focusing on Plato's dialogue, the *Protagoras*, I keep with scholarly opinion that this dialogue most likely depicts the historical Socrates. That is, the early dialogues are generally thought to characterize Socrates as a historical figure, rather than to advance the positive doctrine of Plato. Moreover, I maintain that the dialogue, the *Protagoras*, serves to reveal the views of the *historical* Protagoras. As to the meaning of Protagoras's two famous sayings, it is generally agreed that the "human is the measure" principle is an assertion of extreme relativism, and that the statement regarding the gods is a claim of religious agnosticism. While neither of Protagoras's assertions appears in the *Protagoras* itself, both assertions are nonetheless at work in Protagoras's telling of the Promethean Myth. As Michael Nill states, "If anywhere, it will be in the Great Speech [the Promethean Myth] where the historical Protagoras's views about areté and moral questions will be found" (1985, p. 5).

The events which lead to Protagoras's telling of the myth, and some background to these events, help to place Protagoras's story-telling in context. As the dialogue opens, Protagoras has arrived in Athens and is staying at the home of Callias. Socrates and Hippocrates go to visit Protagoras, but a porter slams the door in their faces, exclaiming, "Ha, sophists" – not knowing that Socrates and Hippocrates, in fact, *are not* sophists. The Athenian suspicion of sophists rested, to a great extent, on the sophists's claim to teach virtue. This claim marked a direct challenge to the traditional Athenian belief that virtue was "a mark of good birth and dependent on natural or divine gifts" (Nill, 1985, pp. 5–6). That Protagoras himself claimed to teach virtue, or practical wisdom, can be seen in the *Protagoras*, when Socrates asks Protagoras what his services will provide the young Hippocrates (who is eager to become his student). Protagoras replies that he will teach "the proper care of [Hippocrates's] personal affairs, so that he may best manage his own household, and also of the State's affairs, so as to become a real power in the city . . ." (319a).[1]

It is noticeable that Socrates warns Hippocrates, at this early juncture in the dialogue, regarding the danger of entrusting his soul to a sophist (312c). This precise warning serves to undermine the 'inconsistency view', for, as I shall demonstrate, Protagoras has a quite consistent set of beliefs. The aim of Socrates's warning is not to guard Hippocrates against a suspected inconsistent belief-set lurking in Protagoras; rather, it is to indicate the peril to Hippocrates's soul if subjected to the particular beliefs of the sophists (and in particular, to those of the sophist Protagoras). Kerferd, for example, ties Protagoras's emphasis on a practical agenda to 'subjective knowledge', 'mastery', and 'humanism' (1981, p. 168), and an analysis of Protagoras's underlying assumptions can be shown to support Kerferd's claim. Protagoras's practical focus can be seen, in other words, to foster these other three 'positions'. Of significance here is the reason why Protagoras conveys his myth in the first place: Protagoras is aware of the general suspicion he faces in Athens, so he attempts to demonstrate – by way of a story – that the virtues he professes to teach are consistent with the values held by the Athenians. Protagoras explains, in his myth, how gods formed the creatures of the earth (320d). Different creatures were allotted various means of survival by the god Epimetheus (320e). Although each type of creature was to be assured a means of preserving itself (321a), Epimetheus ("not a particularly clever person") used all powers available before providing for humans (321c). As the god Prometheus checked Epimetheus's allotments, he found that humans were not well-equipped to survive. Therefore, the gift of "skill in the arts" was stolen from Athena as Prometheus's means of compensating humans.

If Protagoras's assertion, "I know not whether gods exist," is an assertion of religious agnosticism, then the fact that he uses gods in his story can be interpreted as Guthrie suggests: the gods are not to be taken seriously, since in light of Protagoras's agnosticism they cannot be (1981, p. 64). The gods can "be stripped away as adornment to the tale" (p. 64). Guthrie interprets Protagoras's incorporation of gods into his story in a way that makes their use consistent with Protagoras's other beliefs. This is to say, if Protagoras is a religious agnostic, then he cannot truly mean what he says about the gods. The gods, in Guthrie's view, are simply a means of decorating a story whose point could be made without their involvement.

It is critical to note that if Socrates had wanted to expose inconsistency in Protagoras's belief-set, then Protagoras's use of gods in the story – when he had elsewhere professed not to believe in gods – would present an ideal occasion.[2] But Socrates does not do this. Moreover, *contra* Guthrie, it is absolutely crucial that Protagoras does use gods in his myth (i.e. their use cannot be a *sheer* case of adornment). After all, Protagoras wants to put the Athenians at ease; he introduces talk of the gods for precisely this purpose. In turn, it is necessary that Protagoras put the Athenians at ease because he depends upon them for employment. If Protagoras speaks of gods, then he may not appear to be quite the threat that the Athenians had suspected. Protagoras's use of the gods is consistent with his overall project in that, as a sophist, his aim is to persuade. The gods are not merely decoration; rather, they are a device of persuasion. Protagoras's myth, in other words, would not work without them.[3]

Consequently, Protagoras's relativism, agnosticism, and humanism cohere in this way: since Protagoras's aim (as a sophist) is to persuade, and thus, to defeat an opponent, it need not matter which side of the argument one takes – the ultimate aim being simply to win. Truth becomes relativized in accord with the particular position advanced by the arguer. There is no truth to be discovered 'beyond' that centered on human purposes; instead, whatever one deems to be truth is truth for that person. Such a relativized view of truth is possible when truth is solely measured in relation to practical, human interests. That Protagoras does intend the human being to hold this central position is further exemplified when his statement, "The human is the measure of all things," is conjoined with his assertion of agnosticism. These two statements imply that since one cannot have certainty of what may transcend the human realm, the human being must provide the standard for all action and explanation. Through dismissing the gods as indemonstrable, one can proceed solely on the basis of humanistic concerns.

Because Protagoras adheres to a practical or survivalistic conception of the virtues, Socrates is eventually able to demonstrate that Protagoras's basic assumptions are compatible with hedonism (specifically, the view that pleasure is the only good). Since for Protagoras, the human serves as the measure of all things, whatever best secures human survival and well-being is equated with the good. Likewise, since "skill in the arts" plays a crucial role in survival, a hedonistic view supports Protagoras's position as a sophist that "knowledge and wisdom are the most powerful things in human life" (352d). Lacking an objective standard by which to measure virtue (such as independently-existing gods), pleasure can offer itself as the only good to be pursued. As a consequence, and despite some hesitation on his part, Protagoras ends his discussion with Socrates by upholding a hedonistic conception of virtue and knowledge precisely due to its *consistency* with his belief-set.

Keeping in mind that Protagoras primarily advances his views *via* the Promethean myth, the meaning and implication of the term 'promethean', for the present discussion, may be stated as follows:

> The word 'Prometheus' comes from the Greek words *pro* and *mathein*, which together mean 'forethought' and 'thoughtfulness'. 'Promethean', then, may be understood to mean

> the compassionate but thoughtful concern for leading mankind toward greater and greater degrees of self-mastery through understanding and knowledge. Viewed this way, the term promethean aptly describes the ideal form of rationality underlying the institutions of science and technology (Nill, 1985, p. 25).

An emphasis on mastery and control serves to devalue – or aims to eliminate – mere chance. Protagoras's proposal that all are capable of delivering practical wisdom is significant, for whereas the Athenians had traditionally believed that virtue was a matter of luck (i.e. something beyond human control), when virtue is formulated as practical know-how, this element of luck is eliminated. A scientific/technological way of thinking arrives to replace the unpredictability of traditional Athenian luck.

SECTION II: THE ESSENCE OF TECHNOLOGY

Turning now to the opening passage of Heidegger's 'The Question Concerning Technology', Heidegger claims that the essence of technology is nothing technological. In the present context, this means that the case of Protagoras cannot be dismissed as irrelevant to the question concerning technology on grounds that the Greeks had no technological devices or machines (in the modern sense), or because Protagoras does not introduce talk of such equipment within his story. For Heidegger, one overlooks the revealing 'nature' of technology when one conceives it as merely a means to an end – as, for example, in the instrumental conception of technology. Heidegger states, "If we give heed to [the fact that technology is a way of revealing], then another whole realm for the essence of technology will open itself up to us. It is the realm of revealing; i.e. of truth" (1977, p. 12).

By contrast, when one views technology as a means to an end, it is easy to think of modern technology as "something incomparably different from all earlier technologies . . . because it is based on modern physics as an exact science" (1977, p. 14). Lost in the instrumental conception of technology is an essential structure of the *essence* of technology. This structure represents a mode of revealing that underlies and impacts all particularly-held views. Heidegger maintains:

> The word 'technology' stems from the Greek. *Technikon* means that which belongs to *techné* . . . . What is decisive in *techné* does not lie at all in making and manipulating nor in the using of means, but rather in the aforementioned revealing. It is as revealing, and not as manufacturing, that *techné* is a *bringing-forth* . . . . The revealing that holds sway throughout modern technology does not unfold into a bringing-forth in the sense of *poiesis*. The revealing that rules in modern technology is a challenging [*Herausfordern*] . . . (1977, pp. 13–14).

William Lovitt notes that "*Herausfordern* means to challenge, to call forth or summon to action, to demand positively, to provoke" (1977, p. 116; note 13). This is to say that modern technology is characterized by a consistent attitude of 'regulating' and 'securing', challenging and demanding (p. 16).

Thus, the modern age is depicted by Heidegger as one in which life, in all its facets, is organized and planned, predicted and controlled. While Protagoras is precisely not a modern figure, to consider Heidegger's claim concerning technology in this way reveals that Protagoras's worldview nonetheless fits squarely within Heidegger's critique. The desire to eradicate luck – advanced by Protagoras and other sophists – runs counter to Athenian tradition; and it is reflected, for example, in the Hippocratic treatise *On Science*. Here, Hippocrates states, "They did not want to look on the naked face of luck (*tuché*), so they turned themselves over to science (*techné*)" (Wheelright, 1966, p. 194).[4]

This point reintroduces Protagoras's proposal that "all are capable of delivering political wisdom." It is precisely this belief, with its implications of consistency, that Socrates finds disturbing. Socrates objects to this view, in the Protagoras, by stating that when it comes to subjects taught and learned (e.g. architecture and naval design), the Athenians will only accept the advice of experts. However, on general policy, they will allow anyone to give advice – presumably because they do not consider this to be a technical subject calling for training. Protagoras's explanation is that technical sagacity is innate in humans from the beginning. After all, in the Promethean myth, it is bestowed to humans by Prometheus. Political wisdom is only another expression of practical intelligence, which is the first divine gift of Euripides and Aeschylus.

Socrates's scorn for such a conception of political wisdom parallels the following statement by Heidegger. "A strange measure," Heidegger states, "perplexing it would seem to the common notion of mortals, to the cheap omniscience of everyday opinion, which likes to claim that it is the standard for all thinking and reflection" (1971, p. 223). Protagoras would himself seem to be the specific target of Heidegger's charge. While for Protagoras, the human being is the standard and the measure of all things, for Heidegger, the ancient Greeks (prior to Plato) recognized:

> The godhead [as] the 'measure' with which [the humans measure] out [their] dwelling, [their] stay on the earth beneath the sky. Only insofar as [humans take] the measure of [their] dwelling in this way [are they] able to be commensurately with [their] nature (1971, p. 221).

By contrast, Protagoras – in Socrates's eyes – effectively assumes the position of 'godhead'. He attempts to comfort the Athenians by persuading them that he can teach what has hitherto been considered a gift of luck.[5] According to

Heidegger, such a posture issues a way of revealing that forecloses all others. Or, as Guthrie suggests:

> Politics and morals, general theories of human nature, metaphysics, and epistemology cannot be separated. They rest on the assumptions about the nature of reality and the workings of the universe determining [humanity's] position within it . . . (1981, p. 5).

Protagoras's claim of agnosticism, on Guthrie's account, cannot be separated from Protagoras's other views or from his basic manner of approaching the world.

Guthrie's position thereby contrasts with that of Kerferd, who states that the sophistic movement has often been "characterized by a revolt against religion, a profound movement of the human spirit away from a belief in the divine towards a rationalist and humanist outlook on the world" (1981, p. 163). While admitting that sophists and philosophers had differing concerns, Kerferd denounces the above depiction of the sophists as radical. Rather, Kerferd favors a view of the sophists in which they, like the poets and philosophers of the time, were interested in 'reinterpretation' of the gods; that is, in moving from a view of the gods as anthropomorphic cheaters, thieves, and seducers. In Kerferd's view, the sophists were generally one more group engaged in this same sort of reinterpretation. That Sophists, such as Protagoras, were condemned for their atheistic or agnostic positions, according to Kerferd, demonstrates a bias against the sophists that is unjust.

It could be argued, however, that the implications of Protagoras's two famous sayings extend beyond what a mere 'reinterpretation' may suggest. This is precisely, I propose, Socrates's concern. To suggest that Protagoras 'reinterprets' the gods implies that such reappraisal occurs in innocent isolation from other beliefs. While Kerferd views Protagoras's statement about the gods to be a claim about the limitations of human knowledge (i.e. that human beings are simply unable to judge matters concerning the gods), Protagoras's statement actually appears to be grounded in a refusal to acknowledge the possibility of human fallibility.

That Protagoras is not sensitive to human fallibility (i.e. does not truly address the limitations of human knowledge) can be detected in the assertion of knowledge implicit in his statement. That is, if human knowledge were truly appreciated as limited, then human beings would need to proceed with utmost caution so to not assume too much. Human beings would need to act with humility, rather than with *hubris*. Protagoras's statement introduces, however, the idea that by dismissing the gods as being indemonstrable, one can proceed solely on the basis of humanistic concerns. Such humanism places the human being in the central position in the interpretation of reality – a position opposed to a humble recognition of human limitation. It is the danger of this positioning, with its respective consequences, that unsettles and motivates Socrates.

CONCLUSION

Since Protagoras considers virtues such as justice and shame to develop for practical purposes, such virtues are not seen to share in a 'higher' realm. Since Protagoras has no consideration for reality beyond that of human interests, Socrates is ultimately able to show that Protagoras's basic assumptions are compatible with hedonism. Although Socrates's demonstration occupies much of the *Protagoras*, its possibility nonetheless rests on the basic assumptions drawn forth in Protagoras's telling of the Promethean myth. As Cynthia Hampton states:

> The *Protagoras* demonstrates what happens to philosophical analysis when the divine dimension of reality is ignored . . . the door is opened to hedonism once the human rather than the divine is taken as the measure of all (1989, p. 17).

Indications that Socrates disavows [a life of] pleasure can be found in other early dialogues by Plato (e.g. the Euthyphro, *Apology*, *Crito*, *Meno*, and *Phaedo*).[6] For Socrates, pleasures and desires pertain to the realm of the body, whereas virtues such as goodness and justice (or piety) are only achieved by nourishing the soul. The soul, unlike the body, partakes in a 'higher order'. Philosophy, in turn, represents the paradigmatic activity for fostering the soul. The life of bodily pleasures, in Socrates's view, is a lowly life – the ultimate form being a life of hedonism. Inasmuch as a hedonistic attitude denies a higher realm (e.g. the reality of the soul or the gods), such a life is one of arrogance or *hubris*.

For Socrates, Protagoras's position is dangerous not because it lacks coherence, but because it coheres in a way that challenges reality – or a worldview – that Socrates deems genuine. The case of Protagoras reveals that Socrates, in Plato's early dialogues, does not always aim to generate inconsistency within his interlocutor's belief-sets. More significantly, Socrates can be seen to draw forth the implications of beliefs that precisely form a consistent package.

In this respect, Heidegger offers an alternative way of seeing Socrates's activity. On Heidegger's account, such a 'package' constitutes an *approach*, or a way of revealing, that forecloses all others. Moreover, Heidegger maintains that our modern age is one in which the sophistic attitude, as exemplified by Protagoras, reigns.

The modern sophistic attitude is most dangerous when it is construed as simply neutral (i.e. "just the way things are"). Then, it is not seen as *a way* of revealing; rather, it is considered the *only possible* revealing. Likewise, technology, as a way of revealing, marks a consistency of attitude that inhibits other possibilities – perhaps ones that are more original or originary. On this point, Heidegger and Socrates share a common insight, especially as it relates to the figure of Protagoras.

NOTES

I would like to thank the anonymous referees and the editors at *Research in Philosophy and Technology* for their constructive feedback and help in improving the manuscript.

1. All references to Plato's *Protagoras* are to W. K. C. Guthrie's translation of the dialogue, contained in The Collected Dialogues of Plato, edited by Edith Hamilton and Huntington Cairns. References are cited by line number within the body of the paper. The emphasis of the sophists on a practical agenda is often noted in the scholarly literature. For example, Kerferd (1981) states:

> Sophists were concerned with subjective knowledge for practical purposes, to secure mastery over men and over life .... The Sophists were interested in how to succeed in politics and were humanists in that they put man and his values in the central place in the interpretation of the universe (1981, p. 35).

2. This is, after all, precisely the accusation which Socrates levels against Meletus in the *Apology*. Socrates stands accused both of believing in gods (other than those of the State) and for not believing in gods at all.

3. To convince the Athenians that he is able to teach virtue, Protagoras goes on to convey that humans, though possessing the two basic survival skills granted by Prometheus, nonetheless began to meet their demise due to a lack of political skill. This lack of political skill made it difficult for humans to live together congenially in cities. Therefore, Zeus sent Hermes to give to humans a sense of justice and shame as a means of bringing order to the cities (322c). Justice and shame were thereby distributed to all (323a), which made all capable of delivering political wisdom.

4. For a sustained discussion of the role of *tuché* in ancient Greece, the reader should consult Martha Nussbaum's *The Fragility of Goodness* (1986). Traditional Athenian adherence to luck is likewise revealed in Aristotle's discussion of character and action in the *Nicomachean Ethics*. Contrary to common opinion, Aristotle claims that appropriate moral feeling can be cultivated. As Nussbaum indicates:

> The luck view is rejected [by Aristotle] not because it has been found to be at odds with scientific fact about the way things are in the universe . . . but because it strikes a false note .... For we believe that human life is worth the living only if a good life can be secured by effort, and if the relevant sort of effort lies within the capabilities of most people (1986, p. 320).

5. The traditional Athenian gods had likewise been erratic by nature.

6. It should be noted that the longer dialogues, the *Gorgias* and the *Protagoras* (the present focus), are also considered early dialogues, and the Phaedo – which perhaps offers the best example of the connection between pleasures/desires and the body and virtues of the soul – shares features of both Plato's early and middle period dialogues.

REFERENCES

Brickhouse, T., & Smith, N. (1984). Vlastos on the Elenchus. In: J. Annas (Ed.), *Oxford Studies in Ancient Philosophy.* Oxford: Clarendon Press.

Guthrie, W. K. C. (1981). *The Sophists.* Cambridge: Cambridge University Press.

Hampton, C. (1989). Virtue, Knowledge, and the Necessity of Divinity: Plato's Use of the Promethean Myth. Unpublished manuscript.

Heidegger, M. (1977). The Question Concerning Technology. In: W. Lovitt (Ed.), *The Question Concerning Technology and Other Essays.* New York: Harper and Row.

Heidegger, M. (1971). Poetically Man Dwells. In: A. Hofstadter (Ed.), *Poetry, Language, Thought.* New York: Harper and Row.

Kerferd, G. B. (1981). *The Sophistic Movement.* Cambridge: Cambridge University Press.

Nill, M. (1985). *Morality and Self-Interest in Protagoras, Antiphon, and Democritus.* Netherlands: E. J. Brill.

Nussbaum, M. (1986). *The Fragility of Goodness: Luck and Ethics in Greek Tragedy and Philosophy.* Cambridge: Cambridge University Press.

Plato (1985). The Protagoras. In: E. Hamilton & H. Cairns (Eds) and trans. W. K. C. Guthrie, *The Collected Dialogues of Plato.* Princeton, NJ: Princeton University Press.

Vlastos, G. (1984). The Socratic Elenchus. In: J. Annas (Ed.), *Oxford Studies in Ancient Philosophy.* Oxford: Clarendon Press.

Wheelright, P. (Ed.) (1966). *The Presocratics.* New York: Macmillan.

TECHNOLOGY AND EMBODIMENT IN IHDE AND MERLEAU-PONTY

Philip Brey

There is a special class of artifacts, which includes telescopes, probes, hearing aids, and similar items, capable of engaging in 'symbiotic' relationships with the human body. These artifacts are not normally perceived and acted on as objects in one's environment, but instead are used as means through which the environment is experienced and acted on. Glasses and hearing aids, when worn, are not normally encountered as objects in the environment; they are means through which the environment is perceived. Likewise, a blind man's cane or a dentist's probe are means through which the environment is perceived and acted on. Don Ihde (1979) has called the special relations between human beings and such artifacts *embodiment relations*, since such artifacts seem to become part of our embodiment. In contrast, most objects in the environment, including trees, tables, and vases, may well function as objects of perception and action, but do not normally function as objects through which the environment is perceived and acted on.

From an epistemological point of view, embodiment relations are interesting in several ways. First, perception through embodied instruments like microscopes and telescopes raises epistemological questions. Such perception differs from unaided perception in that the original sensory information has been preprocessed (e.g. augmented or distorted) by a technological device. The nature and reliability of technologically mediated perception has already become an important issue in the philosophy of science, in which it is part of

Metaphysics, Epistemology, and Technology, Volume, 19, pages 45–58.

ISBN: 0–7623–0681–5

epistemological studies of scientific instrumentation (e.g. Hacking, 1983; Ihde, 1991; Brown, 1990).

Second, our cognition of embodied artifacts themselves is puzzling. We often do not seem to be consciously aware of them, since after an initial habituation period, they fade into the background of our awareness. Yet the skillful use we make of them reveals that we possess detailed information about them, for instance about their location in space. How is this information gained and represented? This question is especially interesting since it appears that our knowledge of embodied artifacts in some ways seems to resemble our knowledge of body parts. Some philosophers (e.g. Wittgenstein, Anscombe) have argued that we have a special way of knowing about (the position of) our body that is not gained by ordinary sensory perception. An analysis of how embodied artifacts are known may hence also sharpen our understanding of the different ways in which we are acquainted with our own body.[1]

An understanding of embodiment relations may moreover be relevant to philosophical anthropology and the philosophy of action. Embodied artifacts change our relation to the world, sometimes profoundly. The microscope, for example, has forever changed the way we understand our world, and the telephone has had an impact on our awareness of place. Embodiment relations are relevant to the philosophy of action, finally, because many human actions are now mediated by technology, and a proper analysis of technologically mediated action requires an understanding of how technical artifacts that participate in embodiment relations constrain and facilitate action.

So far, few studies of embodiment relations have appeared. Ihde's account of them is probably the most sustained account that currently exists. In this chapter, I will assess Ihde's account of embodiment relations. I will argue that Ihde's account is too limited to serve as a full-blown account of embodiment relations, since it takes as its task the description of recurrent patterns of experience found in embodiment relations, without also explaining how embodiment relations are constituted. This limitation, I will argue, leads Ihde to construe embodiment relations as perceptual, whereas I will argue that they are also frequently behavioral. Moreover, Ihde compounds two kinds of experience found in embodiment relations: experiences of the world as perceived through embodied artifacts, and proprioceptive 'experiences' or representations of the location and orientation of embodied artifacts. The expanded account of embodiment relations that I will end up presenting is directly based on Maurice Merleau-Ponty's original theory of embodiment relations, published in the 1940s, on which Ihde's account is also partially based.

1. IHDE ON EMBODIMENT RELATIONS

Ihde is well known for his theory of human-technology relations (Ihde, 1979, 1983, 1990). The aim of this theory, presented in different versions over the years, is to account for the various ways in which technology plays a role in human experience. As Ihde puts it, the goal of his theory is to "show what is invariable in the way humans experience their technologies" (1993, p. 111). In other words, his theory aims to describe certain patterns of experience that recur in different ways in which humans interact with technology. A taxonomy of recurrent patterns of experience would then amount to a taxonomy of human-technology relationships, as these can be distinguished from a phenomenological point of view.

Ihde (1990) distinguishes different types of human-technology relations. Embodiment relations constitute one important type of human-technology relation, in which artifacts are experienced as means through which the environment is encountered. Ihde defines embodiment relations as uses of technologies in which the technology mediates one's experience of one's environment or world, by being a medium positioned in between individual and world *through* which the individual perceives his or her world. In Ihde's words, in a embodiment relation "I take the technologies *into* my experiencing in a particular way by way of perceiving *through* such technologies and through the reflexive transformation of my perceptual and body sense" (1990, p. 72). In embodiment relations, the embodied technology does not become itself an object of perception – or does so only marginally. Rather, it 'withdraws' and serves as a (partially) transparent means through which one perceives the environment, thus engendering a partial symbiosis of self and technology. As prototypical examples of technologies that are often if not primarily used in embodiment relations, Ihde mentions such items as glasses, telescopes, hearing aids, dentist's probes, and a blind man's cane.

Whereas the previously mentioned artifacts make a good fit with Ihde's definition of an embodiment relation, Ihde further mentions examples of embodiment relations that seem to fit his definition less well. For example, Ihde also identifies the use of a hammer as involving an embodiment relation (1990, p. 80). However, a hammer is clearly not an instrument that is used, like telescopes and dentist's probes, to (better) perceive aspects of one's world. Instead, it is an instrument to act on the world. Ihde may submit, however, that tools like hammers nevertheless transmit perceptual feedback about the world. As the hammer hits a surface, one may feel aspects of that surface through the hammer. This perception may be used, in turn, to better guide the blows of the hammer. So although hammers are not primarily tools for perception, it may be argued that they nevertheless serve to mediate perception and in this sense fit Ihde's definition of an embodiment relation.

Other examples presented by Ihde, however, are still more problematic. Two of these, involving a feathered hat and a car, are quoted by Ihde from Merleau-Ponty's *Phenomenology of Perception.* I will here cite from Merleau-Ponty's text:

> A woman may, without any calculation, keep a safe distance between the feather in her hat and things which might break it off. She feels where the feather is just as we feel where our hand is. If I am in the habit of driving a car, I enter a narrow opening and see that I can 'get through' without comparing the width of the opening with that of the wings, just as I go through a doorway without checking the width of the doorway against that of my body. The hat and the car have ceased to be objects with a size and volume which is established by comparison with other objects. They have become potentialities of volume, the demand for a certain amount of free space (Merleau-Ponty, 1962, p. 143).

In the example involving the feathered hat, the feather is embodied in the sense that the woman has tacit knowledge of the location of her feather, just like she has tacit knowledge of the location of her hands. However, the feather is clearly not a medium through which the woman perceives and recognizes aspects of her environment. It is not used, like a blind man's cane, to extend tactile perception (although it could perhaps be used in this way), since the woman in fact tries to avoid contact between her feather and things in her environment. Similarly, although the driver of a car may have a tacit understanding of the location of the car relative to structures in the environment, it is not normally the case that the driver can feel these structures through the fender of the car. So it appears that these examples do not fit Ihde's definition of an embodiment relation.

There is thus a tension in Ihde's work between different notions of an embodiment relation. It is my aim here to resolve this tension. This tension can be resolved, I will argue, by considering the original account of embodiment relations developed by Merleau-Ponty. Not only was Merleau-Ponty the first author to present a theory of embodiment relations,[2] Merleau-Ponty is also responsible for most of the examples discussed by Ihde that do not seem to fit his definition of an embodiment relation. Hence, a study of the way Merleau-Ponty conceived of embodiment relations is likely to be revealing. My discussion of Merleau-Ponty's views on embodiment relations will result in an extended theory of embodiment relations, one that accommodates embodiment relations in Ihde's sense but also recognizes other kinds of embodiment relations.

2. MERLEAU-PONTY'S THEORY OF EMBODIMENT RELATIONS

Merleau-Ponty's *Phenomenology of Perception* is not, like Ihde's work, primarily concerned with technology, but rather with the nature of the human body and of perception. There are only a few brief passages in Merleau-Ponty's work in which human-artifact relations are discussed.[3] The discussion of human-artifact relation in these passages occurs within the context of a discussion of skills, which in turn is part of Merleau-Ponty's account of the human body. Although brief, these passages are highly suggestive, and constitute, I claim, a proto-theory of embodiment relations. In what follows, I will try to give a brief statement of this proto-theory. Such a statement needs to be preceded, however, by a basic exposition of Merleau-Ponty's account of the human body and its world, and of perceptual and motor skills. Needless to say, Merleau-Ponty's views on these matters cannot be done full justice here. My account of these matters will necessarily be sketchy, and involve a good deal of interpretation.

Perhaps the main idea developed in the *Phenomenology of Perception* is that, experientially, there is a fundamental difference between one's body and objects located outside one's body in the external environment. The external world is experienced as a spatial structure, in which things are, relative to us, remote or near, high or low, to the right or to the left. The body, however, is not normally experienced as a spatial entity of the kind found in one's external environment. The circumference of one's body, as experienced by one, forms a boundary within which one is not normally aware of spatial relations.

There are, admittedly, exceptions. It is certainly possible to experience one's body in spatial terms, and one can, for example, experience one's arms as being positioned above one's legs, or one's hand as being closer than the table. In such cases, one experiences the body as a physical structure just like other physical structures. However, in everyday actions, one does not normally experience the body as a physical, spatial structure in the external world. Instead, one experiences the body as a not expressly spatial means by which the spatial world is engaged. Moreover, although one may be aware of the relations between different parts of one's body, these relations are not normally understood as spatial relations.

This still sounds quite mysterious. So let us consider a hopefully revealing analogy. Suppose, first, that one is wearing a pair of glasses with red lenses. Now one certainly has an experience of these red lenses, since they make the whole world appear red. However, because one looks through them, one does not experience them as located at some definite location in the world that one sees through them (unless, of course, one looks into a mirror). These lenses are hence means by which one engages the world, but not, in one's normal experience, objects that are spatially located in that world!

Suppose, second, that on these glasses, immediately above the pair of red lenses, there has been mounted a pair of green lenses. Moreover, the location of these two pairs of lenses is controlled by a refined mechanism, attached to the muscles surrounding one's eyes, such that when one squints, both pairs of lenses lower two inches, so that one looks through the green lenses. When one stops squinting, the lenses rise again, so that one once more looks through the red lenses. Under such circumstances, the relation between the red lenses and the green lenses is not normally experienced by the wearer of these glasses as a spatial relation (unless he or she looks into the mirror and sees the green lenses being located two inches above the red ones). Instead, the relation between the red and green lenses is experienced in terms of the motor acts of squinting and leaving off squinting.

The analogy is that one's body is, just like the red lenses, a medium through which one engages the world; it is not normally experienced as a spatial entity. Moreover, just as the relation between the red and green lenses is not normally experienced as a spatial relation, the relation between different parts of one's body is not normally experienced as a spatial relation, but instead as a relation defined in terms of motor acts, which are again themselves not experienced as activities located in space. For example, when one is laying on one's right arm, the relation between one's torso and one's right arm is not primordially experienced as one's torso being above one's right arm, but as a dependency relation in which the ability of one's right arm to act is dependent on moving one's torso so as to free one's right arm. As Merleau-Ponty puts it, one's body is a "potentiality of movement," or a "field of possibilities of interaction with the environment": a structure whose parts are understood in terms of their ability to enter into one's projects, rather than in terms of their spatial location.

Still, Merleau-Ponty does not want to say that it makes no sense at all to say that one's body parts are in space. The body is, in a sense, spatial, but the space inside the body is not an ordinary space like the one experienced external to the body. Instead, it is a space in which spatial relations are 'motor relations'. For example, the relation between an outstretched hand and a fist is not understood in terms of ordinary spatial relations, but in terms of the motor primitives of clenching and opening wide. Merleau-Ponty calls the space of the body a "space of situation" which contrasts with the "space of position" in the external world (1962, pp. 98–102).

According to Merleau-Ponty, one's body is immediately known by means of a *body schema* or *body image*.[4] The body schema is an organizing structure contained in one's body that presents one with a unified understanding of one's body, which is experienced as a unified whole or 'Gestalt'. Moreover, the body schema provides one with a pre-reflective, immediate knowledge of the position of one's body parts. This position is again found in what Merleau-Ponty calls the "space of situation" and not the "space of position." When one sees a glass on the table, one may unthinkingly grasp it with one's hand, guided by one's body schema knowledge that one's hand can grasp things, automatically localizes one's hand, and guides it to the glass. Although the body schema makes the positions of one's body parts immediately known, it is not a mere composite of experiences of one's body parts, but rather a structure that precedes such experiences, as the existence of phantom limbs shows.

The body schema is a dynamic structure that reveals the body's orientation not just toward existing but also toward possible tasks. Moreover, the equivalencies between these different orientations, and the trajectories for moving from one orientation to the next, are immediately known through the body schema. For example, one knows through one's body schema how one's limbs when one is in an upright position correspond with one's limbs when one is crouched, and how to move between these two positions. Similarly, one knows by one's body schema whether one is able to walk through a doorpost without bowing, because one's body schema immediately tells one how one's body relates to the doorpost. The body schema does so without assigning a spatial position to the body, because the body does not need to be oriented in space, since it is the anchor relative to which other objects are oriented.

The potentialities of the body schema also determine aspects of the external world, as they are experienced. They determine, for example, whether things count as near or far away, large or small, heavy or light, high or low, passable or impassable, within reach or out of reach, hard or soft, warm or cold, being built up out of two colors or four colors. The space in which the body orients itself is hence not a Euclidean space containing things with objective properties, but a subjective, meaningful space, whose features the body has assigned meanings as possibilities for movement.[5]

Although the body schema contains invariant elements, not all its potentialities are permanent features. The body schema can be changed by the acquisition of new possibilities for movement. A possibility for movement that can be thus acquired, Merleau-Ponty calls a *habit* or *skill*.[6] A skill, when acquired, alters the body schema by altering its potentialities for action. For example, when the child learns to grasp physical objects, or to jump fences, its learns new motor transformations and equivalents that change its body schema.

Note that Merleau-Ponty does not just recognize the existence of *motor skills*, but also recognizes *perceptual skills*, such as being able to discriminate yellow from green.[7] As the body schema changes, experienced properties of one's environment change along with it. For example, a fence that was at first unscalable, or colored yellow-green, may now be scalable, and colored yellow and green.

Most skills that are acquired are perceptual and motor skills by which one orients oneself to aspects of the environment. Such skills relate positions or acts of one's own body to each other, as when one learns to dance, scale fences, or discriminate between blackberries and blueberries. Some skills, however, involve learning to handle objects that are used as tools, or that are otherwise attached to one's body, as when one learns to see through a microscope, to balance a clay jar on one's head, or to cut paper with scissors. According to Merleau-Ponty, when one learns the skill of 'handling' such an object, what happens is that this object becomes incorporated into one's body schema. It is then experienced and used as another means by which one engages the world, rather than as an object in the world that one engages. It becomes a means through which skills are expressed, rather than an object of skilled action. As such, it comes to function as a direct extension of ourselves, as an integral part of our bodies. In this way such artifacts differ fundamentally from artifacts that are not incorporated into the body schema, like the door one opens, the nail one hammers, the light one switches on, and the roof above one's head.

Merleau-Ponty gives at least five examples of such embodiment relations, involving uses of a feathered hat, a car, a blind man's stick, a typewriter, and an organ. The examples of the feathered hat and car were already presented. For the present, I will present one more:

> When the typist performs the necessary movements on the typewriter, these movements are governed by an intention, but the intention does not posit the keys as objective locations. It is literally true that the subject who learns to type incorporates the key-bank space into his bodily space (1962, p. 145).

The skilled typist embodies the typewriter so that it becomes a direct extension of his or her hands. An unskilled typist, in contrast, fails to incorporate the keyboard into his or her bodily space, and has to search for each key in space. Every touch of a key then becomes an encounter with the keyboard, instead of an act that incorporates the keyboard. The boundary between subject and world remains positioned at the end of the subject's hands, and is not extended to incorporate the keyboard.

3. MOTOR AND PERCEPTUAL EMBODIMENT

The three examples of embodiment relations presented by Merleau-Ponty that have been discussed so far – involving a feathered hat, car, and typewriter – are not examples in which the (primary) function of the artifact is to mediate perception. They are not brought up by Merleau-Ponty in a discussion of perception, but in a section on motor skills. Merleau-Ponty's named distinction between perceptual skills and motor skills leads to a distinction between two types of embodiment relations, being two ways in which an artifact may be integrated into the body schema. An artifact may either become a means through which motor skills are expressed or a means through which perception takes place. Combinations are possible and frequent: an artifact may mediate both motor skills and perceptual skills, although in most cases, one type of skill will dominate.

The class of embodiment relations defined by Ihde can, in terms of Merleau-Ponty's theory, be identified as embodiment relations in which an artifact becomes part of the body schema by becoming a medium through which *perceptual skills* are expressed. Perceptual skills such as the skill of being able to discern color or pitch, to identify spatial relations, and to detect motion are then exercised through the artifact. Three of the senses can be mediated in this way: sight, hearing, and touch. Optical instruments such as telescopes, microscopes, and glasses mediate visual perception. Aural instruments such as hearing aids and stethoscopes mediate hearing. Finally, tactile instruments such as probes and the blind man's cane serve to mediate tactile perception.

Artifacts that mediate *motor skills* become part of the body schema by becoming a medium through which motor skills are expressed. There appear to be at least two ways in which artifacts mediate motor skills. First, artifacts may be appendages to one's body with which, or by which, one moves through one's environment. I will call the motor skills this requires *navigational skills*. The feathered hat and the automobile constitute examples of such appendages. They serve to enlarge one's body, and this enlargement has to be taken into account as one navigates through one's environment. Other examples of such artifacts include a worn backpack, a bicycle one rides, a plank one carries on one's shoulder, and any hand-held object that one carries around. Since these artifacts become incorporated into one's body space, one knows tacitly and immediately where they are in relation to objects in one's environment, and one is able to navigate them through the environment. Usually, the goal of navigation is to avoid collision with objects in one's environment as one attempts to reach a desired destination along a chosen path.[8]

Artifacts may also serve as tools for physical interaction with the environment. I shall call the motor skills this requires *interactive skills*. Interactive skills are exercised mostly through hand-held or manually operated tools. Merleau-Ponty's example of the typewriter is an example of an artifact that mediates interactive skills. The pressed key and the typebar activated by one's finger become an extension of fingers through which they make marks on a sheet of paper. Similarly, when one uses a pen, a paintbrush, a vacuum cleaner, a razor, a hammer, a trumpet, a remote control, a knife, or a gun, one performs actions on the environment with a tool that is embodied within one's body schema.

Sometimes an artifact that mediates motor activity is also customarily employed to mediate perception. Merleau-Ponty discusses the example of a blind man using a stick to navigate through the world as an example of both an acquired motor skill and an acquired perceptual skill: "Learning to find one's way among things with a stick, which we gave a little earlier as an example of motor habit, is equally an example of perceptual habit" (1962, p. 152). When learning to use the stick, the blind man acquires motor skills, since he learns to interact with objects in the environment using his stick and to navigate this stick successfully through this environment along with himself, but simultaneously also acquires perceptual skills, since he learns to perceive and locate elements in his environment through the stick. The dentist's probe is another example of an artifact that has both a perceptual and a motor function. In both examples, the *terminus* by which one physically engages the environment – that is, the lower end of the stick and the tip of the dentist's probe – is also the *terminus* where perception begins.

For many artifacts used in motor tasks, their perceptual function is, however, subordinate to their motor function, if it is there at all. A paint brush, for example, is not normally used as an instrument for perception, although it may, like the dentist's probe, be used to perceive the texture of a surface. When it is used in painting, the primary function of the paint brush is not that of a tactile instrument, although the paint brush gives perceptual feedback about the amount of pressure exerted on the canvas, which may be used to adjust one's motor acts. However, when using an airbrush, even this feedback is eliminated, as the contact with the canvas is no longer there. The airbrush is just like the feather on the woman's hat which also does not ordinarily give her perceptual feedback about her environment. Moreover, perceptual feedback does not always aid the execution of motor tasks. When navigating an automobile through a narrow opening, one may feel, through the car, that there are holes in the road, but these perceptions do not help one navigate through the opening more successfully.

Conversely, artifacts that are primarily used as perceptual instruments often, but not invariably, mediate motor skills as well. Instruments for tactile perception, such as probes, are also often used to mediate interactive skills, as when the dentist uses the probe to remove some scale from one's teeth. A modification of navigational skills is required if a perceptual instrument significantly enlarges the bulk of one's body and one moves around while using it.

It may be thought that because embodied artifacts mediate and extend motor and perceptual skills, the primary purpose of embodied artifacts is to enhance skilled action or skilled perception. However, not all embodied artifacts serve to extend action or perception. *Cognitive artifacts* (Norman, 1993) are artifacts that are able to represent, store, retrieve, and manipulate information. They include items such as computers, calculators, abacuses, as well as pencil and paper, and books. Cognitive artifacts extend cognitive abilities, such as abstract thought, memory, problem solving, and language use. Merleau-Ponty believed cognitive abilities, or skills, to be grounded in, but not directly reducible to, sensory-motor skills. When a cognitive artifact, such as a calculator, engages in an embodiment relation with someone, it becomes a medium through which representations (usually in the form of symbols) are perceived and manipulated. In this way, a cognitive task is transformed into a perceptual and motor task.

4. CONCLUSION

To sum up, an embodied technology is a technology that is incorporated into one's body schema, which implies that it becomes part of one's bodily space (Merleau-Ponty's 'space of situation'). It then becomes an integral part of one's repertoire of motor or perceptual skills, and serves as a medium through which motor or perceptual functions, or both, are expressed. Perceptual functions that are mediated are either visual, aural, or tactile functions. Motor functions that are mediated are either navigational functions or interactive functions (or both). As an artifact becomes incorporated into one's repertoire of skills, it often, though not invariably, enhances these skills. That is, it often extends the 'potentialities' of the body schema, and consequently what the body, as mediated by the artifact it incorporates, is able to affect or perceive in its environment. An exception is constituted by objects that merely serve to enlarge the bulk of the body, and complicate navigational tasks without enhancing one's potential, or even serve as limitations.

When Ihde discusses embodiment relations, he discusses them in the context of an account of ways in which human beings experience technology. Given

Ihde's interest in analyzing experiences of technology, it is to be expected that embodiment relations are construed, in his theory, in perceptual terms. Ihde's aim is to give an account of recurrent patterns of experience that embodied artifacts bring with them, whereas Merleau-Ponty is more interested in showing how embodiment relations are constituted. In this sense, the two theories are complementary.

However, when Ihde characterizes the patterns of experience that are correlated with embodiment relations, he only recognizes one pattern of experience, which holds only for artifacts that mediate perception: the experience of one's environment through an artifact. Artifacts that mediate motor tasks were argued also to display a distinct pattern of experience. This is the direct experience of the location of the artifact, relative to objects and places in the environment. Thus, the woman wearing the feathered hat 'feels' where her feather is, just like she 'feels' where her hand is.

In conclusion, my study of Merleau-Ponty's theory of embodiment relations serves to extend Ihde's theory of embodiment relations in two ways. First, it supplements his theory with an account of how embodiment relations are *constituted*. Embodiment relations were shown to be constituted by the incorporation of artifacts into the body schema. Second, it extends Ihde's account of the experience of embodiment relations by distinguishing, next to experiences of one's world through an artifact, direct experiences of the location of an artifact relative to an environment. These are distinct experiences, the first found in perceptually embodied artifacts, the second mainly in motor embodiment.

ACKNOWLEDGMENTS

I would like to thank Hans Achterhuis, Pieter Tijmes, and Peter Paul Verbeek for their comments on an earlier draft.

NOTES

1. See Bermúdez et al. (1995) for an excellent anthology of work on bodily awareness and self-consciousness.

2. Although Merleau-Ponty was arguably the first philosopher to present a theoretical account of embodiment relations, the first observation of the existence of embodiment relations may be found in Heidegger (1962). Heidegger called objects that are only recognized as mere objects present-at-hand (*vorhanden*). Objects that are understood as means to one's ends, he called ready-to-hand (*zuhanden*). In one passage in *Being and Time*, Heidegger appears to distinguish between objects that are merely ready-to-hand, and objects that are ready-to-hand in a special way, which he called 'proximally ready-to-hand' (*zunächst zuhanden*):

> The peculiarity of what is proximally ready-to-hand is that, in its readiness-to-hand, it must, as it were, withdraw [*zurückzuziehen*] in order to be ready-to-hand quite authentically. That with which our everyday dealings proximally dwell is not the tools themselves [*die Werkzeuge selbst*]. On the contrary, that with which we concern ourselves primarily is the work – that which is to be produced at the time; and this is accordingly ready-to-hand too. The work bears with it that referential totality within which the equipment is encountered (1962, p. 99).

Artifacts that are proximally ready-to-hand may now be identified as embodied artifacts.

3. The most relevant passages in *Phenomenology of Perception* are pages 142–147 on motor habits and pages 151–153 on perceptual habits.

4. Merleau-Ponty, 1962, pp. 98ff. Merleau-Ponty's *schéma corporel* is translated as 'body image' in the English translation of *Phenomenology of Perception*, but I, like Tiemersma (1989) prefer the more literal translation 'body schema'. This term is also to be preferred because the body schema is, as Merleau-Pony makes clear, not an image, but a non-representational structure. Moreover, the term 'body image' is now often used in psychology in a different sense, to refer to perceptions of one's own body along social and aesthetic dimensions. Tiemersma (1989) presents a review of the philosophical, psychological, and neuroscientific literature on the notions of body schema and body image.

5. The experienced world is hence subjective, because its features are conditioned by the body schema. However, it is not a mere projection of the body schema. Instead, one may say that an 'objective' world (Kant's *noumena*) and the body schema jointly produce an experienced world, which is therefore not a pure construction of the subject. Because there are invariants in the body schema that all human beings share, there are moreover features in the world that are recognized by every human being, and that are therefore 'objective' in the sense of being independent of any one person's subjective experience. Note, moreover, that according to Merleau-Ponty, Euclidean space is an abstraction of the oriented 'space of position' of the experienced external world (Euclidean space abstracts from 'subjective' orientational features of space such as 'behind', 'near', and 'above'), which is in turn based on the 'space of situation' of the body.

6. Merleau-Ponty's 'habitude' is often translated as 'habit', but I agree with Dreyfus (1996) that 'skill' is a more accurate translation.

7. Merleau-Ponty held that every motor skill has a perceptual aspect, and every perceptual skill has a motor aspect. Nevertheless, he makes a clear distinction between perceptual and motor skills.

8. When considering examples of embodied artifacts that require modifications of ordinary navigational skills, the question is raised how malleable the body schema is, and just how far and how easily it can be extended. A person riding a bike or pulling a cart has to extend body space to include the bike or cart. Similarly, a person flying a radio controlled airplane expresses navigational skills through the airplane. But here, it seems, there is a separation between one's own body and the artifact. The airplane becomes a second reference point, next to one's own body, relative to which objects in the environment are encountered. Similar separations occur when playing video games, or when using virtual reality technology.

REFERENCES

Bermúdez, J. L., Marcel, A. J., & Eilan, N. (Eds). (1995). *The Body and the Self.* Cambridge, MA: MIT Press.

Brown, H. I. (1990). Prospective Realism. *Studies in the History and Philosophy of Science, 21*(2), 211–242.

Dreyfus, H. L., & Dreyfus, S. E. (1999). The Challenge of Merleau-Ponty's Phenomenology of Embodiment for Cognitive Science. In: G. Weiss & H. F. Haber (Eds), *Perspectives on Embodiment: The Intersections of Nature and Culture.* (pp. 103–120). New York: Routledge.

Hacking, I. (1983). *Representing and Intervening: Introductory Topics in the Philosophy of Natural Science.* New York: Cambridge University Press.

Heidegger, M. (1962). *Being and Time.* Trans. John Macquarrie and Edward Robinson. New York: Harper and Row. Originally published in German as *Sein und Zeit* (1927).

Ihde, D. (1979). *Technics and Praxis: A Philosophy of Technology.* Boston: D. Reidel.

Ihde, D. (1983). *Existential Technics.* Albany, NY: SUNY Press.

Ihde, D. (1990). *Technology and the Lifeworld: From Garden to Earth.* Bloomington, IN: Indiana University Press.

Ihde, D. (1991). *Instrumental Realism: The Interface between Philosophy of Science and Philosophy of Technology.* Bloomington, IN: Indiana University Press.

Ihde, D. (1993). *Philosophy of Technology: An Introduction.* New York: Paragon Press.

Merleau-Ponty, M. (1962). *Phenomenology of Perception.* Trans. Colin Smith. New York: Routledge. Originally published in French as *Phénoménologie de la Perception* (1945).

Norman, D. A. (1993). *Things that Make us Smart: Defending Human Attributes in the Age of the Machine.* Reading, MA: Addison Wesley.

Tiemersma, D. (1989). *Body Schema and Body Image: An Interdisciplinaryand Philosophical Study.* Amsterdam, The Netherlands: Swets and Zeitlinger.

THEORIES OF TECHNOLOGY AS EXTENSION OF HUMAN FACULTIES

Philip Brey

The idea that technology is an extension of the human organism is encountered regularly in the history of thought about technology. The idea, in its most basic form, is that technical objects extend the human organism by replicating or amplifying bodily and mental abilities. This idea is usually presented as a key to a better understanding of technology. In this chapter, I will discuss three theories of technology as extension, as developed in book-length studies by Ernst Kapp, Marshall McLuhan, and David Rothenberg. Two authors who have made closely related arguments are Henri Bergson (1911) and Arnold Gehlen (1957) – although neither of their arguments will be considered here. Instead, using the work of McLuhan, Kapp, and Rothenberg my aim is to investigate whether there is a valid and useful sense of 'extension' according to which technology can indeed be analyzed as an extension of human faculties or organs. The outcome of this investigation will be affirmative, and I will end up presenting a unified account of technology as extension of human faculties, that builds on previous accounts.

As I will argue later on, the value of a perspective of technology as extension of human faculties is twofold. First, such a perspective can serve to provide a better understanding of the evolution of technology, and its diverse forms and applications. Second, this perspective can be useful in evaluative analyses of the role of technology in society. The argument that a perspective of technology as an extension has this potential value is made in section five. The bulk of this chapter, however, is devoted to a development of the account of technology as extension of human faculties itself. In the next section, the extension theories

Metaphysics, Epistemology, and Technology, Volume, 19, pages 59–78.

ISBN: 0-7623-0681-5

of McLuhan, Kapp, and Rothenberg are reviewed. In section two, these three theories are evaluated and a unified theory of technology as extension is argued for. Section three further develops and expands this theory, and section four discusses various ways in which technical artifacts, perceived as extensions, may relate to the human organism. (By 'artifact' or 'techn(olog)ical artifact', I mean to refer to human-made tools, machines, utensils, or built structures.)

1. THEORIES OF TECHNOLOGY AS EXTENSION OF HUMAN FACULTIES

In his famous *Understanding Media*, subtitled *The Extensions of Man*, Canadian media theorist Marshall McLuhan depicted technologies as extensions of humanity:

> During the mechanical ages we had extended our bodies in space. Today, after more than a century of electric technology, we have extended our central nervous system itself in a global embrace, abolishing both space and time as far as our planet is concerned. Rapidly, we approach the final phase of the extensions of man – the technological simulation of consciousness, when the creative process will be collectively and corporately extended to the whole of human society, much as we have already extended our senses and our nerves by the various media (McLuhan, 1964, p. 19).

In his book, McLuhan time and again analyzed different technologies as extensions of humanity. However, McLuhan did so in a very informal style, and did not advance a definition of the notion of extension or a clear taxonomy of different types of extensions. Yet it is clear upon careful reading that by an extension (or, as he also calls it, a 'translation', 'repetition', or 'intensification'), McLuhan means an amplification or acceleration of existing human faculties or behaviors. As he puts it,

> It is a persistent theme of this book that all technologies are extensions of our physical and nervous system to increase power and speed. Again, unless there were such increases of power and speed, new extensions of ourselves would not occur or would be discarded (p. 91).

Subsequently, McLuhan claims that increases in speed occur in the interest of increasing power, so that, ultimately, all technologies are created in order to enhance power. In either case, all technologies are analyzed as amplifications or accelerations of functions originally performed by the unaided human organism, that take over or supplement these functions.[1]

McLuhan appears to distinguish two broad classes of extensions of the organism: *extensions of the body* and *extensions of cognitive functions*, which include functions of the senses, central nervous system, and 'consciousness'. By extensions of the body, McLuhan means extensions of parts of the human body that are used for acting on or protecting oneself from the environment, or regulating bodily functions. Specifically, they include the limbs, teeth, and bodily heat-control systems (including the skin). The senses, the central nervous system, and higher cognitive functions are not defined as parts of the body. The basic types of extensions of the body, McLuhan claims, were introduced during the mechanical age. Examples include weapons such as bows, spears, and knives, which are analyzed as extensions of hands, nails, and teeth. Likewise, clothing is seen as an extension of the skin, to extend its function of bodily heat control and protection. The wheel is seen as an extension of 'feet in rotation'. Other mechanical tools and utensils, such as jars, matches, and money, are also analyzed as technologies that extend 'storage and mobility functions'. Mechanisms, generally, are analyzed as specialized, segmented amplifications of 'bodily postures and motions'.

> What makes a mechanism is the separation and extension of separate parts of our body as hand, arm, foot, in pen, hammer, wheel. And the mechanization of a task is done by segmentation of each part of an action in a series of uniform, repeatable, and movable parts (p. 218).

Media are analyzed by McLuhan as extensions of the senses, particularly those of sight and sound. Thus the radio and telephone function as long-distance ears, and visual media, including writing and print, are extensions of the visual function. Electronic media are analyzed, moreover, as an extension of the information processing functions of the central nervous system. That is, electronic media take over functions of information management, storage, and retrieval normally performed by the central nervous system. Consequently, a human being in the electronic age is quite literally, in McLuhan's analysis, "an organism that now wears its brain outside its skull and its nerves outside its hide" (p. 64). 'Consciousness', by which McLuhan seems to refer to creative cognition and higher thought, and which he appears to see as distinct from the central nervous system, is the final form of extension McLuhan envisions. The extension of consciousness is not so much performed by conventional, analog electronic media as it is by digital computers. McLuhan envisioned an era in which human intelligence and creativity would be automated and translated into information functions performed by machines.

Almost a century before McLuhan published *Understanding Media*, German philosopher Ernst Kapp published his *Grundlinien einer Philosophie der Technik* (1877), the first work to present itself as a study in the philosophy of technology. In this work, Kapp makes an extended argument that all technical artifacts are projections of human organs, in that "humans unconsciously transfer form, function and the normal proportions of their body to the works of their hands" (Kapp, 1877, pp. v–vi, my translation). Chapter after chapter, different classes of artifacts are compared with human organs, as Kapp attempts to make plausible his claim that the artifacts under discussion are imitations of human organs. Although Kapp tends to use the term 'projection' (*Projektion*) and not (a German equivalent of) the term 'extension', he does repeatedly refer to artifacts as continuations (*Fortsetzungen*) of biological organs. He also repeatedly states that properties of biological organs are transferred (*verlegt*) to artifacts, and that many tools and machines enhance (steigern) the natural powers of the hands and arms. However, he tends to see artifacts less as supplements than as replacements (*Ersätze*) of human organs.

Perhaps the most important difference between the views of McLuhan and Kapp is that Kapp, unlike McLuhan, argues that the form of technological artifacts imitates the form of human organs. McLuhan only claims that functional properties of human organs are translated, in amplified form, to artifacts. He makes no claim that morphological properties are transferred as well. Kapp, however, claims: "Since the organ whose utility and power is to be increased is the standard, the appropriate form of a tool can only be derived from that organ" (Kapp, 1877, p. 45, my translation). According to Kapp, therefore, form follows function, and for two things to be functionally similar, they must also be morphologically similar. Consequently, artifacts must be morphological analogues of human organs. The bent finger becomes a hook, the hollow of the hand becomes a bowl, the human arm and hand become a rake, oar, or shovel, the human nerves become telegraph cables, lenses in optical instruments imitate the lens in the human eye, railway systems imitate the structure of the vascular system, and the whole human body functions as a model for the construction of machines.

At the time of Kapp's writing, mechanical technologies were still prevalent, and Kapp explicitly states that his focus is on mechanical technology. Hence, his analysis correlates with the mechanical technologies of McLuhan, but treats these much more extensively. McLuhan's analysis, in contrast, places emphasis on electronic media, that were (apart from the telegraph) still to come after Kapp. Kapp's only reference to electronic technology is, indeed, in a chapter devoted to the telegraph. There, Kapp notes, much like McLuhan does for electronic media technologies in general, the profound and 'complete' parallel between the telegraph system and the nervous system. He even claims that the correspondence between a nerve and a telegraph cable is greater than the correspondence between any other organ and its projected technology. As he claims, here the predicates 'gleichsam' (so to speak) and 'gewissermassen' (to some extent) may be dropped: the telegraph cables are quite literally the nerves of humanity, just like nerves are cable installations of the animal body. Kapp also refers to telegraph cables, as McLuhan does to electronic media in general, as 'extended nerves'. Another similarity between Kapp (1877) and McLuhan (1964) is that both see projections or extensions as not limited to tools and machine networks. Language and the state, for example, are also analyzed as extensions or *externa* of human nature.

A second important difference between the views of McLuhan and Kapp is that Kapp holds that the properties that artifacts have in common with human organs are the result of a process of projection. This is to say that human beings use their own faculties as a standard (*Massstab*) whenever they create new artifacts. Kapp holds that this process of the projection of the form and function of organs into artifacts is not normally a conscious process. Instead it is a process executed by the human unconscious, and functional and morphological parallels between organs and artifacts often become apparent only after the fact. However, even if projection is not a conscious process, it is still an intelligent process performed by human beings, which explains the functional and morphological similarities between organs and artifacts. McLuhan, in contrast, does not, or at least not explicitly, hypothesize any process of projection, but merely points out the existence of parallels between artifacts and human faculties because of which the former can be seen as extensions of the latter.

More than a century after Kapp's *Grundlinien*, and almost thirty years after *Understanding Media*, the United States philosopher David Rothenberg's *Hand's End* portrayed technology as an extension of humanity. Rothenberg argues: "Techniques can extend all those human aspects for which we possess a mechanical understanding, that is, we know how it works" (1993, p. 16). These are human faculties that can be conceived of as "a faculty, appendable by a device" (p. 16), such as our eyes, or our hands. What is not extended are faculties that cannot be conceived of as appendable or mechanizable, which are our faculties of judgment, morality, and "the sense of place."

Those human faculties that can be extended by technology, Rothenberg claims, can be categorized into two broad classes: faculties of action, and faculties of thought. Technologies are hence either extensions of action or extensions of thought (or both). Extensions of action roughly correspond to what McLuhan calls extensions of the body. They include, first of all, hand-driven tools and other devices which are "direct extensions of the motions of our body" (p. 31), particularly of manual actions. A second class is formed by technologies for transportation, that are directly controlled by us, that is, vehicles and piloted machines. These are extensions of "our bodily presence if not specifically actions of our arms, legs, hands and feet" (p. 31). A third class is constituted by 'separate machines', which are relatively fixed structures in the environment such as roads, buildings, irrigation ditches, waterwheels, and lighting fixtures, that extend "our restless need for movement" and "our struggle to survive" (p. 33).

Extensions of thought correspond to McLuhan's extensions of the senses, the central nervous system and consciousness. Rothenberg distinguishes three kinds. First, there are artifacts that improve the senses by directly extending perception, such as telescopes, microscopes, telephones, radios, communication satellites, wires, and television. Second, there are 'tools of abstraction', which are instruments that extend cognitive dexterity by extending abstract thought and language functions. These technologies include devices such as computers and calculators, but also 'immaterial' technologies such as natural languages, numerical systems, and formal languages such as those of mathematics, symbolic logic, and computer programming. Finally, there are material extensions of memory, which are material means by which experience can be encoded, stored, and retrieved. Rothenberg's prototypical examples of material memory are not devices that encode information in abstract or symbolical form, but that try to capture human experiences. These devices include storage media such as photographs, video tapes, and sound recordings, and corresponding equipment for recording and retrieval. (Although *Hand's End* contains several references to *Understanding Media*, none of these mentions its theme of technology-as-extension. *Hand's End* also contains no references to Kapp.)

Rothenberg's theory of technology as extension of humanity has a number of similarities to McLuhan's theory. Most importantly, Rothenberg, like McLuhan, construes technological artifacts as entities that extend functional properties of human faculties, but not necessarily morphological properties. Like McLuhan, Rothenberg refrains from claiming that these extensions are the result of a conscious or unconscious process of projection. Rothenberg's typology of extensions also looks a bit like McLuhan's.

There is a major difference between the theories of Rothenberg and McLuhan, however. Whereas McLuhan sees artifacts primarily as extensions of human faculties, Rothenberg sees them only secondarily so. The primary thing that artifacts are an extension of, according to Rothenberg, is human intention. Our intentions, or desires, are normally contained within our own organism. However, as we create technologies, these technologies become carriers of our intentions, and hence extensions of them. Only secondarily can we also recognize technologies as extensions of human faculties. Presumably, although Rothenberg does not say this explicitly, technologies are often extensions of human faculties because these faculties are the original bearers of our intentions. It is only to be expected, then, that as we extend our intentions, the artificial means by which they are extended tend to share functional features with our human faculties.

McLuhan, Kapp, and Rothenberg all develop their accounts of technology as extension of human faculties in order to raise larger questions about human nature, and the social, cultural, and psychological role and implications of technology. Kapp, for one, argued on the basis of his theory that technology only serves the individual and the community when it projects from human faculties. Technology that is instead only developed so as to fit into a technological system designed to serve a higher purpose ultimately limits human freedom, security, and culture.

McLuhan was also concerned with the social, cultural, and psychological implications of technological choices. He claimed that any extension of human faculties implied a 'numbing' of these faculties, or even a 'self-amputation', with great psychological and social consequences. He also believed that by relying too much on our extensions, we were becoming 'servomechanisms' to them. He warned, too, against the commercial exploitation of extensions of our faculties. "Who owns your extended eyes?" he asked, and exclaimed, 'Once we have surrendered our senses and nervous systems to the private manipulation of those who would try to benefit from taking a lease on our eyes and ears and nerves, we don't really have any rights left' (1964, p. 73). Moreover, McLuhan emphasizes the great difference between the 'old' extensions of the mechanical age, which tended to be specialized and differentiated, and the 'new' extensions of the electronic age, which tend to totalize and to translate all social, economic, and cultural processes into the form of information.

For Rothenberg, the purpose of conceiving of technology as extension of humanity is to be able to reevaluate the question of how technology disconnects us from nature, and how it can help reconnect us. He asks what we are extending ourselves for, ultimately, and whether we would be better served with different kinds of extensions. Some artifacts, such as hand-held tools, are more 'transparent' than others, he claims, in that they help us realize our intentions in a natural and effortless way, whereas others, such as roads and buildings, create their own world of order and may in this way take away our autonomy. Rothenberg emphasized that artifacts, particularly those that are not transparent to our intentions, are not neutral means that extend our ability to realize our intentions, but that also serve to modify our intentions and to generate new ones. As new means are added to our existing means, our experience of ourselves and of our world changes, we develop new priorities, and we discern new shortcomings in our existing inventory of artifacts that we resolve by introducing yet other artifacts.

For these authors as well as others, a conception of technology as extension of human faculties enables the asking of novel questions about technology and its consequences, and to treat familiar issues in a novel and unified way. These explorations only have their worth, however, if the claim that technology is an extension of human faculties can withstand scrutiny. Specifically, it has to be shown that there is a sufficiently restrictive sense of 'extension' according to which technical objects can be literally claimed to be extensions of human faculties, with no counter-examples existing. In the next section, I will evaluate the extension theories of McLuhan, Kapp, and Rothenberg according to these criteria. I will end up proposing my own theory of technology as extension of human faculties, a theory that will incorporate ideas from McLuhan and Rothenberg, as well as from some other authors.

2. EXTENSIONS OF THE ORGANISM AS EXTENSIONS OF ITS INVENTORY OF ORIGINAL MEANS

All three of the above theories of technology as extension of human faculties are suggestive, and gain support from many examples. Still, I will claim, they suffer from important flaws that makes them untenable in their current form.

First of all, Kapp's claim that artifacts are morphological projections of human organs does not stand up to closer scrutiny. Admittedly, Kapp presents an impressive range of artifact-organ pairings between which morphological similarities exist, and there are many historical examples of artifacts that were clearly inspired (even consciously) by human organs. However, there are many counterexamples. Fishing nets, books, lighters, telephones, and airplanes, for example, do not seem to have clear morphological similarities to human organs. Of course, anything is similar to anything else in some respect, and so one may try to compare a fishing net to webbed fingers, and airplane wings to outstretched arms, but I take it to be evident that for these and many other artifacts, human organs hardly served as an inspiration.

Second, although McLuhan does not presume that all artifacts are morphologically similar to human organs, he does hold that they are all functionally similar in some respect. Every artifact enhances a functional property displayed by some human organ, such as the ability to move, dig, cut, store, or perceive. An artifact is an extension of a human organ, in McLuhan's sense, in the sense that it has functional abilities also possessed by the organ and adds to, amplifies, or takes over some of its functioning. It can also be demonstrated, however, that not all technological artifacts are functional analogues of human faculties. There are many things that artifacts can do that have little similarity to anything humans do. Electric lighting, for example, is able to give light, but there is no human faculty that even gives a modest amount of light. Explosives, electromagnets, ionizers, and roads are other examples of technological artifacts that have no interesting functional resemblances to human organs or behaviors. It may still be argued that these artifacts nevertheless correspond functionally to human organs in more abstract ways. For example, ionizers may be claimed to share with human hands the ability to 'transform matter'. As one resorts to such abstract and far-fetched properties, however, the idea that artifacts are functional copies of human organs becomes increasingly hollow.

Finally, Rothenberg's theory is ambiguous. Sometimes Rothenberg seems to claim, like McLuhan, that any artifact functionally corresponds to some human organ. This has been shown to be mistaken. More often, however, Rothenberg appears to claim that what is extended by technology are not human faculties, but human intentions. However, if this is what Rothenberg means, then it appears that his theory is really not a theory of technology as extension of human faculties, but a theory of technology as extension of the human will.

It hence appears that neither Kapp, nor McLuhan, nor Rothenberg presents a theory of technology as extension of human faculties that is both coherent and defensible. Should it be concluded, then, that the idea of technology as extension of human faculties has no integrity? No. There is a notion of extension, I will now argue, that is coherent, and to which there are no counterexamples.

My notion of extension has its closest affinities with the notion of extension proposed by Rothenberg. Rothenberg has been seen to claim that all technologies are extensions of the will, or, equivalently, of human desires or intentions. As Rothenberg puts it, "when we make something, we thrust our intentions upon the world" (p. 16). Our intentions, normally only contained within our own organism, become embodied in our artifacts, and are hence extended by them. I have some difficulty making sense of the idea that intentions are things that can be 'extended' by technology or that they can be "thrusted upon the world." Nevertheless, I believe Rothenberg's analysis of technology as extension of human intention contains an important insight.

Rothenberg's idea that technology extends human intentions may be reformulated to say that technology extends the *means* by which human intentions are realized. Thus, it is not human intentions themselves that are extended. Human beings, I claim, are beings who continually try to realize their intentions, whether selfish or altruistic. As a result, they try to change the world so that it conforms to those intentions. To do this, they initially only have their bodily and mental faculties available. These are the original means by which they can realize their intentions. Technological artifacts extend, or add to, these means. This idea may be summed up in the following thesis:

> *Extension thesis:* All artifacts extend the set of naturally given means (i.e. human bodily and mental faculties) by which human intentions are realized.[2]

My extension thesis emphasizes, as do both McLuhan and Rothenberg, that artifacts build on the human body by adding functional features to its functional features. However, it differs in that it makes no claim that these added functional features must resemble, or build on, the functional features of individual organs. In this way, it also accounts for artifacts that have functional abilities not possessed by human faculties, such as the ability to give light, ionize, or magnetize.

Let me expand on this thesis. I am claiming that the human organism harbors a set of mental and bodily faculties that are intentionally employed for the realization of intentions. These faculties include the limbs, controllable muscles, sense organs, those mental faculties that are under conscious control, and any human faculty that can be controlled by the central nervous system. These faculties constitute one's inalienable set of means that can be employed to realize intentions formed by the will.[3] Thus, when one eats an apple, one uses one's arms, hands, and jaws as means, or instruments, by which the apple is eaten, and when one walks, one uses one's legs as instruments for transportation. In this way, human faculties constitute the original 'tool set' that humans use to successfully move around in their world. The body, as Lewis Mumford (1967) put it, is a "primary, all-purpose tool": "man's total equipment for life" (p. 7).[4]

One's original inventory of means, defined by one's mental and bodily faculties, may be extended by other, external means. External means are alienable means, that are not part of the standard inventory defined by our bodily and mental faculties.[5] External means serve to extend the range of actions, or tasks, that someone is capable of, either by enhancing existing capacities, or by adding genuinely novel capacities. They may then help to realize intentions that could not be realized with just one's own faculties. For example, when one is placed under an unclimbable apple tree, with an apple hanging some ten feet above, one cannot, simply by means of one's own faculties, realize an intention of grasping and eating the apple. An instrument such as a long stick, a ladder, or a chainsaw extends one's 'action horizon', and one suddenly finds oneself able to realize one's intention.

3. ELABORATIONS OF THE THEORY

Let me now elaborate this thesis of technology as extensions across three different domains: the social and cultural, the natural, and the individual versus the collective.

Social and Cultural Extensions

Reading Kapp, McLuhan, and Rothenberg, one could almost forget the many social, cultural, and symbolic functions performed by artifacts. All the technological extensions discussed by Kapp, McLuhan, and Rothenberg, concern practical uses of technology, for the extension of physical action, perception, and cognition. This neglect of the social and cultural roles of artifacts may seem to speak against extension theories. However, I will argue that the extension theory proposed in the previous section can easily accommodate the social and cultural dimensions of technology.

The extensions discussed by Kapp, McLuhan, and Rothenberg usually relate to *physical functions* of artifacts: functions that are performed in virtue of their physical capabilities. For example, a knife has the physical function of cutting, which it can only have in virtue of its physical sharpness. Likewise, binoculars enhance perception, in virtue of their optical properties.

John Searle (1995) has argued, however, that artifacts may also have powers and corresponding functions that do not derive from their physical properties but instead derive from collectively imposed *status functions*. Dollar bills, for example, are able to function as a medium of exchange not because they have an inherent physical power to perform this function, but because people have collectively assigned the function of being a medium of exchange to them. Without this assigned function, dollar bills could not function as money. Likewise, a wedding ring can be used to show that one is married only because there is a collective agreement that it has the status of symbolizing someone is married.

Functions of artifacts are hence not just physical functions, but also include status functions, that can be performed because of collective agreements about the status of an artifact that prescribe how the artifact should be interpreted and used. Status functions make it possible for artifacts to perform various social and cultural functions next to their sundry physical functions. These social and cultural functions include symbolic, moral, and religious functions, that are used to satisfy spiritual needs, such as worship, identity formation, and self-expression (e.g. a musical instrument can be used for self-expression, and a religious shrine can be used for worship). They also include social functions such as the marking of social status (e.g. a Rolex watch is not just used to tell the time, but also to signify social status).

Notice, however, that these uses of artifacts are still purposive, even if the purpose is not a practical one. When artifacts are used in their capacity of bearers of status functions, they are hence still means to ends. Thus, a statue that people kneel to and use to get into contact with the divine still serves, in a quite straightforward sense, as a means to an end. It adds to the means available to one for religious worship. Similarly, a musical instrument is a means, next to one's own bodily means, by which one can express oneself. Clothing, likewise, is more than a means to protect the body and to regulate temperature. It can also serve as a means to express one's personality and social status.

Most extension theorists, including McLuhan and Kapp, show insufficient awareness of the social and cultural dimensions of artifacts. Thus, when McLuhan analyzes artifacts as mere means to more power, he appears to stress the practical functions of artifacts and to overlook their cultural roles. Rothenberg displays more sensitivity to the social and cultural role of technology, but still classifies technological artifacts in a functional scheme that only recognizes their role as instruments for knowledge and physical action. Yet it has to be observed that artifacts have both (practical) physical functions and (social-cultural) status functions, both of which serve to extend abilities of the human organism.

It may also be noted, *contra* McLuhan, that not all artifacts designed to perform practical functions are designed to enlarge our natural abilities. Many artifacts are, instead, designed to disburden us, and to relieve us of tasks that we are perfectly well able to perform ourselves (cf. Borgmann, 1984). These artifacts can relieve us without necessarily performing better or faster than our own organs.

Organisms and Natural Objects as Extensions

So far, my emphasis has been on technical artifacts as instances of external means by which one's faculties can be extended. It is a distinguishing feature of technical artifacts that they have been designed to serve human ends. They differ in this from natural objects, such as rocks, sand, sticks, and water, as well as from humans, animals, and plants. However, any object or organism that is capable of being a means to an end can come to extend the human organism. For example, a stick or stone, picked up in the woods, can amplify one's abilities to defend oneself or to modify raw materials. Organisms other than oneself, that is, plants, animals, and other human beings, are also frequently used to extend one's abilities. The distinguishing feature of technical artifacts, as opposed to natural objects, is perhaps, that they have been intentionally designed to function as extensions.

The use of animals and human beings as extensions merits special consideration. Before the large-scale use of water mills in the eleventh century, live animals and human beings were the only reliable means available to perform labor. If people wanted some work to be done that they could not do or did not want to do themselves, their only choice was therefore to enroll the labor force of animals or other human beings. This required them to find a way by which animals or humans could be made to act in accordance to their intentions. Through training, animals were and are made into instruments that perform heavy physical labor for human beings. Animal training is essentially a method of carrots and sticks in which an animal learns to respond to cues with appropriate behavior. It is the way by which human beings impose their will on animals so as to make them useful as an extension of their own inventory of means.

Human beings also use other human beings to perform labor for them. One difference with the use of animals is that human beings may not only be used for 'dumb' tasks that mainly require physical labor, but also for intelligent tasks. They are hence able to serve as extensions of mental faculties as well as of bodily faculties. Suppose, for example, that a landlord wants to know how many pounds of rice he needs to plant in his paddy fields. If he cannot calculate this himself, he may hire an expert to do the calculation for him, someone who then functions as an extension of the landlord's own mental abilities. Subsequently, hired rice planters, functioning as extensions of the landlord's own physical abilities, may do the planting for him.

Perhaps the vastest personal extension that any human being can attain is an extension consisting of a network of human beings, animals, and artifacts that only work for *him* or *her*. For such a network to be manageable for one person, these different actors must be locked into a hierarchical system designed to enable that person to issue commands that are obeyed by it. Lewis Mumford (1967) has noted the existence of such systems in history and has called them *megamachines*. In a megamachine, one individual, an absolute monarch, controls a vast system of human beings and other 'parts' that could be put to work to build pyramids, roads, and temples, to fight wars, or to perform other specialized tasks that require a high degree of organization, in an almost corpselike obedience.

In another example, in his analysis of technical artifacts, Bruno Latour, (1992, 1995; Akrich & Latour, 1992) abolishes the very distinction between human beings and technical artifacts, since maintaining this distinction is held to obscure the fact that technological artifacts and human beings often perform similar functions. Human beings, Latour claims, often seek to 'delegate' tasks to other entities, which he calls 'actants': things that act, or that have a function. Actants may be human or non-human; this does not matter in his theory. For example, the patron of a hotel who wants to keep its front door shut can either close the door behind guests herself, try to discipline guests to perform this task, appoint a porter, or install a door closer. In the last three cases, the patron uses an extension of herself to realize her intentions. It does not matter, for her purposes, whether this extension is human or nonhuman, as long as it does the job.

Individual and Collective Extensions

So far, I have been making the simplifying assumption that artifacts function to further the intentions of single individuals, who are their users. However, many artifacts are used to further collective interests. McLuhan recognized this, and identified, next to extensions of the 'individual organism', extensions of the 'social organism' as well, that is, extensions of social groups or societies. For example, dwellings, in their function as shelter for families or groups, are analyzed by McLuhan as 'collective skins' that function as heat-control systems, and cities as extensions of bodily organs for the use of large groups (1964, pp. 117ff.). Hence, artifacts that are collectively used for a shared purpose do not extend individuals, but collectives.[6]

A single artifact also frequently serves different functions for different individuals, simultaneously extending individuals in different ways. For example, seat belts protect drivers from physical harm, but simultaneously help protect insurers from financial loss. A pair of jeans extends the skin of the wearer, but also furthers the intentions of the manufacturer to sell more of them by means of its prominently displayed logo. Artifacts may hence tie different individuals together in the way that they serve as technological extensions by which these individuals realize their intentions.

4. TECHNOLOGICAL EXTENSIONS AND THEIR RELATION TO THE BODY

At this point it is perhaps useful to consider the ways in which technological extensions may be related to the human body. One may, for instance, distinguish amplificatory/complementary extensions and embodied/separate extensions.

Amplificatory and Complementary Extensions

Not all artifacts, we have seen, functionally correspond to some human faculty. Some artifacts, such as lighters and magnets, have genuinely novel functional properties not found in any human faculties. Nevertheless, there is still a sizable collection of technical artifacts that do serve a function that is similar to that of some human faculty. A distinction may hence be made between artifacts that do not functionally extend some human faculty but introduce new functional features alongside existing ones, which I will call *complementary extensions*, and artifacts that functionally amplify some human faculty, which I will call *amplificatory extensions*. Amplificatory extensions are artifacts that have functions also possessed by some human organ and add to, amplify, or take over some of the functioning of that organ.

McLuhan believed that all technical artifacts are amplificatory extensions, but I have argued that this is not the case. McLuhan is certainly right, however, that many artifacts may be understood as means by which activities of the human organism can be performed better or faster, and that these artifacts have been designed for this reason. Such artifacts range from clothing, which amplifies the function of the skin to protect and regulate bodily heat, to telephones, which amplify hearing abilities of the human ear. The evolution of many artifacts, therefore, can be explained by reference to their amplifying or accelerating function in taking over or extending tasks normally performed by means of human faculties only.

Embodied and Separate Extensions

Amplificatory extensions, artifacts that amplify the functioning of human organs, may maintain three different types of relations with these organs: replacement, supplementation, and enhancement. An artifact may *replace* the functioning of an organ by performing the function of that organ in a way that makes the organ redundant. For example, when driving a car, one's legs are not used as a means for transportation. An artifact may also *supplement* an organ that it extends, by performing a function that the organ in question is also performing. For example, clothing adds to the protective and temperature control functions already performed by the skin.

Still other artifacts *enhance* the functional powers of the organ that they extend, not by independently performing a function that resembles the organ's function, but by cooperating with the organ in a way that enhances its activities, in this way engaging in a *symbiotic relationship* with the organ. For example, a telescope extends visual perception, not by engaging in perceptual tasks independent of the eye (as for example infra-red detectors do) but by teaming up with the eye, thus forming a functional unit consisting of telescope-plus-eye, which is more powerful than the eye alone. When a telescope is used, inception of visual perception no longer takes place when light rays hit the eye, but when light rays hit the front lens of the telescope. Similarly, a megaphone extends the voice by forming a new functional unit, voice-plus-megaphone, that yields a more powerful voice. A megaphone is therefore quite different from a speech synthesizer, which is able to produce speech independent of the human voice.

Not every symbiotic combination of artifact and organ is a functional enhancement of the organ, however. A saw, for example, performs its function in combination with the arm and hand that hold it, but cannot be conceived as an amplification of the arm and hand, because arm and hand are completely incapable of sawing. Similarly, one's hands and a piano constitute a functional unit that produces music, but it is not the case that the piano enhances already existing capacities of the hands to produce music. Saws and pianos are hence not amplificatory artifacts, whereas telescopes and megaphones are.

Artifacts that engage in symbiotic relationships with human organs, whether they amplify a human function, such as telescopes and megaphones, or form a new functional unit, such as saws and pianos, can be identified as artifacts that engage in *embodiment relations*. The notion of an embodiment relation has been introduced into the philosophy of technology by Don Ihde (1979, 1990), and can be traced back to the work of Merleau-Ponty and Heidegger.[7] Embodiment relations are relations human beings have to artifacts or technologies in which the artifact mediates one's experience of, and interaction with, one's environment. They have as a characteristic that they are not themselves objects encountered in the world, but rather, they mediate human encounters with their world. They 'withdraw' and serve as (partially) transparent means through which the world is encountered, thus engendering a partial symbiosis of oneself and it.

A blind man's cane, for example, is an artifact by which a blind man learns to perceive and navigate through his environment. When learning to use the cane, he learns to perceive and locate elements in his environment through the stick. The cane itself 'withdraws' and becomes a medium like other senses, that is not itself experienced, but through which the world is experienced and engaged. As the blind man's cane illustrates, embodied artifacts have a special relationship to the human organism, in that they are tacitly present yet immediately available, just like one's own bodily and mental faculties. They are means through which perceptual and motor abilities are expressed, rather than themselves objects of perception or action. As such, they are experienced as direct extensions of ourselves, as integral parts of our bodies. In this way they differ fundamentally from artifacts that are not embodied, like the door one opens, the nail one hammers, and the roof above one's head.[8]

Symbiosis, and hence embodiment, does not only take place with artifacts that extend action or perception. It also takes place with artifacts that extend cognition, or *cognitive artifacts*, as psychologist Donald Norman (1993) has called them. Cognitive artifacts are artifacts that are able to represent, store, retrieve, and manipulate information. They include items such as computers, calculators, abacuses, as well as pencil and paper, and books. Cognitive artifacts extend cognitive abilities, such as abstract thought, memory, problem solving, and language use. When a cognitive artifact, such as a calculator or computer, engages in an embodiment relation with a user, it becomes a medium through which representations (in the form of symbols or icons) are perceived and manipulated. The user of a calculator, for example, performs cognitive tasks in symbiosis with this device. The calculation is not performed solely by the user (who relies on the calculator), nor by the calculator (as it has to be operated by a skilled user who knows what operations to perform in what order). It is, instead, performed by a new symbiotic unit, a system composed of two mutually dependent functional units, human and machine.

5. CONCLUSION

An analysis of human faculties as means for realizing intentions, and of technical artifact as extensions of the bodily inventory of means provides, I will now argue, a useful perspective from which to explain and evaluate a number of facts about human culture and technology. What this perspective may help explain, to begin with, is the evolutionary trajectory of technology in its diverse forms. Technical artifacts are developed and used to extend the 'natural instruments' of the human body, and when designing artifacts, people will 'build on' these faculties. This implies that artifacts are normally only developed when they are perceived to be a useful addition to, or substitution for, capacities already possessed by the body (along with extensions that are already available). New technologies can therefore be understood by analyzing how they help realize human intentions by extending and building on the existing inventory of human means, including human faculties (at their current state of development) and other extensions already present.

The perspective of technology as extension of human faculties also serves, I have claimed, to put the social, cultural, and psychological role of technology in a new perspective. I have already noted some of the issues that concerned Kapp, McLuhan, and Rothenberg. I will now summarize and expand on some of their main concerns. First, the perspective of technology as extension of humanity raises, for Rothenberg, a question of what we are extending ourselves for. Upon closer analysis, it may be discovered that many artifacts extend us in ways that are incompatible with our most fundamental ideals. So a question raised by the extension perspective is: Do the artifacts that we have created and that we use extend us in a way that is compatible with what we desire most?

A related set of issues, raised by McLuhan, concerns matters of liberty, justice, and social equity. How are the intentions of different members of society represented by the technologies in a society, or to what extent do they fail to be represented? As functions that used to be performed by one's mental and bodily faculties are taken over by machines, what does this imply for one's personal autonomy? To what extent are we becoming 'servo-mechanisms', in McLuhan's terms, that have to subject to the operating conditions imposed by our technologies?

A third set of issues concerns the implications for our own mental and bodily faculties that their 'extension' has. Since many functions that used to be performed by humans are taken over by machines, what functions do our bodily and mental capacities still serve, apart from operating machinery? Is there any truth to McLuhan's claim that new extensions of human faculties have great psychological (and physiological) consequences, and that they put the relation of our faculties to each other in disarray? Ivan Illich (1973) and Albert Borgmann (1984), among others, have argued that contemporary technology reduces human beings to passivity, and does not engage the body. Is this indeed true, and if so, how can technology be designed in a way that does engage the body?

A fourth and final set of issues concerns the consequences for self-understanding of technologies that substitute for human functions. As we "re-create ourselves in technology," how does technology re-create us? As pointed out, Rothenberg holds that technology can only extend human faculties that may be understood in mechanical terms, and not our most fundamental faculties of judgment and morality. So technology only shows us 'mirrors' of quite limited aspects of ourselves. Today one often has a choice about whether to have a task performed by a machine or by a human being. Could this lead to an understanding of human beings in 'technomorphic' terms (and of machines in anthropomorphic terms)?

Finally, I hope to have shown that a conception of technology as the extension of human faculties of the sort outlined is both theoretically and empirically valid, and that such a theory provides a starting point from which many old and new issues concerning technology and its implications can be approached in a unified way.

ACKNOWLEDGMENTS

I would like to thank Hans Achterhuis, Pieter Tijmes, and Peter Paul Verbeek for their comments on an earlier draft.

NOTES

1. Strictly speaking, McLuhan distinguishes two types of extensions: extensions of the 'individual organism' and extensions of the 'social organism'. See below, section 3 of this chapter.

2. Because human intentions are the controlling factor in my analysis of technology as extension of the human organism, my analysis distinguishes itself from a biological, or evolutionary, theory of technology as an extension of the human organism, according to which artifacts are to be explained as biological adaptations. Dawkins (1982), who presents such a theory, analyzes the evolution of both organisms and artifacts as determined by selection processes defined over genes. In Dawkins' theory, genes exercise their 'will' to survive and replicate by means of their host body as well as by means of artifacts and the bodies of other organisms that serve as 'extensions' of the organism, or phenotype.

3. I use the somewhat archaic term 'will' to refer to the mental faculty for deliberate action. The will forms desires or intentions, and tries to implement them by deciding on appropriate courses of action. Thus, it is because of one's will that one chooses to eat a sandwich, to focus one's gaze on some object, to quit one's job, or to execute other voluntary actions. Note that the will does not extend to involuntary actions, such as reflexes, bodily processes such as digestion, and the many cognitive processes, such as processes of recognition and association that are normally executed automatically. There may be no strict boundary, however, between voluntary and involuntary actions.

4. Note that the functional capacities of the 'tools' that are included in one's original 'tool set' are not just determined by their physical properties, but also by the skills possessed by the corresponding tool user.

5. Prostheses replace, rather than add to, instruments in one's original 'tool set'. Prostheses may be (semi-)permanently attached to one's body, in which case they become inalienable instruments. My focus in this essay, however, is not on prostheses but on instruments that remain external to the body.

6. Organizations, conceived of as purposive entities, also use artifacts to further their interests. Organizations are not collectives, because they are more than the sum of their members. Because organizations have no bodies, however, it may make no sense to label such artifacts as extensions of these organizations; the artifacts employed by an organization are merely additions to the set of means by which the organization operates.

7. Martin Heidegger (1962, p. 99) may have been the first to recognize embodiment relations. Merleau-Ponty, 1962 (pp. 142–147 and 151–153) contains a more extended treatment of them.

8. It appears that the distinctions drawn here – those between mere 'stuff' in one's environment, embodied extensions of one's faculties, and extensions that are not embodied – were also recognized by Heidegger (1962). Heidegger called objects that were only recognized as mere objects 'present-at-hand' (*vorhanden*). Objects that are

understood as means to one's ends he called 'ready-to-hand' (*zuhanden*). Objects that are ready to hand so closely that they 'withdraw' (*zurückziehen*) he called 'proximally ready-to-hand' (*zunächst zuhanden*) (1962, p. 99).

REFERENCES

Akrich, M., & Latour, B. (1992). A Summary of a Convenient Vocabulary for the Semiotics of Human and Nonhuman Assemblies. In: W. Bijker & J. Law (Eds), *Shaping Technology/ Building Society: Studies in Sociotechnical Change*. (pp. 259–264). Cambridge, MA: MIT Press.

Bergson, H. (1907). *L'Evolution créatrice*. Paris: F. Alcan. English trans. Arthur Mitchell, Creative Evolution. New York: Henry Holt, 1911.

Borgmann, A. (1984). *Technology and the Character of Contemporary Life*. Chicago: University of Chicago Press.

Dawkins, R. (1982). *The Extended Phenotype: The Gene as the Unit of Selection*. Oxford, England: W. H. Freeman.

Gehlen, A. (1957). *Die Seele im Technischen Zeitalter: Sozialpsychologische Probleme in der Industriellen Gesellschaft*. Hamburg: Rowohlt. English trans. Patricia Lipscomb, *Man in the Age of Technology*. New York: Columbia University Press, 1980.

Heidegger, M. (1962). *Being and Time*. Trans. John Macquarrie and Edward Robinson. New York: Harper and Row. Originally published as *Sein und Zeit* (1927).

Ihde, D. (1979). *Technics and Praxis: A Philosophy of Technology*. Boston: D. Reidel.

Ihde, D. (1990). *Technology and the Lifeworld: From Garden to Earth*. Bloomington, IN: Indiana University Press.

Illich, I. (1973). *Tools for Conviviality*. New York: Harper and Row.

Kapp, E. (1877). *Grundlinien einer Philosophie der Technik: Zur Entstehungsgeschichte der Cultur aus Neuen Gesichtpunkten*. Braunschweig, Germany: Westermann.

Latour, B. (1992). Where Are the Missing Masses? The Sociology of a Few Mundane Artifacts. In: W. Bijker & J. Law (Eds), *Shaping Technology/Building Society: Studies in Sociotechnical Change*. (pp. 225–258). Cambridge, MA: MIT Press.

Latour, B. (1995). A Door Must Be Either Open or Shut: A LittlePhilosophy of Techniques. In: A. Feenberg & A. Hannay (Eds) and trans. C. Cussins, *Technology and the Politics of Knowledge*. (pp. 272–281). Bloomington, IN: Indiana University Press.

McLuhan, M. (1964). *Understanding Media: The Extensions of Man*. New York: McGraw-Hill.

Merleau-Ponty, M. (1962). *Phenomenology of Perception*. Trans. Colin Smith. New York: Routledge. Originally published as Phénoménologie de la Perception (1945).

Mumford, L. (1967). *Technics and Human Development: The Myth of the Machine*. Vol. 1. New York: Harcourt Brace Jovanovich.

Norman, D. A. (1993). *Things That Make Us Smart: Defending Human Attributes in the Age of the Machine*. Reading, MA: Addison Wesley.

Rothenberg, D. (1993). *Hand's End: Technology and the Limits of Nature*. Berkeley, CA: University of California Press.

Searle, J. (1995). *The Construction of Social Reality*. Cambridge, MA: MIT Press/Bradford Books.

HERMENEUTICS AND THE PHENOMENON OF INFORMATION

Rafael Capurro

In the preface to *Philosophy and Technology II: Information Technology and Computers in Theory and Practice*, Robert Cohen and Marx Wartofsky consider two questions to be at the heart of philosophical thinking about modern information technology: What is information? and What is the relation between computer calculation and human reasoning? They also point toward a third issue – one which in my opinion is central and underlies the other two – namely, the interaction between society and information systems.[1]

My contribution to the first question is a tentative proposal utilizing recent discussions of the concept of 'modernity'. I take a restrictive view of the second question by analyzing the hermeneutical components of the interaction between an inquirer or user and data bases. In regard to the last issue, I contrast the whole field of the technological instrumentation of information with a non-instrumental but at the same time highly 'productive' view of human 'logos'. I will argue that the hermeneutical dimension of language allows not only a critical view of what I call the 'information Ge-stell', but also gives us some hints about the fundamental potentialities of language which remain hidden when the information phenomenon blinds our eyes.

INFORMATION AND MODERNITY

What is information? Information is the shape of knowledge at the end of modernity. Some characteristics of the end of modernity are: (a) abandonment of the primacy of rational or scientific thought as qualitatively superior to all other types of discourse; (b) abandonment of the idea of human subjectivity as

Metaphysics, Epistemology, and Technology, Volume, 19, pages 79–85.

ISBN: 0–7623–0681–5

opposed to objectivity, in which intersubjectivity and contextuality play only minor roles; and (c) abandonment of the (Platonic) idea of human knowledge as something separate from the knower.

All these positions, especially the third, were core discoveries of Edmund Husserl. They are also fundamental characteristics of the phenomenon of information, so that information can be described as the shape of knowledge at the end of modernity.

First, with regard to the abandonment of the primacy of scientific rationality, information is admitted to be fragmentary, to come in pieces. The *fragmentation* is two-fold: in reducing knowledge to pieces, the original contextuality disappears or becomes tacit. Knowledge becomes, literally, partial, dependent on prejudices or on the knower's frame of reference. This relativity of knowledge to a changing horizon of interpretation also brings to the fore of epistemology a new category: that of truth as now, at the end of modernity, inseparable from that of relevance.

Second, with regard to the abandonment of the subjectivity-objectivity opposition, information is described as having a certain *commonness.* Information is something basically human which, in principle, should be accessible to everyone. Modern knowledge is something common, shared by a community, in the first instance by the scientific community.

Third, with regard to the abandonment of the idea that knowledge is something separate from the knower, there is the notion of *mediation.* Modern information technology disseminates all kinds of knowledge all the time to everyone in a way prefigured by printing. Information becomes part and parcel of media, becomes a medium.

In sum, the characteristics of fragmentation, commonness, and mediation point toward the nature of information, its present shape at the end of modernity.

THE HERMENEUTICAL PROCESS OF INFORMATION STORAGE AND RETRIEVAL

In setting up (say) a bibliographic database, the fragmentation of information forces us to create the conditions of possibility for the retrieval of the pieces because their common context remains tacit. The partialization opens the possibility for different perspectives of interpretation.

This situation can be described in terms similar to those used by Martin Heidegger to analyze the structure of understanding: the general conceptual background (*Vorhabe*), the specific viewpoint (*Vorsicht*), and the corresponding terminology (*Vorgriff*). In the same way, a database creator must pre-define the field of knowledge, which is usually thematized through a classification scheme. The terminology of the field is objectivized in a thesaurus. Bibliographic description, abstract, index terms (or descriptors), and classification codes are the document surrogates that can be searched, taking into account the objectivized pre-understanding (thesaurus, classification scheme, etc.) of a user community. Once the database is implemented in the system we have, on the one side, the objectivized pre-understanding and, on the other, the interpreter or inquirer.

According to existential hermeneutics, a human being is not an isolated inquirer trying to reach others or the outside world from his or her encapsulated mind/brain, but is already sharing the world with others. Within this world of open possibilities, a person meets others and shares things and concerns, developing in its variety and complexity what Hannah Arendt calls the "web of human relationships" (Arendt, 1970, pp. 183–184). With the fixation of a part of the community background in some database, the inquirer or searcher can match his or her questions against it. Here Gadamer's "fusion of horizons" (Gadamer, 1975) describes not only the situation of text interpretation in general but also and even more accurately the dynamic process occurring during online dialog.

We can consider the process of the storage and retrieval of information hermeneutically as the articulation of the relationship between the existential world-openness of the inquirer, his/her different open and socially shared horizons of pre-understanding, and the fixed horizon of the system. The information-seeking process is basically an interpretation process having to do with the (life-) context and the background of the inquirer, and with that of the people who store different kinds of linguistic expressions having a meaning within fixed contexts of understanding (as, for instance, thesauri, keywords, and classification schemes).

With the fixation of a part of a community background in a system, the inquirer can match his/her questions and backgrounds of pre-understanding against it. The online dialogue is a specific form of the *hermeneutic circle*. In contrast to a *vicious circle*, the concept of the hermeneutic circle refers to the process of understanding as a dynamic process of *fusion of horizons* (Gadamer, 1975) in which statements are considered as answers to questions. Questions arise within a pre-understanding which is itself the result of having asked questions, and so on. Thus the dynamic of the interpretation process has its roots in a Socratic attitude of questioning. This attitude is existentially grounded in the fact that we are embedded into an already structured world (historical situation, culture, language, and so on), being confronted at the same time with an open field of possibilities, i.e. with an open frame or a horizon.

The inquirer's pre-understanding is embedded into a community's pre-understanding, which is itself part of the web of interrelations of things or matters that in their openness and finitude arise within the shared world-openness itself. If we call this world-openness *Pre-Understanding* (PU), as the ontological condition of possibility to the ontic *pre-understandings (pu)*, then the basic difference stated by this theory is that *PU* cannot be reduced to *pu*, which is what mentalism in fact postulates. Of course, the difference between *PU* and *pu* does not preclude the possibility of integrating dynamic learning patterns into retrieval systems (Doszkocs, Reggia & Lin, 1990). This can be visualized as follows:

(open horizons) *INQUIRER* < – – – > *SYSTEM* (fixed horizons)

Some consequences with regard to the understanding and design of information systems are:

- In setting up, say, a bibliographic database, the fragmentation of information forces us to create the conditions of possibility for the retrieval of the pieces. We need conceptual backgrounds, for instance the scope of a data base, specific viewpoints (classification schemes), and a terminology. The result is an objectivized or fixed pre-understanding. These backgrounds belong to historical, cultural, linguistic . . . situations. There is no *knowledge in itself.*
- Users are not isolated minds with cognitive structures, but are bodily human beings sharing a theoretical and practical pre-understanding with, for instance, professional communities. There are no *users in general.*
- The question of relevance relates to the different horizons of pre-understanding. The hermeneutical paradigm offers a framework for the foundation of different relevance criteria such as systems relevance and individual relevance or pertinence (Lancaster, 1979; Salton & McGill, 1983). In fact, this distinction is not enough. According to T. J. Froehlich (1994), hermeneutics can provide a more productive framework for modelling systems and user criteria. This framework should include a hermeneutic of users, of the information collection, and of the mediation through the system. This is in accordance with what I proposed in my book *Hermeneutik der Fachinformation* (Capurro, 1986).

Information systems are embedded in different cultural contexts. The study of information processes includes rhetorical, ethical, and political questions. Information science can be conceived as a rhetorical discipline (Capurro, 1992). Information scientists such as N. J. Belkin, Robert N. Oddy, H. M. Brooks, and Peter Ingwersen (Ingwersen, 1992) have developed a cognitive paradigm that conceives the information retrieval process as an interpretation process, where the requester's knowledge structures actively interact with the system (Capurro, 1985; Allen, 1991). Bernd Frohmann (1990, 1992) and Blair (1990) have made different objections to this paradigm. Existential hermeneutics can also provide an antidote to mentalism in information science.

I agree with Don Swanson's postulate on the "future of an illusion" as he states:

> An information need cannot be fully expressed as a search request that is independent of innumerable presuppositions of context – context that itself is impossible to describe fully, for it includes among other things the requester's own background of knowledge (Swanson, 1988, p. 95).

It is, of course, not only his/her background of knowledge that cannot be described fully, but the very fact of the requester's *being-in-the-world*. What we do when we retrieve information is, in fact, to interpret it not only intellectually but existentially. Information retrieval is, as Swanson remarks, a misleading metaphor. As Karen Sparck Jones (1991) points out, the role of artificial intelligence in information retrieval in the foreseeable future will not be "to replace humans by machines" (Sparck Jones, 1991, p. 563), but to support different kinds of natural language manipulation, operating only at the linguistic level, without being based directly on 'world knowledge' or, as we would hermeneutically say, without operating on an existential basis. 'Information-as-process' as well as 'information-as-thing' (documents) are, in other words, situational (Buckland, 1991, p. 50; Cornelius, 1996).

LANGUAGE, KNOWLEDGE, AND THE INFORMATION GE-STELL

Information technology is able to help us become more human if we make joint efforts to investigate its presuppositions in all their complexity. This historical reflection in its philosophical dimensions is the task of hermeneutic phenomenology. Let me now try to illuminate this topic, reflecting on the potentialities of human logos.

According to Heidegger, modern technology is two-sided: as a *techne* it partakes of *poesis* and brings something forth into unconcealment, but at the same time it crystallizes into the instrumental structure of the *Ge-stell*.[2] Instrumentality is good, provided it does not degenerate into a totalitarian or one-sided view. From this perspective, the development of information technology at the end of modernity is the creation of an *information Ge-stell*.

Whereas, on the one hand, we bring forth linguistically mediated knowledge in a new shape, on the other, we transform language into a mere instrument.

Yet even when this happens, as I have argued in the previous section, the process of interpretation is needed for the constitution of meaning. In fact, written as well as spoken *logos* never comes to an end, can never be definitively fixed once and for all. It conceals itself in its re-presentations. Modern subjectivity does not pay attention to this concealment while transforming the event of information, its weakness or dependence on interpretation, into an information and/or knowledge *establishment*. In this way it gives up its ethical responsibility, hoping to rest on a strong or fixed structure (Capurro, 1996).

Nevertheless the *information Ge-stell* is an opportunity for modernity to recuperate in one of its characteristic formations the hidden dimension of language. The *information Ge-stell* can become a voice within the polyphonic nature of human *logos* – if and only if it is interrelated to the whole range of its hidden potentialities. If it is not, then we will have no more than an information society. The key issue in today's knowledge society is our relation to what we do not know in and through what we believe to know. To do this in a digital environment is one of the major challenges of today's networked environment, where the partiality of knowledge is the strength of a decentralized, non-totalitarian, and opaque structure we call the Internet. What we get is not a fully enlightened or transparent society, but an opaque one, where the perspectives are continuously undermined by chaos and creativity (Vattimo, 1989; Capurro, 1995).[3]

NOTES

1. Robert S. Cohen & Marx W. Wartofsky, 'Editorial Preface', in Mitcham & Huning, eds., 1986, pp. v–vi. Paraphrasing from p. vi.

2. Heidegger, 1967, 'Die Frage nach der Technik', pp. 5–36.

3. The origin of this paper goes back to the International Conference 'Phenomenology and Technology' held at the Philosophy and Technology Studies Center, Polytechnic University (New York), October 2–4, 1986, which was organized by Wolfgang Schirmacher and Carl Mitcham. After thirteen years, obviously, things have changed and I have done some further work, too. My book *Hermeneutik der Fachinformation* was published in 1986, and since then I have written some articles on this subject, as well as another book, *Leben im Informationszeitalter* (1995). Some of these articles, as well as a list of publications, can be found in my homepage (http://www.capurro.de). The present text is an enriched version of the original one. I have added some later insights without changing the basic ideas, which I still think are valuable and can also be of help when reflecting, for instance, about the nature of communicating and searching for information in the Internet.

REFERENCES

Allen, B. L. (1991). Cognitive Research in Information Science: Implications for Design. *Annual Review of Information Science and Technology, 26*, 1–37.

Arendt, H. (1970). *The Human Condition* (6th ed). Chicago: University of Chicago Press.

Blair, D. C. (1990). *Language and Representation in Information Retrieval*. Amsterdam, Netherlands: Elsevier.

Buckland, M. K. (1991). *Information and Information Systems*. New York: Greenwood Press.

Capurro, R. (1985). *Epistemology and Information Science*. Stockholm, Sweden: Royal Institute of Technology Library. Report TRITA-LIB–6023.

Capurro, R. (1986). *Hermeneutik der Fachinformation*. Freiburg, Germany: Alber.

Capurro, R. (1992). What Is Information Science For? A Philosophical Reflection. In: P. Vakkari & B. Cronin (Eds), *Conceptions of Library and Information Science: Historical, Empirical, and Theoretical Perspectives*. (pp. 82–96). London, England: Taylor Graham.

Capurro, R. (1995). *Leben im Informationszeitalter*. Berlin: Akademie.

Capurro, R. (1996). Information Technology and Technologies of the Self. *Journal of Information Ethics, 5*(2), 19–28.

Cornelius, I. (1996). Information and Interpretation. In: P. Ingwersen & N. O. Pors (Eds), *CoLIS 2: Second International Conference on Conceptions of Library and Information Science: Integration in Perspective, October 13–16, 1996*. (pp. 11–21). Copenhagen, Denmark: Royal School of Librarianship,

Doszkocs, T. E., Reggia, T., & Lin, X. (1990). Connectionist Models and Information Retrieval. *Annual Review of Information Science and Technology, 25*, 209–260.

Froehlich, T. J. (1994). Relevance Reconsidered – Towards an Agenda for the 21st Century: Introduction to Special Topic Issue on Relevance Research. *Journal of the American Society for Information Science, 45*(3), 124–134.

Frohmann, B. (1990). Rules of Indexing: A Critique of Mentalism in Information Retrieval Theory'. *Journal of Documentation, 46*(2), 81–101.

Frohmann, B. (1992). Knowledge and Power in Library and Information Science. In: P. Vakkari & B. Cronin (Eds), *Conceptions of Library and Information Science: Historical, Empirical, and Theoretical Perspectives*. (pp. 135–149). London, England: Taylor Graham.

Gadamer, H.-G. (1975). *Wahrheit und Methode: Grundzüge einer philosophischen Hermeneutik*. Tübingen, Germany: Mohr.

Heidegger, M. (1967). *Vorträge und Aufsätze* (3rd ed.). Pfüllingen, Germany: Neske.

Ingwersen, P. (1992). *Information Retrieval Interaction*. London, England: Taylor Graham.

Lancaster, F. W. (1979). *Information Retrieval Systems*. New York: Wiley.

Mitcham, C., & Huning, A. (Eds) (1986). *Philosophy and Technology II: Information Technology and Computers in Theory and Practice*. Studies in the Philosophy of Science, Vol. 90. Boston: D. Reidel.

Salton, G., & McGill, M. J. (1983). *Introduction to Modern Information Retrieval*. New York: McGraw-Hill.

Sparck J. K. (1991). The Role of Artificial Intelligence in Information Retrieval. *Journal of the American Society for Information Science, 42*(8), 558–565.

Swanson, D. R. (1988). Historical Note: Information Retrieval and the Future of an Illusion. *Journal of the American Society for Information Science, 39*(2), 92–98.

Vattimo, G. (1989). *La societá trasparente*. Milano, Italy: Garzanti.

TOWARD A PHILOSOPHICAL ANALYSIS OF EFFICIENCY

José Manuel de Cózar

This chapter attempts to pursue a philosophical analysis of efficiency. Beginning with a distinction between two discourses about efficiency, it identifies the epistemological and methodological problems of unicity and applicability, respectively. After an extended examination of each of these problems in turn, the paper argues for a general reconsideration of efficiency.

TWO PROBLEMS OF EFFICIENCY

Efficiency, the technological goal par excellence, is a dominant value in contemporary societies. Any analysis of efficiency reveals a basic dichotomy between: (a) standard engineering discourse involving problem definition (delimitation) and solution, and (b) a philosophical approach based on 'problem widening' and choice. Regarding the practical, often dramatic consequences, of technological problems, analyzing the idea of efficiency is not only of philosophical interest, but also a useful way of addressing questions concerning techno-economic policy.

In view of this, consider the following case. A mine generates a large amount of toxic waste. An engineer 'solves' the problem by designing a dam where the waste is dumped as mud. From the engineer's perspective, the dam is the right solution because the problem has been set up in a very circumscribed way (pollution within the mining boundaries) – which separates it, metaphorically and literally, from the wider context, and sets the criteria for an optimal solution: to get rid of the waste by confining it in a reduced area and with the lowest economic cost.

Metaphysics, Epistemology, and Technology, Volume, 19, pages 87–100.

ISBN: 0–7623–0681–5

When the matter is examined from a wider point of view, however, it is possible to raise doubts about the problem setting and the problem solution. The engineer, from this new approach, has not solved the problem, but has put in danger the surrounding environment and its inhabitants – human and non-human – together with the local economy. The problem needs to be widened, since the environment around the mine has an important ecological value. What if the dam one day breaks, causing a completely uncontrolled dumping of waste? This would bring about a cascade of serious, far-reaching, heterogeneous effects, both spacial and temporal.

What the engineer did, together with the company managers and the bureaucrats or politicians who issued the licenses, was to make a decision by which the complexity and diversity of the risks to the security, health, and preservation of the natural environment and the local economy were ignored or, at most, drastically circumscribed. The problem setting prevented the consideration – or led to the discarding of – other alternatives, such as waste treatment neutralization, or even closing down the mine. In the end, the most efficient solution could turn out to be the worst.

This hypothetical case illustrates fairly well the dilemma of efficiency contextuality.[1] The dilemma is not always present in public discussions concerning technology, but it is real, and becomes more difficult to deal with as the contextual complexities of a techno-scientific problem grow. At the very core of the technical we find an attempt to neatly specify conditions in order to work out a model and, in so doing, to obtain an optimal solution. What is revealed is that for virtually every relevant social problem faced technologically, the solution is in some way being chosen in advance, by delimiting the initial conditions and criteria. In practice, it is often impossible to claim seriously that the most efficient solution has been found. The *modus operandi* of technological 'problem solving' involves both considerable power but latent weakness as well.

There are two major questions involved in the analysis of efficiency from a philosophical perspective. First, is efficiency an univocal concept, despite its various formulations (in the natural sciences, the different branches of technology, economic theory and methods, management and policy)? Second, is its application in search of solutions and in evaluating artifacts, projects, and situations subject to any epistemological and methodological obstacles? In what follows, these questions will be referred to as the unicity problem and the applicability problem, respectively.

The problems are related, thus preventing any completely isolated understanding of them. If the concept of efficiency is univocal, some general conditions must be met in every case, and the fulfilling of the conditions gives rise to difficulties affecting the success of the methods employed to solve a problem efficiently. Conversely, the several available methods can be regarded as different concretions, as alternative readings, of the supposed single concept. In fact, what we frequently meet is a kind of dialectic of mutual reinforcement between concept and methods in every successful application.[2]

THE UNICITY PROBLEM

The most general way of approaching the concept of efficiency is to define it as a ratio of results (already obtained or estimated) to resources (used or usable). In systems terminology, the relation is said to hold between 'outputs' and 'inputs'. This ratio is maximally efficient if, given some parametric or stipulated conditions,

(1) the maximal amount of result(s) from a given resource obtains, or if
(2) given a previously specified result (objective), the minimal amount of resource(s) to reach the result is used.

This way of defining efficiency has its roots in what is, historically speaking, the first technical concept of efficiency: engine performance or, more broadly considered, thermodynamic efficiency. The performance of a thermal engine was defined as a ratio of the useful work obtained to the energy (heat) used. At best, the maximal amount of energy obtained would be the same as the energy consumed by a technological system.

Now if we subscribe to a classic analytical approach, we could claim that thermodynamic efficiency is just a variety of efficiency, a consequence of meeting two conditions. First, results and resources must be expressed in magnitudes referring to the amount of energy or translatable into it. Second, and more interesting, the quantity of the result must be, as a maximum, equal to the quantity of the resource. In other words, the ratio of results to resources cannot be more than one, the interval ranging between 0 and 1. This holds for every kind of efficiency, because, as a matter of convention, maximal efficiency is set as 1. However, from a strictly mathematical point of view, the interval goes from 0 to the infinite. A physical constriction (the second law of thermodynamics) makes it impossible to surpass 1 in energy efficiency.

Two comments:

The first comment is historical. The concept of thermodynamic efficiency was originally restricted to engine performance. From there, it was appropriated by economic theory and other academic fields, eventually moving into the political and public domains. Economists saw themselves as engineers devoted to promoting social welfare, thanks to improvements in the management of scarce resources, in the same way as engineers did their best to find economical solutions. Influential economists and engineers disseminated the idea of efficiency as a central value of universal application.[3] Does this mean that the concept of efficiency has remained the same during its spread into wider areas? We will address this question later, comparing the use of the term 'efficiency' in economic, technological, and social settings.

The second comment is more analytic. The fact that in thermodynamics one never obtains more output than input allows, in this context, an interchangeable use of the words 'efficiency' and 'efficacy' (or even 'effectiveness'). An object, action, or system is efficacious (effective) insofar as the fixed objectives are obtained. It is a ratio of attained results to aimed objectives (R/O). Now, from an energy point of view, the aimed objective coincides with the input, that is, the rendered maximal amount of energy is equal to the energy taken up.[4]

Nevertheless, in the present analysis effectiveness and efficiency are not necessarily associated, because what we find in the former is a comparison of results with objectives, whereas efficiency compares results with resources. It is possible to reach a predetermined objective, but by squandering resources. The contrary also holds: despite an efficient management of resources, a failure to attain some level required to achieve given objectives. In addition, in the ratio expressing efficiency, the numerator is valid for objectives, meaning estimated results, as well as for achieved results. As far as the denominator is concerned, if the resources are fixed, they can yield results not explicitly foreseen.[5]

There are interesting differences between a purely mathematical formulation of efficiency and its evaluation in real-world situations. Efficiency is directly proportional to results and inversely proportional to resources. This means that efficiency tends to the infinite when the results tend to the infinite and resources tend to zero. For evident reasons, it does not make a lot of sense to set out the concept of efficiency in this way. Every measurement procedure has to be empirically reasonable. In practice, an infinite resource or an infinite result is not conceivable.

Posing the question in a slightly different way, efficiency is a concept expressing a relation, and is usable as a criterion to evaluate objects, actions, and systems. It is important to notice, however, that the concept is associated with a set of methods to improve the efficiency of such objects, actions, and systems. For this reason, it does not take us very far to conceptualize efficiency as the search for the maximal quantity of result using the minimal quantity of resources. This would lead to the search for an infinite result without using any means at all. There are two options: to calculate (and search for) the efficiency of some resources to a set result, or to estimate (and search for) the efficiency in the attainment of results from the available resources. These options are called 'output efficiency' and 'input efficiency', respectively. Even if the mathematical model is the same, what is undoubtedly different – as we shall see later – is both the understanding of the involved concepts and the practical consequences of their application.

We have yet to determine whether there is a univocal concept of efficiency with several variations or if what we really find are different concepts. A quick survey of appearances in databases and the media of the term 'efficiency' shows a wide array of uses, involving several heterogeneous fields: nuclear physics, energy technology, agriculture, health, business management, public administration, and even academic and personal performance. We encounter a tension between the disparity of uses and a clear 'family resemblance'. On the one hand, a pragmatically inspired approach might argue that the meaning of a technical concept should be found in its paradigmatic applications, the procedures associated with its uses or, at least, the consequences of its use for experience. In other words, there would be different concepts of efficiency. On the other, one might propose the satisfaction of a few basic conditions – say, a ratio of results (inputs) to resources (outputs), and a scale to consistently order the values so obtained, and thus be able to compare them. In other words, there would be a single concept of efficiency.

There is no hard and fast solution to the unicity problem. It depends on choosing between pragmatic and analytic criteria. The point, nevertheless, is that defending a univocal concept of efficiency makes sense only to the extent that the concept is defined in a highly abstract, quasi-vacuous sense. Virtually everything may be conceptualized as a result or as a resource – as output or as input. (Not everything can be meaningfully translated into a numerical value and mathematically manipulated, but there are sophisticated attempts to do so.) In any case, in real life situations, the general definition of efficiency must be illuminated by a careful scrutiny of its applications. If not, it is always possible to take advantage of an abstract and quasi-vacuous definition to promote the concealing impression of unanimous agreement among all individuals and perspectives involved in a technological project.

THE APPLICABILITY PROBLEM

Efficiency is basically a concept with a comparative role. It allows us to compare objects (designs, situations, etc.), no matter whether they actually exist, are related to the past or future, or are ideal. The ability to compare has not only a descriptive, but also an evaluative dimension.

Comparison demands a proper scale to set in order (of preference) the considered solutions. To preserve consistency and avoid serious application problems, a well-established comparative function is needed. Otherwise, we shall have to face indeterminate or incomplete situations.[6] (A third kind of problem, uncertainty, can emerge even though the 'utility function' is correctly established.)[7]

One reason why the concept of efficiency is so attractive to so many people has to do with the fact that the mathematical solutions obtained are arranged in a scale ranging from 0 to 1. In this way, comparison between alternatives seems *prima facie automatic*. The calculation of efficiency is contemplated as a maximization calculation. Usually, this implies the aggregation of several results, on the one hand, and several resources, on the other. Therefore, numerical units expressing quantities are needed. In other words, although comparability does not depend on quantitative measures, efficiency determination turns almost immediately to a question involving quantitative calculations.

Among the well-known difficulties discussed in the technical literature are the problems raised by metrization, the task of translating the relevant elements into measurement.[8] These difficulties do not arise when the problem setting concerns well-established physical magnitudes, such as mass or speed. Thermodynamic performance falls into this group, because measurements in this field are carried out using energy units. Compare this with cases of metrization in the social sciences such as, for example, intelligence determination. This example and others in social engineering vividly remind us of the practical consequences of ill conceived problem setting.[9]

Setting aside the aforementioned formal problems concerning the mathematical expression of efficiency – and there are other factors that can affect the making of comparisons, especially non-linearity – the key to understanding both its definition and the scope of application of the concept is contextuality. The concept of efficiency is a contextual one because to decide whether something is efficient or not depends heavily on the surrounding (relevant) conditions. But contextuality is far from being a recognizable, assumed element in the dominant approach to efficiency, even though it is central to finding out the real scope of many supposed efficient solutions. Contextuality also becomes a pivotal issue in understanding the conflict between the modern technological project and critical attitudes toward it on the part of many philosophers of technology – a conflict that set the framework for the present discussion.

The context in which an intervention in social reality may be assessed as improving the efficiency of some aspect is not always something already given.

Context is not just a set of preexisting conditions. It is constructed in an epistemological and methodological sense by deciding what has to be taken into account as pertaining to the problem conditions and what is simply left out (in a process often called 'modeling'). In addition, context is constructed, frequently, in a very physical way because technological innovation and/or economic innovation actively models (shapes) the social and physical environment in order to facilitate the suitability and eventual success of the innovation. Against the image of a preexisting context, with clear boundaries separating what belongs to the system and what is external, it is possible to oppose the image of a net-like frame incessantly woven and unwoven.[10]

Contextuality problems have several aspects. The most important are:

- *Translation* of efficiency into more concrete goals, such as performance, order, profitability, productivity, and so on. Efficiency becomes a sort of meta-value, formally invariant, while its applications change dramatically according to the putting into practice of more specific criteria. This is perhaps the most general and visible sense in which efficiency is contextual.
- *Problem definition.* Although this point may seem like a platitude, it is useful to remember that efficiency depends on the definition of what has to count as relevant conditions, as an objective, as a resource. It also bears heavily on the relevant effects of a technological action. For example, it is common to calculate agricultural efficiency by taking as inputs the labor hours expended directly on agriculture and as outputs the bushels of grain. However, when inputs are arable land, healthy soil and fresh water, and outputs are human beings adequately fed, traditional, sustainable agricultural systems may well be more efficient than mechanized, large-scale production agriculture.[11]
- *Physical, biological, psychological or social conditions.* Context dependence may appear in the form of regularities and laws of physical, biological, psychological, and sociological realities. Sometimes, they are simply not taken into account. On other occasions, they are not well known, or even completely unknown. Uncertainty appears in efficiency calculations when it is impossible, due to lack of information, to properly set the order of the different solutions to be able to compare them.
- *Geographical limits.* Other forms of contextual dependence are literally geographical. Drawing up arbitrary boundaries is a very common strategy when dealing with problems involving natural resources such as water or air.
- *Temporal vector.* Clearly, what was considered efficient in the past may not be so in the present or in the future. Efficiency assessment relies on the level

of technological development. There is also a more interesting kind of temporal contextuality, which is related to predictions about future events. Uncertainty appears again, especially when trying to forecast the evolution of social and environmental systems. Finally, the estimations of economic efficiency become much more difficult as a temporal lapse increases.

- *'Side effects'*. In technology assessment literature and applied economics there are many technical terms to describe the effects of techno-economic interventions in a natural or social system. 'Externalities', 'impacts' ('direct' and 'higher order'), 'colateral effects' and 'side effects' – just to mention the most widely used – are expressions that reflect the strained relation between what has to be counted in the analysis as outputs and what is ill considered as *other consequences* or even left out – deliberately or because of the difficulties in dealing with them. On occasion, especially when the physical space involved is not very extensive, local knowledge (frequently neglected by experts) shows itself to be superior to sophisticated technical approaches.
- *The institutional structure* also has a crucial role in most cases of efficiency improvement that are connected with the development of economic and technological systems. Institutional structure does not remain unchanged across time, as is supposed in many efficiency approaches. Big companies know this very well, and have as a priority to take some control of the institutions, trying to modify and reshape legal and other institutional frameworks in order to improve their penetration and benefits.[12]
- *Strategic decisions*. In these cases, the efficiency calculus depends on success in predicting another's actions, a kind of uncertainty that can never be completely eliminated. Moreover, since the efficiency of a project will probably decrease to the extent that people's actions move in opposite ways, efficiency valuation cannot be any longer independent of relevant individuals and groups.

Any of these factors may dramatically change the assessment of the efficiency of a technological project. In many cases, there is a significant degree of arbitrariness in the decisions taken in order to set the problem conditions. In other cases, the level of uncertainty is so high as to make naive the assumption that all the relevant information can be gathered. In short, the above-mentioned problems often make it useless to appeal to efficiency as a predictive tool, rather turning it into an *ex post facto* justification for actions already chosen. The development of 'new technologies', such as genetic engineering and computers, are paradigmatic in this respect.

EFFICIENCY: TECHNOLOGICAL, ECONOMIC, SOCIAL

Instead of delving into the technical details concerning contextual dimensions, the rest of this chapter will be devoted to emphasizing the conflicting views of efficiency that appear in real life situations.

These situations are likely to demand the reconciliation of opposing goals such as, for example, speed and safety in transportation – or, to refer to a different field, profitability and general access to education. In such cases, it is rare to find a solution that unequivocally combines the opposite demands in the right proportion. Value judgements have to be made, which can be extraordinarily complicated when fundamental values intervene. For example, how can one decide, by employing an exclusively technical approach, the optimal proportion of plantations to forests?

Even if they share the same academic and professional background, experts in the same field of expertise may not agree on the problem formulation and the proposed solutions to a concrete problem. It is likely that technical experts from different fields will disagree in many cases. Particularly manifest can be the contrast between purely technological conceptions of efficiency and economic ones.[13]

To an engineer or technologist, the most efficient solution is that which makes the best use of available resources to accomplish an intended goal, given a set of well-defined conditions of physical and technical feasibility, regardless of the cost of the project and of the obtained solution. In real life situations, however, economic constrictions and profitability demands must be taken into account by the engineer, hindering in most cases purely technological efficiency approaches. Dominant economic thinking, by contrast, usually does not worry about any constriction economists call an 'externality', unless forced to translate it into monetary terms. They count on having a result (understood as profit) always exceeding invested resources. Moreover, the economic approach to efficiency tends to stress available resources over pre-determined objectives. Its point of departure is resource scarcity and the efficient allocation of resources. For this reason, its customary option lies in fixing resources and seeking the maximization of results, a view which is obviously close to the concept of profitability. But even productive efficiency, understood as obtaining a specific product from the combination of the minimal amount of inputs is clearly subordinated to the final goal of getting as many benefits as possible.

Things become even more complicated when we recall that efficiency also has a social dimension, which is – or should be – above all recognized by policymakers. Here, the common good may be in conflict with private goods. Economists, playing now the role of political assessors, resort to a cost-benefit analysis that is supposed to meet the Pareto principle. A so-called Pareto Optimum is reached when it is not possible, by any reallocation of resources, to make any person better off without making at least someone else worse off without being compensated. One important social difficulty with this is that compensation is almost never given. From a strictly conceptual point of view, however, it is possible to cast doubt on agreements of economic efficiency, solely conceived in terms of income, while ignoring the social perception of efficiency and the pursuit of social well being.[14]

Cost-benefit methods, guided only by economic criteria, may favor the withdrawal of resources from socially valuable objectives in order to allocate them to a single objective, the most profitable, but one perhaps less valuable or even negative according to social perceptions. In other words, economic efficiency may yield counterintuitive results. In addition, for laypeople, demands for the efficient management of public resources express themselves as a wish, sometimes diffuse, for "things to work better" or "without wasteful spending." While people are conscious of an economic dimension to these matters in terms of the taxes they have to pay, they are frequently unable to clearly figure how to assign prices to goods and services or, indeed, they actively oppose them.[15]

People's intuitions about efficiency may not only differ from those of economists, but also from technology professionals. Admittedly, they can all agree on the need for an 'instrumental rationality' approach in order to solve many social problems. But "the most appropriate way of problem solving," being a commonly shared slogan, is so vague an expression that it lends itself to several interpretations when put into practice. Technological efficiency seeks to be strongly 'objectivist', aiming to propose numerical solutions as the result of methods rigorously applied. These methods would not allow disagreement – experts perhaps excepted – and would be completely neutral in regard to social values. As already stated, the methods' strength is proportional to delimiting strategies in problem formulation and modeling. In contrast, public concerns about things working well, together with the philosophical analysis of technology, offer a view of diffused contours, conflicting valuations, complex risk perceptions, all irreducible to a single and uncontroversial solution.[16]

A complementary and interesting way of approaching the analysis of efficiency is to regard it as a 'standardized package'. According to the definition of this concept proposed by Joan Fujimura, a 'standardized package' is a combination of a scientific theory and a standardized set of technologies which is adopted by the members of multiple social worlds in order to stabilize definitions, simultaneously facilitating interactions and cooperative work between these different groups.[17] Adapting Fujimura's proposal to our object, we may think of efficiency as a 'standardized package' consisting, basically, of a general concept and several methods to evaluate and improve the efficiency of an object, system, practice, or situation. Conceptualization and methods define, restrict, and reinforce each other, narrowing the range of possible actions without wholly defining them.[18] Together, they provide people from heterogeneous professional and social environments with a homogeneous and conventionalized way of dealing with efficiency problems. But this can be done at the price of forcing the application of pernicious strategies supported by an illusive, shared understanding of the relevant elements at stake and by each apparent success of the methods.

EFFICIENCY RECONSIDERED

The analysis pursued in this paper has something of a preliminary nature. In addition to a general refinement of the analysis, two important tasks remain – one 'historical', the other 'sociological'. The former would consist of tracking the genealogy of the concept of efficiency, from its origin in thermodynamics up to its widespread acceptance by society at large. The latter would require the study, in detail, of some particular technological controversy. Its aim would be to show the disparity of interpretations between experts of the same field, between them and other professionals from different fields of expertise, and also between all of them and the social groups involved in a project to improve the efficiency of an object or system. Both tasks would improve our understanding of efficiency 'at work'.

Much is at stake in the analysis of the concept of efficiency. We have to bear in mind that the dangerous discourse about technological neutrality is based on an exclusive commitment to internal technical values, efficiency being the most prominent.[19]

Engineering and the economic and public framing of an efficiency problem has frequently led to contradictory conclusions. For this reason, references to a general concept, commonly shared, are useless or even confusing. But what are less justified in cases of disagreement are accusations of having an irrational attitude, accusations that are often raised by technical experts against non-professionals.

A good deal of the problem we have been discussing may rest on the ambivalence of the concept of efficiency, on being in part descriptive, in part prescriptive. The technological approach seems to justify prescriptions from a dispassionate description of objects and situations. The prescriptive dimension of efficiency, however, has to be found in the value judgments on 'means' and 'ends' that have to be made – before, during, and after the calculus. In many cases in real world situations, the prescriptive demands of us an essentially non-technical estimation of their axiological weight. To the technocratic faith in a (more or less) benevolent and paternalist reorientation of what is conceived as a confusing social aspiration it would be sound to oppose the idea of making explicit the technical limitations and value judgements involved in interventions to improve efficiency in social life.

To sum up, even if addressing some practical problems from an instrumental rationality point of view attracts wide agreement, efficiency is only one interpretation among others when trying to state what the proper means are to reach a goal.

NOTES

1. As a matter of fact, something similar to what has been related happened in Southern Spain, some years ago. Anyhow, the example has been inspired by other one described in Gary Bowden's paper, 'Coming of Age in STS. Some Methodological Musings', in *Handbook of Science and Technology Studies*, ed. by S. Jasanoff, et al. (Thousand Oaks: Sage, 1995), p. 70. In his paper Bowden clearly depicts the above-mentioned confrontation between the engineering discourse of problem definition and solution and the philosophical approach that inspired the birth of STS, based upon problem understanding and choice.

2. At this point it is appropriate to reference the pioneering paper by Stanley R. Carpenter, 'Alternative Technology and the Norm of Efficiency', *Research in Philosophy and Technology*, vol. 6 (1983), pp. 65–76. As Carpenter rightly claims, "a failure to appreciate the varied ways the term 'efficiency' is used" creates problems in debates on the different ways of understanding technological practice. "One would wish for a little more clarity and precision in these discussions. This I take to be an important and appropriate task for a philosopher of technology" (p. 66).

3. See Carl Mitcham, *Thinking through Technology. The Path between Engineering and Philosophy* (Chicago: University of Chicago Press, 1994), pp. 225 ff.

4. Drawing on Ivan Illich's phenomenology of tools and machines, Mitcham points out that the efficacy of a tool can be known by a user immediately and through direct experience. The user is deprived of this knowledge when tools are replaced by machines. To compensate for this a science of mechanics was developed. "The quantification of efficacy by the input-output calculus of efficiency in turn gives birth to new constructions of artifice, the world of machines." Cf. Mitcham, *Thinking trough Technology*, p. 184. See also Illich classic essay, *Tools for Conviviality* (New York: Harper and Row, 1973).

5. Alternatives accounts of efficiency can be found in Miguel Angel Quintanilla, *Tecnología: Un enfoque filosófico* (Madrid: Fundesco, 1989), pp. 102–104, and Mario Bunge, *Philosophy of Science and Technology* (Dordrecht: Kluwer, 1985), vol.2.

6. In this context, a situation is indeterminated if it is not possible to reach a single solution or at least equivalent solutions. A situation is incomplete if the scale order is incomplete or disconnected. The order of preferences must satisfy the following

conditions: reflexivity, completeness, transitivity and continuity. See, e.g. S. P. Hargreaves Heap and Y. Varoufakis, *Game Theory* (London: Routledge, 1995).

7. Some conditions under which uncertainty dramatically increases will be specified later on in this paper. For a review of the crucial role of uncertainty in science policy, see Silvio Funtowicz and Jerome Ravetz, *Uncertainty and Quality in Science for Policy* (Dordrecht: Kluwer, 1990). For a classical discussion of the concept of optimizing – and satisfacing as an alternative – and their applicability problems see Herbert Simon, *The Sciences of the Artificial* (2nd ed.) rev. (Cambridge, MA: MIT Press, 1981). See also Jon Elster, ed., *Rational Choice* (New York: New York University Press, 1986), and Elster, *Solomonic Judgements: Studies in the Limitations of Rationality* (Cambridge, MA: Cambridge University Press, 1989).

8. See, e.g. the classical work by Carl Hempel, *Fundamentals of Concept Formation in Empirical Science* (Chicago: University of Chicago Press, 1952).

9. See, for example, Steven Jay Gould, *The Mismeasurement of Man* (New York: Norton, 1981).

10. The well-known French STS scholar Bruno Latour, one of the advocates of the actor-network approach, stresses this point in a forceful way. See, among others works by Latour, *Aramis, or the Love of Technology* (Cambridge, MA.: Harvard University Press, 1996). See also his recent book, *Pandora's Hope: Essays in the Reality of Science Studies* (Cambridge, MA.: Harvard University Press, 1999).

11. This argument is developed by Peter S. Wenz, 'Pragmatism in Practice: The Efficiency of Sustainable Agriculture', *Environmental Ethics*, vol. 21, no. 4 (Winter, 1999), pp. 390–410. As Wenz notes, "when rating practices for their efficiency, we generally concentrate on inputs that are not only necessary, but in relatively short supply" (p. 398).

12. For a good exemplification of this reshaping of the institutional framework by companies see Thomas P. Hughes, *Networks of Power* (Baltimore: John Hopkins University Press, 1983).

13. Although in both cases the activity involved consists of optimizing, constrains and objectives often diverge. In several works, the economist Nathan Rosenberg has dealed with these matters in regard to innovation. See, for a general presentation, Stephen J. Kline & Nathan Rosenberg, 'An Overview of Innovation', in Ralph Landau and Nathan Rosenberg, eds., *The Positive Sum Strategy* (Washington, DC: National Academy Press, 1986).

14. A comprehensive criticism of conventional economic efficiency analysis can be found in the writings by Nobel Prize recipient Amartya Sen. For an overview of the various conceptions of economic methods vis-à-vis human well-being, see *The Quality of Life*, Martha Nussbaum and Amartya Sen, eds. (Oxford: Clarendon Press, 1993). See also D. W. Bromley, *Economic Interests and Institutions* (Oxford: Basil Blackwell, 1989). For a more specific criticism of economic methodology in regard to the impacts of science on society, see Kristin S. Shrader-Frechette, *Science Policy, Ethics, and Economic Methodology* (Boston: D. Reidel, 1984).

15. See, e.g. J. O'Neill, 'Cost-Benefit Analysis, Rationality and the Plurality of Values', *Ecologist*, vol. 26, no. 3 (1996), pp. 98–103.

16. See Kristin S. Shrader-Frechette, *Risk and Rationality* (Berkeley, CA: University of California Press, 1991). Constructivist approaches to the technological phenomenon have done some interesting empirical and historical work in order to show the failures of a reductionistic efficiency explanation of technological change. See, for example,

Wiebe Bijker, Thomas Hughes & Trevor Pinch, eds., *The Social Construction of Technological Systems* (Cambridge: MIT Press, 1987), and Wiebe Bijker & John Law, eds., *Shaping Technology/Building Society* (Cambridge: MIT Press, 1992). 'Constructive Technology Assesment' aims at making explicit values and decisions that configure the development of technological systems, so as to allow wider social participation in the process. See Arie Rip, Thomas Misa, and Jan Schot, eds., *Managing Technology in Society* (London: Pinter, 1995).

17. John Fujimura, 'Crafting Science', in *Science as Practice and Culture*, Andrew Pickering, ed. (Chicago: University of Chicago Press, 1992), pp. 168–211.

18. Fujimura, 'Crafting Science', p. 169.

19. According to the well-known definition of Jacques Ellul, technology is 'the totality of methods rationally arrived at and having [aspiring to] absolute efficiency (for a given stage of development) in every field of human activity'. 'Note to the Reader', in J. Ellul, *The Technological Society* (New York: Knopf, 1964) p. xxv. One of the main aims of Ellul's work is to denounce the 'fixed end of technique', by which the multiplicity of means and ends are reduced to one: the most efficient. Thus "reason appears clearly in the guise of technique." See Ellul, *The Technological Society* (New York: Knopf, 1964), p. 21.

THE FATE OF SKILLS IN THE INFORMATION AGE

Bill Hook

The information age has changed the breadth, depth, and distribution of skill in contemporary culture. I offer a detailed analysis of photographic, programming, and musical skills that will yield a number of helpful distinctions and expose the role, value, and evolution of skills. The distinctions that I will draw parse focal, device, and service skills from their broader contexts, as well as delineate direct from indirect deskilling. In conclusion, I will propose two avenues for a reform, one that will forestall the loss of skill locally and another that will reclaim it broadly.

PHOTOGRAPHIC SKILL

In my photography bag rests a Pentax 6 × 7 medium format camera. This is a remarkable instrument of photography, not one that is comfortable in the hands of the unskilled. It is fully manual in the sense that the photographer must set the aperture, shutter speed, and focus. It does make one concession to the electronics revolution: it has an electronically controlled shutter curtain that increases the accuracy of its movement. This though is transparent to the user so long as the battery does not run down. Contrast this to its 35 mm counterparts today. They are fully automatic in practically every way. The photographer is not required to set the aperture, shutter speed, focus, or even fuss with the details of loading film and rewinding. A sensor in the eyepiece will detect where your eye is looking and will focus the lens on that object. Its electronics and software are sophisticated enough to discern all manner of exposure situations such as back lighting, red eye, and flash fill. These cameras

Metaphysics, Epistemology, and Technology, Volume, 19, pages 101–123.
2000 by Elsevier Science Inc.
ISBN: 0–7623–0681–5

rest easily in the hands of the novice and expert alike. Filmless cameras are just becoming good enough that $30,000 may buy a model that rivals a $500 film camera in some situations. Who among us, though, is not sufficiently cognizant of the technological paradigm to have any doubt that technology will eventually render film obsolete?

The probability of a fine exposure coming from my 6 × 7 camera in unskilled hands is too poor to calculate. The camera has no light meter, so the novice is going to have to guess the exposure. Having no understanding of exposure in any event, he or she is probably not going to set the aperture or shutter speed at all. The novice will easily figure out how to focus the camera and will hand-hold its bulk unsteadily while pressing the shutter release. Even in the unlikely situation that the exposure happened to be right by chance, the image will be badly blurred because one simply cannot hand-hold a medium format camera in most situations. As either film size or lens magnification is increased, sensitivity to vibration is also increased. The 6 × 7 camera has a film size that is four times the area of 35 mm film.

To make a proper exposure with my 6 × 7, you need a handheld incident light meter. The one I have, the Sekonic L-508, measures incident light, reflected light, flash, 1°–4° spot, EV, and makes numerous calculations such as averaging several spot sources. This meter takes hours to learn to use and is sufficiently complicated that, if you do not use it often, you will forget its operation. You also need a good tripod to steady the camera as well as a cable release to trip the camera's shutter release without shaking the camera. Because the mirror on this camera is much larger than that found on 35 mm cameras, you need to lock the mirror out of the way before tripping the shutter release, otherwise the vibration caused by the mirror flipping up and out of the way of the line of exposure will spoil the sharpness of the image. These things take time to learn well. And none of these technical details address the skill needed in the darkroom to transform an image on film successfully to photographic paper. The skill in the darkroom is much deeper.

In the darkroom you have to know how to mix chemicals for different films and exposure problems, how to keep the temperature of the chemicals within exacting tolerances, and how to filter light in an enlarger for differing contrasts, dodging, burning, and color balance. Not even dozens of books on darkroom technique could transform a novice into an expert. It takes years of practice and study. Todd Walker (as quoted in Schaefer, ed., 1998) tells us that

> [i]n years past the techniques I used to print photographic images with ink were based upon a complex photographic separation of the information in a negative. I was able to create from six to fourteen litho positives that corresponded to different density values in the starting negative. These positives were contact printed onto litho film to generate a set of

> litho negatives. This procedure captured in tonally separated negative form the entire content of the original negative. By accurately registering and progressively exposing in sequence a sheet of photographic paper to these negatives, I could emphasize any tonal region I chose . . . (Schaefer, ed., 1998, p. 302).

Walker describes a process that took days to make a single satisfactory print, a process so complex he admits that he "often lost all interest in the image [he] was trying to print along the way!" (Schaefer, ed., 1998, p. 302).

PHOTOGRAPHIC DESKILLING

For $7,000 I could purchase a medium format film scanner that would transfer information from a negative or transparency to my computer at 4,000 lines per inch optical resolution and 36 bit color depth: sufficient detail to make preservation of the film no longer necessary. Instead I have a document scanner that can do low resolution film scanning at 600 lines per inch at 24 bit color. I scanned some of my transparencies and negatives just to understand its capabilities. I used photo-editing software to display the scanned image and make color corrections, contrast adjustments, and even alter at the smallest level the objects in the image. I was able to repair a damaged image completely and remove an entire object in another image. If I had a photo quality printer, which would cost about $15,000, I could print any of these images in minutes. In time the costs of this equipment will decrease and the functionality and quality will increase.

In my lifetime the Pentax 6×7 in my camera bag will become a historical artifact, as too will darkrooms the world over. Indeed, the darkroom has already been rendered obsolete for Walker:

> When a computer became available I thought about using the procedures I had worked out photographically with this new tool. I first used an Apple II, which offered very low-image resolution and only white/black, orange/blue, and magenta/green colors. I learned a lot using this computer, but I now have a far more powerful PC, capable of working with highly resolved images and a broad spectrum of color values. Decisions and procedures that once took weeks of work in the darkroom can now be done in hours on a computer.
>
> On the computer I can separate my black-and-white negatives into 256 separate tones of the gray scale, whereas with photographic methods I was limited to 14 or fewer. With these separations I can do everything that I have been able to imagine . . .
>
> The computer allows me to revisit the world I photographed in, using old negatives to construct and reinvent the entire scene. Working with a computer is like making a photograph in a perfect studio that has available at your fingertips every option you can imagine (Schaefer, ed., 1998, p. 303–304).

John P. Schaefer agrees and explains that "digital imaging offers creative possibilities well beyond the capabilities of the experienced photographer and the most sophisticated darkroom facility" (Schaefer, ed., 1998, p. 326). Without digital imaging, so argues William J. Mitchell, there is no doubt

> that extensive reworking of photographic images to produce seamless transformations and combinations is technically difficult, time-consuming, and outside the mainstream of photographic practice. When we look at photographs we presume, unless we have some clear indications to the contrary, that they have not been reworked (Mitchell, 1992, p. 7).

In the early days of digital photographic manipulation, many photographers felt that anyone who used a computer to alter the image exposed on film was somehow cheating. People fail to be so impressed when they discover that something that appears to be amazing is actually trivial. This is evident in wildlife photography and cinematography where animals that appeared to be wild as recorded on film are discovered to be captive. The element of truth to reality in photography is rarely expected any longer. In fact, says Mitchell, digital images are

> more susceptible to alteration than photographs, drawings, paintings, or any other kinds of images. So the art of the digital image cannot adequately be understood as primarily a matter of capture and printing . . . intermediate processing of images plays a central role (Mitchell, 1992, p. 7).

Mitchell fails to give proper credit to intermediate processing in analog photography. If all analog photography ever amounted to was a true recording of reality, then good riddance to it. But it was always about manipulation at the point of exposure and then again in the darkroom. That manipulation has not disappeared; in fact it has been enhanced significantly by the digital era. Ansel Adams' analog work was celebrated not because his subjects or compositions were so much better than those of his contemporaries but because his zone system showed a mastery of manipulation that was without equal in his day. Already the level of manipulation that he studied a lifetime to achieve is available to the comparatively unskilled digital photographer. Schaefer explains that

> [f]or the vast majority of digitized images, the computer is only a sophisticated enlarger that can carry out operations identical to those that can be accomplished in the darkroom — but with superior results and much less effort . . .
>
> What kinds of traditional darkroom activities are easily performed at a computer workstation? Dodging and burning-in can be accomplished over a broad area of a photograph or on a microscopic portion of it. Scratches, pinholes, and other negative defects are easily fixed; image contrast can be altered at will; images can be combined to create a montage; colors can be changed to any one of millions of distinct possibilities; four-color-separation negatives are available at the touch of a button; negatives can be enlarged or copied without loss or change of contrast, highlight, or shadow details; faded images can be restored to their original densities and tonal values, and so forth. With access to the appropriate equipment and software, any and all of these transformations can be

> accomplished with results equal to or better than those that can be achieved in the best photographic laboratories and darkrooms by highly skilled professionals (Schaefer, ed. 1998, pp. 325–326).

Learning in a few days or months to do on a computer what Adams studied decades to master in a substantial darkroom is both a deskilling and a reskilling. It is a reskilling in that a computer skill set is acquired to replace a darkroom skill set. It is a deskilling because the computer skill set is much smaller, as evidenced by the rapid acquisition of the knowledge necessary for mastery. Even when we flex our new digital muscles powerfully and produce work that is far beyond the realm of analog possibility, we employ a skill set that is significantly attenuated over its analog counterpart.

I have rarely used my camera in recent years largely because of this deskilling. There is increasingly less that is remarkable about a composition that is well exposed and printed. Is it the rule rather than the exception that when a skill is appropriated by technology its devotees either lose interest or embrace the deskilling? Not entirely, but I think that if they do not, they eventually become marginalized.

Once weaving was an impressive and useful skill. At the weaver's shop in Colonial Williamsburg you get a glimpse of the skill of production and the beauty of product that was once had. The people who held these skills did not learn them quickly. As young apprentices to the weavers, they acquired the techniques over years so that they would some day work alongside and eventually replace the artisans at the looms. But to see people today weaving the fabric for the clothes that they wear would be odd. Certainly these people exist, even outside of colonies like the Amish that reject much of modern technology, but they are marginalized by society.

Similarly we never see a photographer using the photographic plates of a century ago in box cameras. Those skills were completely appropriated by technology and anyone still practicing them would have to be a photographic historian or a few sandwiches short of a picnic. Skills are being subsumed by what some hold to be a symbiotic relationship between the military, scientists, and corporate management, and what others, who are more world appropriating, hold to be the paradigmatic face of technology (Borgmann, 1984).

David Noble, in *Forces of Production*, argues that deskilling serves a more nefarious goal: the quest for power.

> It is a common confusion, especially on the part of those trained in or unduly influenced by formal economics (liberal and Marxist alike), that capitalism is a system of profit-motivated, efficient production. This is not true, nor has it ever been . . . . The goal has always been domination (and the power and privileges that go with it) and the preservation of domination (Noble, 1984, p. 321).

This quest for power, according to Noble, came to fruition after World War II as

> three sets of complementary ideas reinforced one another and converged in the postwar period. And this intellectual climate was sustained and institutionalized by the power of these three communities: that of the military to subsidize and shape technological development, that of the technical community to lend scientific sanction and prestige to the chosen course of development, and that of management to decide how the new technology would be used and to impose this decision upon the work force (Noble, 1984, p. 192).

To see deskilling, however, solely as an issue of economic efficiency and labor costs, or the greed for power and control by institutions, misses its salient features.

DEVICE PARADIGM

This deskilling of photographic processes is but one instantiation of deskilling in general, which is itself part of a broader paradigm that began unfolding in the modern period. This paradigm is what Albert Borgmann calls the 'device paradigm'. To understand the device paradigm we must understand the difference between a thing and a device. For Borgmann a thing

> is inseparable from its context, namely, its world, and from our commerce with the thing and its world, namely, engagement. The experience of a thing is always and also a bodily and social engagement with the thing's world. In calling forth a manifold engagement, a thing necessarily provides more than one commodity. Thus a stove used to furnish more than mere warmth. It was a focus, a hearth, a place that gathered the work and leisure of a family and gave the house a center .... Physical engagement is not simply physical contact but the experience of the world through the manifold sensibility of the body. That sensibility is sharpened and strengthened in skill .... Limitations of skill confine any one person's primary engagement with the world to a small area. With the other areas one is mediately engaged through one's acquaintance with the characteristic demeanor and habits of the practitioners of the other skills. That acquaintance is importantly enriched through one's use of their products and the observation of their working (Borgmann, 1984, pp. 41–42).

A device for Borgmann is very different from a thing:

> A device such as a central heating plant procures mere warmth and disburdens us of all other elements. These are taken over by the machinery of the device. The machinery makes no demands on our skill, strength, or attention, and it is less demanding the less it makes its presence felt. In the progress of technology, the machinery of a device has therefore a tendency to become concealed or to shrink .... [T]hose aspects or properties of a device that provide the answer to "What is the device for?" constitute its commodity, and they remain relatively fixed. The other properties are changeable and are changed, normally on the basis of scientific insight and engineering ingenuity, to make the commodity still more available. Hence every device has functional equivalents, and equivalent devices may be

> physically and structurally very dissimilar from one another (Borgmann, 1984, pp. 42–43).

When the notion of availability is added we see the paradigm take shape:

> [L]et us note that the notions of liberation and enrichment are joined in that of availability. Goods that are available to us enrich our lives and, if they are technologically available, they do so without imposing burdens on us. Something is available in this sense if it has been rendered instantaneous, ubiquitous, safe, and easy (Borgmann, 1984, p. 41).

The elements of the device paradigm critical to the deskilling of photography are instantaneity, ubiquity, and ease of use. We will assume that safety is no longer a great concern in photographic excellence. The critical element, though, is ease of use. Once ease of use has been guaranteed, instantaneity and ubiquity naturally follow. Cameras are now instantaneous in that once a subject for exposure is decided upon, it takes only a moment to make the exposure. They are ubiquitous because ease of use accentuated the commodity and concealed the substructure.

Were the skills involved in photography subsumed by technology in the form of electronics and software, or were they appropriated by electronic engineers and software developers? Many millions of people were involved in the development of the technology that subsumed photographic skills over the past few decades. The breakthroughs that resulted in autofocusing came as a result of research into lasers for use by science, industry, the military, and medicine. A beam of monochromatic light is reflected off the photographer's retina, which is read and calculated to be centered on a particular point in the viewfinder's grid. The engineers of cameras used and maybe even added additional refinements to the technology, but they did not invent it. The same is true for much of the electronic and software wizardry that makes a camera so easy to use. So it was not really the thousands of engineers and programmers working for Nikon, Cannon, Olympus, and Pentax that deskilled photography, rather it was millions of people building up complex and impenetrable layers of technology that were gradually applied to cameras. You cannot point to any particular set of people who acquired the subsumed skills that photographers once had and say that there was simply a transfer of skill from one group of people to another. It is right, I believe, to say that the skill was appropriated by electronics and software, not by engineers and programmers.

Daniel Dennett speaks to this point indirectly in *Consciousness Explained*, where he shows that consciousness is the sum of billions of participants in the brain working together and separately to process information. He argues that

> the exclusive attention to specific subsystems of the mind/brain often causes a sort of theoretical myopia that prevents theorists from seeing that their models still presuppose that

somewhere, conveniently hidden in the obscure 'center' of the mind/brain, there is a Cartesian Theater, a place where "it all comes together" and consciousness happens (Dennett, 1991, p. 39).

The Cartesian Theater of consciousness is what we feel we have because we have an urge to identify our consciousness in terms that are not so diffuse. With Dennett's concept of consciousness, we are unable to point to any few or even hundreds or thousands of parts and say these are consciousness. Consciousness is diffuse in the extreme. The electronics and software in the camera are to the programmers and engineers as the theater of consciousness is to the brain. The electronics and software is something sufficiently finite that I can point to and say this is where the skill went. It is a mistake to point to a software engineer in charge of base class objects at Borland and say she now holds the skill, or even a part of it, that photographers once held. She probably knows nothing about photography at all, has never worked even remotely for a camera company, and would deny any culpability for the deskilling even though she will admit to being a part of the software infrastructure. Similarly sound engineers who understood that the music of an entire orchestra could be recorded as a single waveform and played back by reproducing that waveform did not subsume the skill of the musicians. The skill of the sound engineers has nothing to do with the skill of playing instruments and making music.

More broadly it is the device paradigm that deskilled these people, not engineers and programmers. Engineers and programmers serve the device paradigm, but they do not now hold the skill that was subsumed by the paradigm. I do not want to simply wave these engineers and programmers away; theirs is an important role.

SERVICE SKILL, DEVICE SKILL, AND FOCAL SKILL

Let us call the skill that serves the paradigm *service skill* and the skill that is deskilled by the paradigm either *focal skill* or *device skill*. Focal skill is skill involved in what Borgmann calls a focal practice. Focal skills center on eminent things that eminently engage people's skills and disclose the world they live in. Playing a violin and skiing are focal practices. Device skill is skill involved in the manipulation of devices. Using a computer and driving an automobile are device skills.

Focal skills are a complement to things, and device skills are contingent upon devices. Things engage us and broaden our cultural and natural faculties most powerfully when we entertain skills that employ them. Skills that operate on devices do so on the commodity aspect of the devices and not on their technological substructures. As a result, device skills typically occlude more than they disclose, allowing of course that their substructures are more substantial than their commodities.

Making music by playing a violin is an eminently cultural activity. It maintains a multifaceted commodity: it engages the musician, his or her mentor, and audience musically and socially. Tremendous skill is demanded that takes many years to acquire after a great deal of practice and dedication from both student and teacher. The substructure of the violin is mostly natural, and where it is not, it is fashioned through intelligible processes that conceal little. The substructure has no functional equivalents. Replacing the gut strings or the wood body with a metal matrix or a carbon fiber composite makes a significant and unacceptable difference such that the changes are immediately recognized by musician and audience, both visually and acoustically. An accomplished violinist is unlikely to translate that focal skill to a piano without a great deal of practice and learning. Certainly the musical skill acquired by learning the violin will make learning the piano easier, but it will still require great dedication to become accomplished on another instrument.

Driving a car competently is a device skill that takes a few days or weeks of practice, certainly a tiny fraction of that required to learn to play a violin. The substructure has so many functional equivalents that auto mechanics have difficulty keeping track of them. The commodity aspect of the car remains either relatively or completely unchanged by differences in substructure. This is evidenced daily at the counters of car rental business at airports where travelers arrange to competently drive cars of various makes and models. Not only is most of the substructure concealed, but so is most of the environment you travel through. The car insulates you from differences in temperature, steepness of terrain, and danger from bears, mountain lions, and deranged mathematicians (if you live in Montana), or muggers if you live in a metropolitan area.

Failing to understand the difference between focal and device skill leads to the kind of words we find in Malcolm McCullough's *Abstracting Craft: The Practiced Digital Hand* where he writes:

> If computers continue to become smaller and more numerous at the same mind-boggling pace, they may conceivably pervade everything we own and operate, from office doors to garage workbenches. When it accommodates not only touch but also sound, physical context, social community, and accumulated experience, this ubiquitous technology may actually seem quite pleasant – especially to the more skillful or sensitive person (McCullough, 1996, pp. 18–19).

To the contrary, these devices, when ubiquitous, will seem quite pleasant to the least skillful and insensitive person, of which there will be many.

Curiously he follows that thought with:

> One thing is clear. We have reached a point in the history of technology where it is especially important to take pride in human abilities. We must not only defend against further deskillings, but also direct inevitable technological change in a more human-centric direction (McCullough, 1996, p. 19).

McCullough is able to make these seemingly dissonant statements because he ties skill closely to hand-eye dexterity:

> The way of hands is personal, contextual, indescribable. Little can surpass the hands in showing that we know more than we can say. Psychologists and social scientists have studied this inarticulable knowledge extensively, and they have many names for it: operative, action-centered, enactive, reflection-in-action, know-how. The most common word is skill (McCullough, 1996, p. 3).
>
> In these many different facets and guises, the computer has become a visual medium. It has subsumed many older visual technologies such as photography and video as well .... Visual computing has provided a new form of hand-guided, continuous processes .... Increasingly computing shows promise of becoming the medium that could reunite visual thinking with manual dexterity and practiced knowledge (McCullough, 1996, p. 50).
>
> [C]an there be a reunion of manual and mental work? May the coordinated hand, eye, and tool at last benefit from what has traditionally been the most powerful means of abstraction – processed symbols? And conversely, may the work of symbolic processing be made more human by the ways of the eye and the hand? (McCullough, 1996, p. 111).

By allowing skill to be embodied principally in the conjunction of visual thinking and hand-to-eye coordination, one is pushed toward the position that Kasparov exercises no chess skill if someone else moves the pieces for him. Similarly, playing a video game should demonstrate more skill than literacy. Certainly there is much to be applauded in McCullough's attention to the importance of eyes, hands, tools, and mind in the embodiment of skill, but it can lead to absurdities and worse.

Borgmann too sees bodily engagement as being tied to skills. However, where McCullough has us bodily engage the commodity, Borgmann demands "bodily and social engagement with the thing's world" (Borgmann, 1984, p. 41). Kasparov's skill is not merely comprehending chess abstractions or moving the pieces on the board, but is also understanding the nuances of an opponent, his/her/its prior games, and conducting himself gracefully when he loses a match. Well, one is not always skillful in every way. Reading too is a skill that draws most powerfully on the world of experiences with things and social interactions. One is challenged by literature to take austere symbols on a page and make them come alive with meaning that connects to the world at large.

The recovery of skill that McCullough champions in the digital age occurs through the virtualizing of reality. When a significant portion of reality has finally been virtualized, how then are skills like literacy to survive in any meaningful way? These austere symbols will have less and less to refer to in the real world. The virtual world has few if any moorings on which to anchor a stable community and robust culture, both of which are necessary for sustaining a rich literacy. Sven Birkerts, considering the effect multimedia and hypertext are having on historical literacy, writes that "we may start producing generations who know a great deal of ìnformation' about the past but who have no purchase on pastness itself" (Birkets, 1994, p. 137). Likewise, in a highly virtual world, we may well have abundant information and virtual skill, but little purchase on the real reality that subtends and supports the virtual one.

WHERE THE SKILL GOES

The important question then remains as to whether an equal amount of skill, though entirely different in nature, is transferred from focal skill and device skill to service skill, or whether there is a net loss of skill. If it is conserved, then is the transferred skill of equal value?

Empirical evidence to determine whether there is a loss of skill when deskilling occurs may be hard to come by and have poor precision in any event. Our intuitions, I believe, tell us that there is a net loss of skill, that service skills are not created in the same proportions in which device and focal skills are lost due to efficiencies.

Noble argues that numeric control and its subsequent deskilling represents a second industrial revolution. He does not mince words when he writes that

> in the meantime, the compulsion to automate (and to dominate) – fueled by newly inflamed competitive fears – continues apace (and resistance grows). As a result, we see, not the revitalization of the nation's industrial base but its further erosion; not the enlargement of resources but their depletion; not the replenishing of irreplaceable human skills but their final disappearance; not the greater wealth of the nation but its steady impoverishment; not an extension of democracy and equality but a concentration of power, a tightening of control, a strengthening of privilege; not the hopeful hymns of progress but the somber sounds of despair, and disquiet (Noble, 1984, p. 353).

Noble can be forgiven for failing, in 1984, to anticipate post-modernism and how information processing would sweep through technological cultures, creating whole industries of information processing and management.

It is clearly difficult to determine whether a net amount of skill is lost through deskilling because in many cases a device conceals what, if any, focal and device skills mutate into service skills as they become subsumed by the paradigm. One would have to calculate the total skill in existence both before and after the production of devices, taking into consideration population growth, production, unemployment, etc. Noble is adamant, and I think part of his claim can be tested empirically. He argues that efficiencies through numerical control deskill those who remain at the job and put the rest out of work. If automation really reduced the workforce as he claims, then we should see a very high rate of unemployment. But we do not. Device and service skills, as well as the growth of production, account for some of this.

As Systems Manager for a federal district's law offices throughout most the 1990s, I helped to deskill both support staff and attorneys. In 1990, before they used computers in any significant way, attorneys used the office library for much of their research, secretaries spent most of their day typing on inflexible word processors that amounted to typewriters with screens, cases were tracked by paper files and files by index cards, and collection activity was organized by hand in rows of filing cabinets. All of this was automated by computers and software in the intervening years. Now attorneys rarely venture into the library; their research is on-line from their desks. Secretaries spend less time word processing and more time doing legal research. Cases and collections are tracked via client-server databases. Every employee's job has been transformed to the extent that the old-timers occasionally reminisce about how they accomplished tasks in the days before computers.

Regardless of whether a process is performed manually by using paper or automatically by using software, business rules determine how to proceed. When the office was organized using paper, people embodied all of the business rules. Those who had twenty years of experience were highly valued for their detailed knowledge of all of the business rules. When the office migrated to software, many of the business rules became concealed in binary code and database schemas. As the memories of those who predate the software begin to fade, those rules will become translucent and eventually will drain into the digital machinery and disappear. In exchange for this deskilling, command and control of information increased. Prior to computerization, determining the number of inactive and active criminal cases brought in the last five years that involved drugs and the use of a gun involved lengthy handwork by attorneys, paralegals, and secretaries. The case count would be highly inaccurate because some of the investigation would rely upon the memories of those who worked the cases. After computerization, that determination would be made less by those who worked the cases and more by the computer support staff. Instead of taking days, it would take hours. The computer nerds would analyze the request, compare it against the database schema map, and run queries against the database to arrive at the wrong answer. After consultation with a paralegal who could eyeball the report and find omissions, the request was reevaluated against the database schema, the query adjusted and run again, and the correct wrong answer would finally be achieved. It would be correct in that it accurately captured the data, but wrong because the integrity of the data input is not perfect.

As the databases and software continue to be refined and the business rules clarified and abstracted further in the database schema, such determinations will be cut from hours to minutes and will not involve the memories of those who worked the cases. With enough abstraction and refinement, the report will not even require a highly skilled database nerd; an entry-level drone would be able to satisfy the request. A sufficient number of business rules remain outside of software at the present so that those crusty veterans of the office remain valuable for their experience, but if the current trend toward abstraction and concealment holds, their worth will soon begin to erode noticeably.

VIRTUAL DEVICE SKILL

I said that in many cases a device conceals the transformation of device and focal skills to service skills. In the case of photography, the device is an automated camera. The skill of the photographer is lost and the skills of engineers and programmers are increased. The transformation of skill from one to the other cannot be clearly followed because the device blocks attempts to trace the skill's path. But in some cases a device does not obtain, and the path of transformation can be laid bare. This has happened in programming.

Computer programmers have a very definite skill that takes years of practice and study to become competent with. The skill is so deep that most never become competent. The skill they have does not have its place in the natural world but in the virtual world.

The history of programming compilers is replete with deskilling. Years ago assembly language was one of the principal languages a programmer learned. Assembly has a one-to-one correlation with the binary instructions that direct the CPU to action. There is a definite skill involved in assembly programming and it is tedious work in the extreme. It takes a long time to write even a simple program when you have to hand code every instruction to the CPU.

An assembly programmer has to understand the operating system well and be familiar with its thousands of function calls. Assembly programs call these functions; and data, in the form of arguments, are passed back and forth between the program and operating system. These functions are often slightly different, and sometimes radically different, between operating systems. An assembly program written for Windows NT is unlikely to run under Windows 98 unless care was taken to avoid specific calls to functions in Windows NT that are not a part of Windows 98 and vice versa. There is of course much more the assembly programmer must understand, but we see that the programmer is highly skilled.

The C language was written by Brian W. Kernighan and Dennis M. Ritchie and published by Bell Labs in 1978. It became the dominant computer language by the mid-1980s. With C there is no longer a relation between one line of code and one binary instruction. One line of C code may result in a single binary instruction, or it may result in a hundred. It depends upon the line of C code and how the authors of the C compiler implemented the compilation. This has the advantage of hiding the underlying complexity of assembler from the programmer. A C programmer can write a much bigger program, having more processor instructions, with fewer lines of code. There are other languages that are more or less functionally equivalent to C, though not as widely accepted.

C++, a superset of C, was created by Bjarne Stroustrup at Bell Labs and was published in 1985. It contains everything C has but includes an object-oriented system of class objects. Other object-oriented languages have been developed with various levels of acceptance. Class objects conceal additional complexity from the programmer by combining functions and variables together in such a way that they may be inherited by other class objects in a possibly long lineage. An inheritance model of a class object may reveal a number of ancestors that conceal possibly hundreds or thousands of lines of code from the programmer. Suppose class object X stores strings of characters. Let class object Y inherit all of the functionality of class X and in addition create a dialog box and place the strings of characters in the box. Allow class object Z to inherit class Y and add the functionality that will accept user input in order to alter the strings of characters. This allows a programmer to build a complex hierarchy of code that at every step conceals complexity. The programmer may never directly deal with class objects X and Y, and in fact may even be unaware of the existence of X and Y. Class object Z may be the only thing the programmer needs to know how to use.

A compiler called Delphi arrived from Borland in the mid 1990s. The language the compiler utilizes is Object Pascal. Like C++, it is object oriented. The difference here is that the Delphi compiler is so thoroughly object oriented that virtually everything complex is an object. There are usually six levels of ancestry in a typical class object, which results in enormous concealment. On top of this class hierarchy rest the elements of rapid application development (RAD). A RAD compiler allows the programmer to build the visual interface and much of the non-visual code by dragging an icon off a toolbar and dropping it on a form, thereby instantiating hundreds or thousands of lines of code. This saves an enormous amount of development time when programming for a graphical operating system like Windows NT. I have written programs using Delphi that compiled to twenty million binary instructions. Clearly I could never have written such a program in assembler. With this powerful compiler I can write a complex program in better than one-thousandth the time it would take using assembly language to write the same program.

The visual aspects of this compiler conceal a great deal of detail not just behind object classes, but behind visual objects that are drop and drag. McCullough appears to understand this concealment when he argues that work has begun on making our

> data constructions coincide with our abstractive powers of visual thinking. Visual computing has expanded our capacity to visualize abstract symbolic structures as physical images. It lets ideas become things .... Visual production demands, and visual computing increasingly supports, a cultivated practice of abstraction based on direct manipulation of graphical symbols (McCullough, 1996, p. 52).

When using Delphi the programmer rarely calls any operating system functions directly. When operating system functions get called, it is by base class objects, or distant ancestors, that the programmer is unaware of or is unconcerned with. The programmer's skills in these areas have been subsumed by Delphi. But Delphi presents a whole new level of skill that must be mastered. There are thousands of objects now in place of thousands of operating system function calls. These objects allow the exploitation of power on a level much higher than was previously possible.

However, a person is able to make intelligible only so much organizational complexity. There is more complexity in assembly and Delphi than any one person can fully articulate. When complexity is concealed, as with Delphi, it is not the case that the programmer is now rendered unskilled, but rather is now free to assume complexities at a higher and more abstract level. This abstraction allows for additional power that was previously impossible.

Software compilers are not devices in the sense that Borgmann describes devices. Devices for him are concrete objects. A software compiler is an instruction for manipulating information and does not exist as a material object. Allow me to call it a *virtual device*. It parallels a device in many important respects.

Virtual devices may help us see more clearly how skill is transformed during deskilling without it severing service skill from device skill, or in this case, virtual device skill. The virtual device I am working with here, a compiler, differs from a device in that those who service the virtual device are small in number and their skill is well contained and defined apart from the usual layers of technology that produce devices. It is not the case that millions of people were involved in the technology that constitutes Delphi. Delphi does not utilize lasers, composite alloys, plastics, or any of the other materials of technology. It is all virtual material, built wholly from the ground up by the minds at Borland.

Do we see a net loss of skill as a result of greater efficiency? So far the answer has been no. In fact the net amount of skill has increased. The level of device skill remains the same while additional skill is added by the creation of service skill. The result of this is that the same number of programmers are able to produce more software that is more powerful and easier to use. But there must be a limit to how much software will be acquired by our culture. Information processing is a rapidly growing area at the present. Allow fifty or a hundred years to pass and imagine that most aspects of life and business that can be computerized have been computerized. Additional growth in information processing is then more difficult. If compilers in that era encapsulate far more complexity, will there be a need for so many programmers? It seems to me that we will need a lot fewer. Taken to its natural limits, the compilers will encapsulate so much complexity that there will be insufficient complexity remaining to fully occupy a programmer. Ordinary people could then direct the compiler to produce a needed program. I am hesitant to admit that the evolution of the programmer to higher levels of abstraction is finished even at that futuristic point. Perhaps the evolution will turn programmers from coders into artificial intelligence designers. Thinking out to these limits, though, is just so much hand waving.

Let us return our focus from virtual devices to that of actual devices. Why was it that photography became so deskilled while programming has not? An examination of music will bear fruit here.

THE ROLE OF ARTIFICIAL INTELLIGENCE

Has the making of music been deskilled via the device paradigm? It has been in some ways, and not at all in others. Nothing has occurred to deskill Emma Kirkby or her vocal counterparts, Andras Schiff and his instrumental counterparts, or Christopher Hogwood and his conducting counterparts. Their skills appear untouchable by technology, at least at the present. There are electronic devices that can cheaply simulate various instruments and even whole orchestras, but we are not fooled by these base imitations. I once attended what was billed as a baroque ensemble at Cisel Hall in Billings where the music of Handel was promised. Understandably I was excited to hear live music and equally understandably I was chagrined when I took my seat and noticed that in place of a harpsichord, they had an electronic keyboard programmed to simulate a harpsichord. The other instruments in the ensemble were real, though not period instruments.

But let us not underestimate the power of the device paradigm, as Hubert Dreyfus did in 1979 when he said of chess programs that "we can in every area where expertise is based on experience continue to find some experts who outperform even the most sophisticated programs" (Dreyfus, 1979, p. 32). It took but eighteen years from Dreyfus' pronouncement for Deep Blue to beat Kasparov. But this does not represent artificial intelligence at work as IBM verifies:

> Does Deep Blue use artificial intelligence? The short answer is 'no'. Earlier computer designs that tried to mimic human thinking weren't very good at it. No formula exists for intuition. So Deep Blue's designers have gone "back to the future." Deep Blue relies more on computational power and a simpler search and evaluation function (Frequently Asked Questions: Deep Blue, at http://www.chess.ibm.com/meet/html/d.3.3a.html#ai, March 30, 1999).

Perhaps eventually sound engineers and programmers will make the computerized instrument capable of all the nuances of a real instrument, and perhaps even of a period instrument, such that the most sensitive ears and knowing minds will not be able to discern the difference. At that point, instruments will be supplanted by electronics and the process of performance will change. Some deskilling is certain to result.

A period oboe must be played differently from a contemporary model. It has subtle differences that require special skills and care when playing. The electronic oboe will not have these. There will be a program in the device that will alter its sound for period or contemporary output. The device will be more forgiving in hands that have fewer years of practice. Still, the musician will have to obtain skill to play the device. Let us then advance many years further and allow software the level of musical articulation that the musician has, such that the musician is no longer needed. At that point, the performance is completely subsumed.

To get to the point of completely deskilling the musician, you need very strong artificial intelligence. Presently that is for science fiction. But what we see from this examination of music is that skills with the most depth cannot be appropriated by devices without serious artificial intelligence being a part of the device. Skills that are easily subsumed by devices are those that do not make powerful claims on us. Focusing a camera does not make a powerful claim on me. I just look through the lens and turn a ring until the focus appears sharp. The autofocusing mechanism takes this skill and refines it by making focusing faster and more accurate. Simple light metering and setting the exposure calibration of the camera makes no strong claim on me. I could go on, but the pattern has emerged already – devices appropriate skills that have the least depth.

We see now that the most technical and mundane aspects of photography have been deskilled. None of the artistic decisions have been threatened by the device; to the contrary they have been magnified. If there is not enough left of photography after the current round of deskilling, then it must mean that too much of photography always rested on something that was never all that interesting. Or does it?

DISCLOSING THE WORLD

There are greater concerns directly related to deskilling that have not yet been touched upon here. Borgmann is principally concerned with the loss of focal things and practices in our lives. Devices disburden us and insulate us from reality in as much as they are consumer and not scientific devices. Focal things engage us and disclose the world we live in. Disclosure is the opening up of reality where we obtain information about and for the construction of reality that makes the world more perspicuous and prosperous. Information about reality assumes the position that there is a reality independent of our frameworks of understanding. Information for reality is the recognition that we in turn shape reality materially, culturally, and technologically.

If focal things, practices, and skills are to have meaning, reality must exist independently of intelligent beings and be structured in a particular way. The power of focal things lies in the disclosure of a structured reality. Various ontological positions and arguments are already well established and defended, so I will offer only two brief arguments that I have found persuasive for focal reality.

If reality, minus intelligent frameworks, is unstructured, then intelligent frameworks may not be a part of reality. More simply, if reality has no shape independent of my understanding, then I cannot be a part of reality, otherwise I will be unstructured and have no understanding and no frameworks. If I exist outside of the reality that I project structure onto, then my projection of structure onto reality must be organizing reality from pieces of something. Those pieces have to exist and be already structured, ordered, and lawful in some way. If the building blocks of reality are already structured in some way, then Occam's Razor instructs me to choose the simplest explanation: reality in its entirety is already structured in a particular way.

Focal realism is the recognition that reality is shaped in a particular way and that focal things reveal this shape to us through focal practices and skills. Manual or automatic cameras and assembly or Delphi compilers are devices or virtual devices. They are not focal things and never have been. They do engage people's skills, but do they disclose the world that we live in? Clearly Delphi, being a virtual device, does not disclose the natural world. In the virtual world, though, Delphi not only fails to disclose it, Delphi conceals it most powerfully. Its strength and utility lie in the fact that it conceals rather than discloses the virtual world. Delphi does add complexity to the virtual world, but in so doing it conceals more than it adds, otherwise there would be no gain in the power of manipulation.

The camera as a device in use by a photographer has some world disclosing properties, but as the device becomes more technologically refined, that disclosing power fades. When using my 6×7 camera, I know that I will need about fifteen minutes to properly expose a composition that catches my eye. Therefore there is no rush. I study the scene carefully, consider it from different angles and may even decide to return when the sun is in a different location of the sky, or on a cloudy day if the contrast is too high. If it is windy and the object moves easily in the wind, then I may have to wait for a day that is calm if the object must be still, or windy if motion is needed. I have to examine the sky to make sure that the sun is likely to remain either blocked or unblocked by clouds for the duration of my preparation. I will use my handheld light meter to take both incident and spot readings of the scene. I am thinking about the photographic processes involved. Do I want a long or slow shutter speed? A wide or narrow aperture? Do I want the entire scene in focus? I focus the lens on the nearest object that must be sharp. I look at the focal distance on the lens barrel. Then I focus on the furthest object that must be sharp and again check the focal distance. Splitting the distance and focusing on the central point between the distances does not work because there is more depth of field behind the point of focus than in front of it. So instead I use the calibration marks on the lens to determine exactly what distance the lens should be focused for. The shutter release bulb is then attached. I lock the mirror up and set the lens iris closed and check the lens front for dust. The shutter fails to trip. Oops. I forgot to advance the film. I do so and then trip the shutter. By now I am thinking that a different angle would be better so I adjust the camera and tripod and start over.

Compare this with what most people do with a state-of-the-art 35 mm camera. They see something interesting and they walk up and snap an exposure and are off in less than a minute. So much skill is concealed by the technology of the camera that the photographer moves too quickly, fails to engage the scene, sun, clouds, and wind. The photographer allows the camera's software to make decisions about depth of field, exposure, motion, and focus. He or she can switch the camera to manual operation, in order to make all of these decisions personally, but will not. I have never seen anyone in my many years of working with other photographers switch an automatic camera to manual mode. Undoubtedly it does happen, but it is the exception. Digital cameras, as they become more refined, will increasingly overcome the limitations that inattentive photographers place on the image. More manipulation will be possible in the computer to save the image from neglect at the time of the exposure.

As the device subsumes ever more skill from the person, less of reality is disclosed by using the device. Borgmann claims that it "is the consonance of cosmic context and focal things that makes the world semantically coherent and allows references to emerge and submerge" (Borgmann, 1999, p. 20). My 6×7, in all its manual glory, conceals reality in a way that oil and canvas never did. Artists may spend hours or even days painting a scene. The wind, sun position, clouds, and weather all play a much greater role in oil painting. The texture of all the objects in the scene are observed and considered. Reality, for the oil painter, is not captured as in a photograph, it is understood and made real again.

Earlier I claimed that skills that are easily subsumed by devices are those that do not make powerful claims on us. This is true. But what we now see is that those subsumed skills, while not by themselves making powerful claims on us, mediate powerful claims on behalf of reality. They disclose the world in ways that, taken individually, do not seem substantial, but when taken together, reality emerges vigorously in our lives.

DIRECT AND INDIRECT DESKILLING

Consider now how the device paradigm has changed the face of music in Missoula, Montana. In the early days you could not have music easily, instantaneously, or just anywhere. It required a place for the performance, the organizing of skilled musicians, and the arranging of social agreements and promises to attend. Today music is easy, instantaneous, and ubiquitous. Students may easily listen to Handel's *Carmelite Vespers* on a Diskman while waiting for the professor to arrive at the seminar room for the day's lecture.

Having music is not the same as making music. The musicians have a skill, that of making music, but the having of music involves social arrangements and promises. For Borgmann the making of music would be concurrently a labor and leisure, a means and an end. The having of music is a leisure, an end. When music is had through social arrangements and promises, it reaffirms the skill of the musicians. When it is had by way of a Diskman, it is a consumption that makes no affirmation of skill. It takes only a few gifted superstars of instrumental and vocal production to make music that may be heard by millions at any time and in any place. To say that technology has done nothing to threaten Emma Kirkby with deskilling is true but says too little. Kirkby is a masterful superstar. What of the thousands of other baroque vocalists whose skills fall short of her performance? The device paradigm deskills them not by subsuming their skill, but by making their skill irrelevant. When the Diskman makes Kirkby ubiquitous, we are less inclined to affirm the skill of the soprano in the local music department by making the necessary social arrangements and promises to attend.

This is different from the kind of deskilling that happened when the computer made the darkroom obsolete. The skills of the darkroom photographer were in some sense made irrelevant by the computer, but the computer replaces the darkroom as a refinement of the darkroom. The Diskman, however, does not represent a refinement of live music, but is a substitute for it. This distinction appears at first blush to be splitting hairs, but it is a crucial one.

Allow me to call deskilling arising from a refinement of a device *direct* deskilling and deskilling that occurs from making skills irrelevant *indirect* deskilling. Direct deskilling threatens only those skills that make the least claims on us. Skills that remain safe, at least until strong artificial intelligence is practical, are the ones that make powerful claims on us. With indirect deskilling nothing is impervious. Skills that make the strongest claims on us are vulnerable to irrelevance. Indirect deskilling can have a tight symmetry, as the Diskman does to the performance of live music, or it can have no symmetry, as television does to the reading of classics, the writing of poetry, or the attention a parent provides a child. For direct deskilling, the simpler skills of the craftsperson will be assimilated; resistance is futile.

Indirect deskilling takes two shapes: it confronts us as consumers, and it confronts us as citizens. As consumers, indirect deskilling can be resisted, but rarely is. Borgmann claims that

> [o]nce a television set is in the house, the daily decision whether to read a book, or write a letter, or play a game, or tell stories, or go for a walk, or sit down to dinner, or watch television no longer really ranges over seven possibilities. The presence of the television has compressed all alternatives to one whose subalternatives are contained in the question: What are we going to watch tonight? (Borgmann, 1992, p. 112).

Borgmann explains further how we have lapsed into this predicament:

> The segregation of doing from making, of morality from production, goes back to Aristotle and has been carried forward into the modern era by Immanuel Kant .... [T]he modern developments that entirely upset and destroyed traditional production left morality intact, if not in its particulars, then, at any rate, in its general cast as the theory and practice of human conduct.

> What remains unexamined all the while is the power of products, of the material results of production, to shape our conduct profoundly. Any moral theory that thinks of the material setting of society as an essentially neutral stage is profoundly flawed and unhelpful; so, in fact, is most of modern and contemporary ethics (Borgmann, 1992, p. 110).

CONCLUSION

Taking responsibility for the important material decisions helps us reclaim our lives, but, as long as culturally our community has failed to take responsibility, only within narrow personal limits. Deciding not to purchase a television and instead to purchase a fine chess set, which will be located centrally in the living space, does nothing to encourage the playing of chess when all of your acquaintances own televisions. You can use your reclaimed time to engage in reality, but not with others who fail to engage in focal practices. Similarly, studying voice and practicing to be a good soprano falls short if people do not make the commitment to gather and listen attentively.

As citizens we have come to believe that we have left open all possibilities, that each of us is free to choose the type of life he or she wants to engage in. When we build highways, the roads do not and can not lead us everywhere; they lead us in particular directions. As citizens we have denied that we are deciding the boundary conditions of material consumption. We strike down additional funding for our schools, libraries, and the arts, but make sure that cable television service is affordable and available for every family. We have made definite material decisions as citizens that will channel our material decisions as consumers. To again find players at the chessboard, to have Shakespeare in the park, to keep our promises to the soprano in the music department, to read aloud Faust for family and friends, to avail ourselves of the culture of the table, the recognition and preservation of focal things and practices need to be undertaken by us as citizens.

REFERENCES

Birkerts, S. (1994). *The Gutenberg Elegies: The Fate of Reading in an Electronic Age*. New York: Fawcett Columbine.

Borgmann, A. (1984). *Technology and the Character of Contemporary Life*. Chicago: University of Chicago Press.

Borgmann, A. (1992). *Crossing the Postmodern Divide*. Chicago: University of Chicago Press.

Borgmann, A. (1999). *Holding On to Reality: The Nature of Information at the Turn of the Millennium*. Chicago: University of Chicago Press.

Dennett, D. (1991). *Consciousness Explained*. Boston: Little, Brown, and Company.

Dreyfus, H. L. (1979). *What Computers Can't Do: The Limits of Artificial Intelligence*. New York: Harper Colophon.

McCullough, M. (1996). *Abstracting Craft: The Practiced Digital Hand*. Cambridge, MA: MIT Press.

Mitchell, W. J. (1992). *The Reconfigured Eye: Visual Truth in the Post-Photographic Era*. Cambridge, MA: MIT Press.

Noble, D. F. (1984). *Forces of Production: A Social History of Industrial Automation*. New York: Oxford University Press.

Schaefer, J. P. (Ed.) (1998). *Basic Techniques of Photography, Book 2: The Ansel Adams Guide*. Boston: Little, Brown, and Company.

TOWARD NEW FOUNDATIONS IN PHILOSOPHY OF TECHNOLOGY: MITCHAM AND WITTGENSTEIN ON DESCRIPTIONS

Andrew Light and David Roberts

Over the last twenty-five years, philosophy of technology has become a recognizable sub-discipline in the Americas and Europe. There are journals, societies, and international meetings devoted to the subject.[1] But in some sense the field suffers from the lack of a common ground on which to base questions that might define it as a philosophical discipline, central questions whose resolution will drive the discipline forward. Certainly there are many views now on the social effects of technology and how we are to evaluate those effects, but the field nonetheless lacks a critical intradisciplinary discussion of those competing views of the kind that characterizes most philosophical sub-fields.

Something similar to this worry was voiced over fifteen years ago by Elizabeth Ströker in one of the first anthologies assembled on philosophy of technology. Ströker even came up with a name for this problem; she called it the "paradox of continual beginning":

> [O]ne is still confronted at present [in philosophy of technology] only with a multitude of detailed philosophical studies each of which can be characterized as merely a sketch or an attempt . . . . Furthermore, although it occurs quite often that authors refer to each others' work – even if they disagree in their views – it is only in rare cases that a critical discussion of the same subject is carried out . . . . [S]ystematic elaborations and assessments have yet to be entered upon (Ströker, 1983, p. 324).

Metaphysics, Epistemology, and Technology, Volume, 19, pages 125–147.

ISBN: 0–7623–0681–5

While the level of philosophical sophistication in the field has gone well beyond a series of mere "sketches," the lack of critical discussion among the robust theories in the field remains a problem. Indeed, it seems that each new entry rebuilds philosophy of technology from scratch rather than moving forward through critical interaction. Ströker's solution to this problem involves the adoption of a more rigorous interdisciplinary connection between philosophers studying technology and scientists involved in the production of new technologies. While we are a bit wary of her prescription, we share the motivation which drives her analysis: "only as a systematic discipline can [philosophy of technology] take on the rank among other philosophical disciplines that it deserves (today more than ever) if one judges by the importance of its tasks" (Ströker, 1983, p. 334).

Our view is that philosophy of technology as a form of applied philosophy has not made very much progress since Ströker diagnosed the paradox of continual beginning. Our intuitions are that the lack of self-criticism and healthy philosophical debate is linked to the absence of common questions in the field around which it might systematize itself. Joseph Pitt, a past president of the Society for Philosophy and Technology, indicated a similar worry at a symposium in late, 1995. Pitt argued that the more substantive challenge to the health of the discipline involved its overwhelming focus on the social consequences of technology: "a lot of philosophers working in the philosophy of technology see the philosophical job exclusively in terms of identifying and bemoaning the impact of technology on society."[2] Later in the paper he asks, "Is it really true that the only interesting philosophical problems about technology are the problems associated with social impact?" (Pitt, 1995, paragraph 11).

We believe this to be an excellent question, one which we will try to answer here. If Ströker and Pitt are right, what we find in philosophy of technology is a field out of balance: there is little substantive philosophical exchange, and what mostly makes up the substance of the field is social critique. But while our worry is similar to Pitt's, it is ultimately broader. While we agree that philosophy of technology is too exclusively focused on normative social claims, the problem with these claims is not that they are exclusively negative, as Pitt maintains. It could simply be the case that negative social assessments are closer to the truth about the social effects of technology. Given that possibility, the success of Pitt's argument awaits an explicit resolution of the social-normative questions.

In this chapter we will attempt to bring together the concerns represented by Ströker and Pitt. We will argue that one of the biggest problems in philosophy of technology is the predominance of social-normative assessments of technology as the almost exclusive focus of the field.[3] We argue that Ströker's worries about the lack of critical discussion could in part be solved through an increased role for philosophical descriptions of technology which, while connected to social-normative arguments, can be separated as the basis for a common ground of philosophical inquiry. We believe that philosophy of technology becomes a distinct field of applied philosophy first in its articulation of philosophical descriptions of technology itself; this amounts to saying that the field should be built from the bottom up, not suspended from higher level (and thus more vulnerable) social-normative judgments. We are not opposed to social-normative analysis; on the contrary, we believe such analysis is rightfully viewed as a vital part of the field. We simply believe that a common descriptive apparatus would serve to cohere those analyses, to give them ground on which to engage one another. Because social-normative categories come from wildly differing philosophical schools of thought (Marxist, Heideggerian, Dewian, etc.), it is no wonder that philosophers of technology often do not engage with each other's assessments of the social effects of technology. Attempting to restart the field at the level of descriptions of technology may provide a more common ground for philosophical debate.

After reviewing the general state of the field, we will turn to an analysis of the work of Carl Mitcham, one philosopher of technology who has focused on a more descriptive approach to identifying the subject of his philosophical endeavors. We will then sketch an alternative descriptive approach to Mitcham's grounded in Wittgenstein's descriptive strategies, in part to articulate our own account of improving the descriptive base of the field and in part to show how one form of descriptivism (ours) can critically interact with another (Mitcham's).

1. THE SOCIAL-NORMATIVE CONTENT OF PHILOSOPHY OF TECHNOLOGY

Early philosophy of technology is marked by a tendency to classify types of technology strictly in terms of their social-normative effects. In other words, most of the philosophical forerunners of what we call philosophy of technology today – for example, Heidegger and Ellul – run together their descriptive classification of technology with a normative argument.[4] At first such arguments were very crude. For example, Ellul claims that all of modern technology (which he calls *technique*) shares an overarching ontological-normative dimension, one that is predominantly harmful to humans. But considering the period in which such views were expressed, the one-sided vehemence of such accounts of technology is understandable. The prevailing view against which such accounts were directed was an instrumental description of technology, one that characterized it as a normatively neutral set of artifacts.[5] By claiming that technology is normatively substantive, thinkers such as Ellul and Heidegger made a case for something other than a simple instrumental view of technology. By making these claims as strong as they did – for example, by arguing that the effects of technology on society were pernicious and determinant – they made their case more compelling. In the field's natal stages, then, a strong but defensible normative thesis was needed to counter the dominant view that technology had no substantive theoretical content at all. If Ellul and Heidegger did nothing else, they established that technology is a philosophical issue worthy of our most careful attention.

In the years since the initial articulation of this view, the general form of such arguments has changed and the normative claims of philosophers of technology have become more balanced. On the whole, however, the structure of argumentation in the field has not advanced far beyond such social-normative typologies of technology. For example, one of the most influential writers in the field today, the neo-Heideggerian Albert Borgmann, divides technologies up into "devices" and "focal things."[6] Without attempting a thorough account, we may simply point out that Borgmann defines focal things as those objects which engage us in a beneficial social context and are inseparable from that context; he defines devices, characteristically the artifacts of modern technology, as objects that are context-independent. Devices may be used anywhere at any time. In removing the need for the skills and social interaction required by focal things, devices leave us "disburdened" and disassociated from our social milieu.

Borgmann's paradigmatic example of a focal thing is the wood stove found in the North American frontier prairie home. Each family member fulfilled a special function in relation to the stove; some gathered wood, others stoked the stove, and others cooked (Borgmann, 1984, p. 42). The stove was a medium for family life in at least three ways: it provided temporal regularity for the household, marked the different seasons, and served as a locus for familial responsibility and individual identity.

In contrast, the device is the foreground of modern mass technologies for which the world of things is the rapidly receding contrasting background. As Borgmann explains:

> A device such as a central heating plant procures mere warmth and disburdens us of all other elements. These are taken over by the machinery of the device. The machinery makes no demands on our skill, strength, or attention, and it is less demanding the less it makes its presence felt (Borgmann, 1984, p. 42).

As artifacts that need no particular context, devices diminish the quality of social relations, relations once centered around focal things and their accompanying practices. It is because of this focus on the pernicious effects of devices on our social relationships that Borgmann calls his theory the "device paradigm."

Borgmann's account, while certainly more substantive and interesting than Ellul's, shares with the latter's at least one quality: the ontological typology of technology is reducible to purely social-normative dimensions. An object is classified ontologically according to its social-normative consequences; this means that an objection to Borgmann's social-normative criteria counts as an objection to his ontology, which is merely a reification of those criteria.

And, unsurprisingly, it is precisely Borgmann's social-normative criteria that are the most controversial part of his theory. Andrew Feenberg, for example, has argued that Borgmann's views are romantic, resting on a naive picture of an older, less intrusive form of technology.[7] Regardless of whether this charge – or any of the others directed at the device paradigm – is true, a problem emerges: Borgmann's typology of different kinds of technology would not withstand a rejection of his social-normative analysis. This leaves the entire theory in a state of perpetual vulnerability, something which could be avoided if Borgmann's ontology could be separated, even theoretically, from his social critique.

The lack of critical interaction in philosophy of technology comes into focus when we consider that this absence of a distinction between the formulation of an ontology of technology and an argument about its normative properties has remained largely unchallenged in the field. Even as strong a critic of other philosophical accounts of technology as Feenberg has failed to raise this particular issue.[8] The lack of attention to this problem is odd, given its stunting effect on the field's growth and maturation. Today, philosophy of technology is in an awkward situation: any of its major theories, if rejected on social-normative grounds, will be forced to start anew from the ground up.

As was mentioned earlier, it can be argued that this fragile dependence of ontologies on social-normative judgments explains the rarity of direct engagement among major figures in the field. Each socially normative view of technology imports a fully formed social theory which helps to make sense of the social implications of technology. Thus, to reject these socially normative views of technology one must be prepared to take on the entire social critique implied by the analysis. But of course, this is quite a challenge. It would be easier simply to offer an alternative social theory rather than to engage with another on its own ground. But if substantive descriptions of technology could be divorced from a theorist's social-normative views, then there could be another level on which the theory could be engaged. This level, describing the object of philosophy of technology, might not rely on any particular social theory, even though it would surely rely on some kind of imported framework of analysis. A separation of descriptions of technology from its social-normative consequences (and the social theories that such analyses rely on) could help to establish a new ground on which different schools of philosophy of technology could engage one another. At the very least, encouraging an account of descriptions independent of social analysis would be a point of entry into a critical discussion among different views.

Let us call the claim that philosophers ought to embrace the importance of descriptions in the philosophical study of technology the "descriptivist thesis." It does not follow from this thesis that higher level social-normative claims cannot be built on top of descriptions of technologies. It simply means that descriptions of technology ought to stand alone as philosophical claims, rather than finding their philosophical import solely in their social-normative implications.

2. MITCHAM'S DESCRIPTIVISM

Although he does not articulate it in the same context, there is at least one figure in the field who seemingly agrees with the descriptivist thesis. Carl Mitcham, in his book *Thinking Through Technology*, provides what is arguably the most thorough descriptivist account of technology yet produced.[9] The question we engage here is whether Mitcham's analysis actually avoids the problem we are identifying or ultimately instantiates it.

The core of Mitcham's book is his elaboration of a distinction between "engineering philosophy of technology" (EPT) and "humanities philosophy of technology" (HPT). EPT "takes technological thought and action as the models for all human thought and action and attempts to explain or reformulate all apparently non-technological thought and action in technological terms." HPT, on the other hand, "argues that technological thought and action are only one aspect or dimension of human thought and action, and seeks to delimit or restrict the technological within a more expansive framework" (Mitcham, 1995a, paragraph 3). Mitcham elaborates on this distinction in great detail, arguing that it is grounded in a history of these largely separate pursuits, a history that he exhaustively documents in the book. In accord with our analysis, Mitcham also suggests that HPT is concerned primarily with normative social issues: "humanist philosophers [of technology] seek to define and delimit technology out of a concern that the expansion of technology can in many instances undermine human social and cultural experience" (Mitcham & Waks,

1995, p. 236). Again, in an encyclopedia article that draws heavily on the account in his book, Mitcham remarks that, "Not theoretical analysis, however, but ethical and political concerns predominate in philosophy of technology" (Mitcham, 1995b, p. 2480).

In the second half of his book, Mitcham turns to prescription. In order to become a rigorous academic discipline, he argues, HPT must be more inclusive of EPT. This primarily involves an increased commitment to technical understanding, the purview of engineers. "Humanities *philosophy* of technology" will need to become "humanities philosophy of *technology*." The synthesis of the two, with the change in emphasis, becomes "philosophy of technology simpliciter" (p. 143, author's emphasis). On its face, this turn appears hopeful to the descriptivist; greater technical understanding would surely lead philosophers of technology to enrich their descriptions of their subject. But in what way, and to what end? Our overall concern with Mitcham's recommendation is twofold. Problem 1 is that it is at best uncertain whether engineering detail, and the descriptions it engenders, carries any philosophical weight; if it is, strictly speaking, aphilosophical, then there seems little reason to expect that its inclusion in philosophy of technology will have any impact on the discipline's philosophical practices.[10] Problem 2 is that if the preceding is accurate, then the larger descriptivist goal – establishing a shared ground on which opposing philosophers can meet and engage – will not be importantly served by the turn to EPT.

On problem 1, the fact that philosophers of technology in an expanded HPT would be more sophisticated in their understanding of, as Mitcham puts it, the "internal" workings as opposed to "external" consequences of technology does not mean that they would thereby become good descriptivists. Further attention to the elaboration of Mitcham's argument will make this point clear. Mitcham explains that his new philosophy of technology simpliciter would first undertake an analysis of the characterization of technology itself. Careful attention must be given to this step, because it is the same one in which figures such as Borgmann have moved directly into normative characterizations. Mitcham puts forth two divergent characterizations of technology, mirroring the HPT-EPT distinction: the "engineering" and the "social science" senses of technology. The engineering view is limited, reserved "for more direct involvement with material construction and the manipulation of artifacts" (p. 147). The social science view is broader, including 'all of what the engineer calls technology, along with engineering itself' (p. 149). Of course Mitcham takes the engineering view very seriously, so when he worries that the tension between these two senses of "technology" cannot easily be resolved, we ought to pay attention.

Mitcham identifies three responses to this tension concerning the meaning of technology: (1) stipulate around the problem; (2) use subscripts to distinguish one usage from another; and (3) "provisionally adopt the extensive [social science] meaning, with the intention of gradually formulating distinctions within it by whatever means become available and appear appropriate during the course of a deeper analysis" (p. 152). Mitcham chooses the third alternative and then summarizes his approach accordingly:

> The thesis is that "technology" is not a universal term; it does not mean exactly the same thing in all contexts. It is often, and in significant ways, context dependent – both in speech and in the world. But neither is it a pure equivocal such as "date," which can refer to wholly unrelated things on a calendar or a palm tree. There is a primacy of reference to the making of material artifacts, especially since this making has been modified and influenced by modern science, and from this is derived a loose, analogous set of other references. An initial need in the philosophy of technology is for some mapping out or clarification of this conceptual one and many, a conceptual one and many that can be assumed to reflect a real diversity of types of technologies with various interrelations and levels of unity (p. 152).

Mitcham is on the right descriptivist path; thus far in his characterization, there have been none of the customary sweeping social-normative distinctions. All that has been established is Mitcham's central intuition, reflected in his provisional preference for the social science sense of technology, that technology is "pivotally engaged with the human" (p. 159). Proceeding with his characterization, Mitcham argues that if technology has something to do with human activity, loosely speaking, then a "philosophical anthropology" is needed to divide up the different forms of life associated with technology. Mitcham accordingly draws distinctions between technology's "manifestations in the mind, through bodily activities, and as independent objects that take their place in physical and social worlds" (p. 159). From these manifestations, Mitcham divides technology into its forms as knowledge, activity, and object. He adds "technology as volition" to this list to distinguish technology as a concern of the "will" as separable from the sphere of cognition (p. 159).

This development in Mitcham's account is very encouraging to the descriptivist. Mitcham certainly realizes that his new ontology of technology is unique in the field, in part because it is developed independently of social-normative views. He suggests for example that the "framework at hand can support either a technological determinism (in which objects or ideas exercised a controlling influence on human activity) or a theory of human freedom (in which individual volition or creative knowledge plays a dominant role)" (p. 160). His framework apparently gives Mitcham the clear advantage of retaining a philosophically interesting account of technology without resorting to an articulation of more divisive social-normative issues. After surveying many of the definitions of technology made by a variety of philosophers (e.g. Frederick Ferré, Mario Bunge, and Stanley Carpenter), Mitcham concludes with a statement of purpose that distinctly echoes our descriptivist goal: "It remains then to propose and develop a typology that can encourage an active dialogue with such previous attempts, protecting and ordering the insights they contain" (p. 157).

However, the next few chapters of Mitcham's book (those between the chapter just discussed and the summary conclusion) do not deliver on the descriptivist promise. In these chapters, Mitcham expands on each of the four aspects of technology outlined above: technology as knowledge, activity, object, and volition. The format of each chapter is the same, with an opening etymological discussion followed by a thorough description, leading into discussion of the relevant literature on the topic. There are two problems with these chapters from the descriptivist perspective: one involves the philosophical neutrality of the detail in Mitcham's discussion; the second is the drift in the chapters toward invoking social-normative issues to justify the descriptive discussion. We will elaborate on these problems through a discussion of the first of these chapters, "Types of Technology as Object."

As one proceeds through the chapter and the descriptions of technology as objects get more detailed, a worry arises over whether the philosophical content of Mitcham's descriptions has completely fallen out of his account. For example, early in the chapter Mitcham pauses to argue that the six classical machines of antiquity – lever, wedge, wheel and axle, pulley, screw, and inclined plane – are really all variations on two: the lever and inclined plane. But there is no explanation why this classification of technology as object is of philosophical importance or why it is the product of a philosophical line of inquiry. Even more confounding is Mitcham's distinction between power tools and machine tools:

> Power tools and machine tools are not always carefully distinguished. Power tools are power-driven hand tools such as the electric drill and saw or the air-driven jackhammer or even kitchen and house-hold appliances. A rotary electric handsaw is a lot different from the electric table or radial-arm saw, which are stationary shop power tools. Machine tools, by contrast, are power tools for metal cutting – machines used to make machines. Some basic classes of machine tools, moving from the early and general to more recent and specialized are lathes, shapers and planners, drilling machines, milling machines, grinding machines, power saws, presses, turret lathes, production millers, gear-cutting machines, and broaching machines. Recent innovations in machine tools include automatic control by

> tracing or record-playback techniques and numerical control, as well as chipless methods of removing metal by plasma arc and laser-beam machining (pp. 167–168).

The description here is perplexing, though certainly not because it is unclear. It is nothing if not clear. It is confusing because, again, there is no clear philosophical rationale for its articulation. Later in the chapter power tools appear again, but only, to put it crudely, for their "de-skilling" and hence social-normative effects. However, Mitcham does not explain why this exhaustive description of the difference between power and machine tools is philosophically important *as a description*. Later in the chapter, it becomes clear that the point of distinguishing power tools from other tools is clearly and exclusively to assess their social-normative implications. Importantly, the power tools / machine tools discussion just cited is no tangent in Mitcham's text. It is offered as part of a serious philosophical discussion and as the beginning of an approximation of what a reworked philosophy of technology would include. We can assume, then, that this paragraph is only a prefatory indication of what a full fledged and appropriate description of power tools would look like on Mitcham's descriptivist theory.

But is it philosophical? After many of these intricate descriptions, Mitcham says: "Two basic philosophical questions about artifacts are, What are they, both conceptually and ontologically? and How are their distinctions and operations to be understood?" (p. 172). It is clear that the first question is consistent with the descriptivist thesis and also securely philosophical. The second question, however, which seems to have elicited the power tools analysis, is not obviously or necessarily philosophical. Simply describing the mechanical workings of an artifact would not count as a philosophical description and hence would not be the sort of activity called for by the descriptivist thesis.[11]

Mitcham's discussion certainly does not stay at this level, nor is it exhaustively concerned with non-philosophical (or at best philosophically marginal) descriptions. There are some incredibly interesting discussions in this chapter that fall under the general rubric of his first question concerning the ontology of technology. But as the chapter progresses, problem 2 concerning the implications of Mitcham's descriptivism emerges. By page 179, Mitcham has turned to a discussion of the social effects of technology. The only descriptions of technology carried forward through the chapter serve this social-normative discussion.

A good deal of this part of the chapter is an extended discussion of the "artifactist" thought of Ivan Illich, Lewis Mumford, and others. This discussion falls ostensibly under a subsection on the "phenomenology of artifacts"; in it we learn that the point of "artifactology" is the advancement of the thesis that "artifacts have consequences" (p. 182). Nevertheless, as mentioned above, it is encouraging here that Mitcham uses this discussion to criticize those artifactists who rely on the most general "external" or social descriptions of technology as the point of their philosophy. On a cursory reading, Mitcham could appear to be applying a strict descriptivist criteria here. According to Mitcham:

> In none of the cases listed, however, do the authors provide extended and detailed analysis of the inner structures of artifacts and how such structures give artifacts inherent tendencies toward specific types of human engagement and use. Their focus remains largely at the macro and in one case symbolic level, stressing external relations (p. 182).

And specifically, concerning Ellul: "his characterology applies to technology and tools more as societal institutions than as material objects" (p. 182). Even Mumford, one of the more technically adept early figures in the field, is also criticized for failing to follow Mitcham's version of descriptivism: "when Mumford analyzes the influence of machines on human affairs, as with the case of the mechanical clock, he does not relate this influence to structurally distinct properties in artifacts" (p. 183).

But even with this criticism, it becomes clear that for Mitcham the point of relating social-normative influences to structural properties of artifacts is to bolster the credibility of what remains an essentially social-normative analysis. This point is clearest, again, when Mitcham finishes his critiques and provides some examples of the sort of analysis of technology he would like to see. Approvingly echoing the more substantial descriptions of Illich (and clearly Borgmann), Mitcham compares the social consequences of hand tools and power tools: "The latter less and less allow end-users to introduce their personal intentions into the world, to leave traces of themselves in those rich constructs of traditional artifacts that have served for millennia as the dwelling places of humanity" (p. 184).

We need go no further to see that our original fears about Mitcham have been confirmed. While he began with descriptivist intuitions, Mitcham's purpose in shoring up the descriptions of HPT with the analytical tools of EPT is ultimately to provide a more solid base for judgments about the social implications of technology, without necessarily legitimating the descriptions themselves as philosophical activities. In this case, power tools, like Borgmann's devices, cut us off from some social connection that we would otherwise have. This explains why power tools are fit subjects for philosophical analysis. The philosophical propensity to build arguments around a narrow focus on social-normativity has not been challenged, for the added elements of

Mitcham's new philosophy of technology are not philosophical but mechanical.

3. TOWARD A WITTGENSTEINIAN DESCRIPTIVISM

It is important to note that we are arguing not merely for the importance of descriptions in philosophical accounts of technology, but for the importance of *philosophical* descriptions in philosophical accounts of technology. The incorporation of descriptions from Mitcham's EPT is not necessarily conducive to the end of establishing a firm, philosophically self-sufficient descriptive base for the philosophy of technology. While we acknowledge that a healthy engineering acumen is important to philosophers in the field and that such acumen will prevent embarrassing errors and encourage clarity, our concerns are not allayed by the prospect of philosophers of technology learning the difference between, say, a lathe and a level. What we think the field needs is a rudimentary ground on which a set of descriptions can be philosophically debated as interesting in their own right, a set which will serve as a touchstone around which contrasting accounts, normative and otherwise, may be centered. This may allow for substantive debate in the field rather than today's collection of largely unconnected normative frameworks. We now turn to sketching an alternative to Mitcham's account of developing a descriptivist foundation for the field.

When we consider the role of descriptions in philosophy, Wittgenstein's later work comes naturally to mind. In it, he makes the celebrated claim that description is the sole task of the philosopher: "We must do away with all explanation, and description alone must take its place."[12] We will examine in a moment whether Wittgenstein's descriptive strategies are separable from this claim. First, however, we should explain what we interpret Wittgenstein to be saying in his discussions of description.

Wittgenstein's most sustained treatment of description is offered in §106–§130 of the *Philosophical Investigations*. A selection of quotes will show the general character of the passages:

> Here it is difficult as it were to keep our heads up – to see that we must stick to the subjects of our every-day thinking, and not go astray and imagine that we have to describe extreme subtleties, which in turn we are after all quite unable to describe the means at our disposal (§106).

> These are, of course, not empirical problems; they are solved, rather, by looking into the workings of our language, and that in such a way as to make us recognize those workings: in spite of an urge to misunderstand them. The problems are solved, not by giving new information, but by arranging what we have always known (§109).

> Philosophy simply puts everything before us, and neither explains nor deduces anything .... Since everything lies open to view there is nothing to explain. For what is hidden, for example, is of no interest to us (§126).

> The work of the philosopher consists in assembling reminders for a particular purpose (§127).

In general, commentators have acknowledged that Wittgenstein stressed the priority of description over explanation, but little has been done to explore just what type of descriptions he had in mind. The tendency has been simply to conclude that he intends us to describe *language*. This is undoubtedly true, but what aspects of language? Given Wittgenstein's general emphasis on meaning-as-use, it is often assumed that what he had in mind was description of specific terms in a language-game, or perhaps of language-games themselves. These interpretations are certainly in line with the passage above regarding the "workings of our language," as well as other passages throughout the work. However, we are inclined to agree with Heather Gert that the passages in question do not determine these interpretations, at least not exclusively.[13] Wittgenstein is not forthcoming on the issue. We are inclined to decide the question by looking at what Wittgenstein himself chooses to describe.

When Wittgenstein considers descriptions, they rarely turn out to be of the use of particular terms or language-games. Rather, the descriptions he most often considers are descriptions of *phenomena* – colors, paintings, facial expressions, music, sensations, etc. He attempts to present these descriptions as he finds them, eschewing (as much as possible) the characteristic philosophical tendency to sweep them too quickly into generalizations. He envisioned himself as a painter inventing a new style, a style that clearly represents what is in itself fuzzy: "For I want to keep telling myself: 'Make sure you really do paint only what you see!'" (Wittgenstein, 1980, p. 68e). These passages are scattered throughout his work; what they have in common, Gert points out, is that Wittgenstein seems to be concerned with *describing these descriptions*.

For what purpose? What is gained by describing descriptions of everyday phenomena? In a sense, there are as many answers to this question as there are phenomena to describe. Here we only need to establish that understanding such descriptions is important to a clear understanding of the phenomena themselves. Wittgenstein forcefully expresses this sentiment in §122:

> A main source of our failure to understand is that we do not *command a clear view* of the use of our words .... Our grammar is lacking in this sort of perspicuity. A perspicuous representation produces just that understanding which consists in "seeing connexions ..." The concept of a perspicuous representation is of fundamental significance for us. It earmarks the form of account we give, the way we look at things.

As Wittgenstein says in §109, we have an "urge to misunderstand" the workings of language. This is due to "pictures" that are woven into our grammar. Certain language-games urge pictures on us, and we produce confusion and error when we literalize these pictures or apply them to other language-games that are not always closely related.[14]

A celebrated example of Wittgenstein's struggle against such a picture is his treatment of the "problem" of sensation. He stressed that our confusion with regard to sensations is a function of a picture woven into our language, one which models sensations as objects that must be spatially located. This is not the place for an extended discussion of Wittgenstein's treatment of the problem; what is important to see is Wittgenstein's goal: he believes that by clearly describing the language that we use to describe sensations – that is, describing our descriptions, including the network of similarities and differences connecting them to surrounding, related language-games – we can come to understand the *phenomena* of sensations. As Wittgenstein wryly points out in §314: "It shows a fundamental misunderstanding, if I am inclined to study the headache I have now in order to get clear about the philosophical problem of sensation."

When Mitcham turns to EPT as a source for a new HPT, when he asks after the "inner structures of artifacts and how such structures give artifacts inherent tendencies toward specific types of human engagement and use," is he studying his headache? That is to say: is the structural investigation of various technologies conducted by EPT the best way to understand the *philosophical* problems of describing technology? Mitcham thinks that part of the explanation of "human engagement" with technology is hidden in the "inner structures" of various particular kinds of technologies. The problem with this focus on engineering detail is that it damagingly limits the very philosophical anthropology that Mitcham urges. If technology is "pivotally engaged with the human," if it is woven into a fabric of diverse practices and activities, then it follows that engagement and use will be influenced by a whole network of factors. The 'inner structure' described by EPT is but one of those factors, and it remains for the descriptivist philosopher of technology to clarify its place in the fabric.

We seek a descriptive methodology for the philosophy of technology, but we think there is a better way to produce it than looking beneath our engagement for structural explanations. As Wittgenstein says, "it is difficult as it were to keep our heads up – to see that we must stick to the subjects of our every-day thinking" (§106). To turn to EPT is to attempt to explain our engagement, to deduce what causes it, how it is possible. But Wittgenstein warns us away from this tactic; he would have us focus on describing that engagement with technology as it stands. From a Wittgensteinian point of view, human engagement with technology does not take on meaning or significance by virtue of the physical structure of technologies any more than sensation-talk takes on meaning by virtue of the physical structure of our brains. The philosophical problems manifest themselves on a different level. We want to suggest that Wittgenstein's method of description is a more effective way of exposing and clarifying these problems than Mitcham's attempt, though we do not suggest that this is the only valid form of descriptivist philosophy of technology.[15]

So how might Wittgenstein's insights play out in the philosophy of technology? It would involve, for one thing, a claim that the descriptions of technology present in natural language already contain the distinctions and judgments necessary to provide a self-sufficient base for the philosophy of technology.[16] We would use as our starting point philosophical anthropology (along with Mitcham). We would investigate the forms-of-life and descriptive grammars bound up with various families of technologies. It is true, no doubt, that many of the distinctions present in natural language will be normative in character. For instance, we often distinguish those technologies that "make our lives easier" (e.g. ATM machines) from those that serve as orienting rituals, binding our immediate community (e.g. the equipment used to homebrew beer). This distinction lends itself to the normative conclusions drawn by Borgmann and others. But what is important is that we need not begin with these normative conclusions, interpreting everyday language solely in terms of such frameworks. We need not relate to or speak of technology solely in terms of its positive or negative effects on the social fabric any more than we relate to it solely as a set of inert artifacts.

For example, the rise of the information age and its attendant information technologies presents an interesting collection of case studies for the descriptivist. Furthermore, information technology is a perspicuous example of the need for philosophers of technology to pause and fully describe before they pass social judgment. A univocal condemnation or celebration of information technology's social-normative effects is at this point, to put it mildly, premature. What is needed is a flexible and inclusive attempt to descriptively map the complex and shifting web of "engagement and use" of which these technologies are a part; to focus immediately on social-normative effects is to sacrifice the messy complexity of the subject matter to a rush to judgment.

The rubber of the descriptivist thesis hits the road, so to speak, in such concrete events as the U.S. government's recent attempts to regulate the content and use of the Internet. Congress unwittingly mimicked the dominant maneuver in philosophy of technology by quickly moving, in an apparently defensive reaction to the daunting complexity of the subject matter, into an exchange of strident and narrow social-normative judgments. The debate was characterized by confusion about how exactly to characterize the proposed subject of legislation. Perhaps, given today's Congress, this was not surprising, but in this case the general ignorance seems less contemptible than usual, if only because internet experts and casual users alike are in the grip of a similar confusion. What is the Internet?

In a recent book that admirably enacts the spirit of philosophical anthropology called for by the descriptivist thesis, Mark Stefik, a scientist working on information technology at the Xerox Palo Alto Research Center, has collected excerpts from a wide variety of sources to illustrate the varied metaphors used to describe the Internet.[17] Is the Internet a vast library, a storehouse for cultural information? Is it a medium for communication along the lines of the postal service or telephone systems? Is it a global marketplace for unmediated exchange of goods and services? Or is it a digital world in itself, a gateway to experiences like groupware and virtual reality? These are the four primary metaphors that Stefik uses to organize his book, but it is not difficult to think of others. Much recent work at the crossroads of philosophy and geography has explored the consequences of viewing the Internet as a place, with all the accompanying associations that attend our notions of place. Those interested in evolutionary theory might view the internet as Teilhard de Chardin's "noosphere," the evolving realm of ideas, given material manifestation. And, of course, let us not slight U.S. Vice President Al Gore by passing over the popular metaphor that he is believed to have coined: the information super highway, a network of connections that will, like the interstate road system before it, shrink the distances between us.

These are merely gestures, and inadequate ones at that. We do not claim to be philosophical anthropologists on par with Wittgenstein (or Mitcham), nor is it our purpose to attempt such a feat here. We simply want to establish that Wittgenstein's method, describing descriptions, is one useful way of uncovering philosophically interesting distinctions and classifications that can serve as a groundwork for a robust philosophy of technology.

Something remains to be said about Mitcham's provisional adoption of the social science usage of the word "technology," quoted earlier. Mitcham acknowledges that the word is "often, and in significant ways, context dependent" (p. 152). However, he proceeds from a "primary reference" – the making of material artifacts – and then suggests in investigation how this "one" relates to the "many" context-dependent uses. We are concerned that Mitcham has moved from a narrow, univocal definition of technology, that of EPT, to an only slightly less univocal definition for use in HPT. It is the attachment to technology's purported "primary reference" that has stymied attempts to give a full accounting of technology's role in contemporary life.

Wittgenstein often warned against a craving for generality which leads to a "contemptuous attitude towards the particular case" (Wittgenstein, 1958b, p. 18). He might point out that one factor involved in the problem of exclusively social-normative theories in philosophy of technology is the influence of the assumption, echoed in Mitcham's words above, that technology must be "one" as well as "many." When technology (or any phenomenon) is conceived in this way, it is the "one" that tends to get the attention. Particularly in analytic philosophy, we are tempted to proceed directly to theories and frameworks that purport to explain technology as such. This is one influence which urges philosophers of technology toward social-normativity, since little but a normative framework could subsume all the multiplicitous varieties of technology.

Wittgenstein can offer us a helpful heuristic, expressed in his discussion of the notion of a "game":

> Consider for example the proceedings that we call "games" .... What is common to them all? ... Don't say: "There *must* be something common, or they would not be called 'games' " – but *look and see* whether there is anything common to all .... For if you look at them you will not see something that is common to all, but similarities, relationships, and a whole series of them at that. To repeat: don't think, but look!
>
> And the result of this examination is: we see a complicated network of similarities overlapping and criss-crossing: sometimes overall similarities, sometimes similarities of detail (Wittgenstein, 1958a, §66).

The philosophical import of technology lies in its engagement with the human, as Mitcham rightly points out, but this engagement is diverse, like the many things to which we refer as "games." When we talk about technology, what we refer to is not always derivable from, or even connected to in any non-trivial way, the making of artifacts. Again, information technology is a telling example: as hardware shrinks and software – i.e. strings of information – grows, it becomes more and more unhelpful to think of this process in terms of artifacts and their production. It may be that we can view rearranged information (computer programs) as artifacts, but as those programs become more able to rearrange themselves and interact with users in more intuitive ways, this way of talking about software will enter into tension with the reality of human engagement and will either end or impede that engagement.

For instance, Netscape posts various types of software on a network, which allows the subscriber to use these programs through a network connection rather than buying the software for herself. Is "access" an artifact? Is a string of programming code? Presumably this would fit under Mitcham's category of technology as knowledge (programming information), or perhaps technology as activity (access). Nevertheless, one could make the argument that a good deal of our confusion with regard to information technology lies precisely in our tendency to assume this "primacy of reference," that is, to assume that we must understand all transactions on the model of artifact production, exchange, and use.

If we adopt Wittgenstein's strategy, when we consider technology we first *look*. Our first step is not the assumption of a systematic conceptual unity, but rather the illumination and clarification of diversity. On this model, the primary goal of philosophy of technology, still a relatively young philosophical sub-discipline, ought to be the development of a shared descriptive foundation that reveals these "overlapping and criss-crossing" similarities.

But simply telling philosophers to "look" is not much help. Look for what? In the passage that follows the one quoted above, Wittgenstein begins to give his heuristic some substance:

> I can think of no better expression to characterize these similarities than "family resemblances"; for the various resemblances between members of a family: build, features, colour of eyes, gait, temperament, etc., etc., overlap and criss-cross in the same way .... And I shall say: "games" form a family (Wittgenstein, 1958a, §67).

To get a feel for what this heuristic would look like in the hands of a philosopher of technology, we have made some suggestive substitutions in brackets through the remainder of this passage:

> And for instance the kinds of [technology] form a family in the same way. Why do we call something ["technology"]? Well, perhaps because it has a – direct – relationship with several things that have hitherto been called [technology]; and this can be said to give it an indirect relationship to other things we call by the same name. And we extend our concept of [technology] as in spinning a thread we twist fibre on fibre. And the strength of the thread does not reside in the fact that some one fibre runs through its whole length, but in the overlapping of many fibres (Wittgenstein, 1958a, §67).

On a Wittgensteinian model, descriptivist philosophers of technology make it their primary goal to describe these fibers in detail and to elucidate the ways in which they interconnect and overlap. Importantly, such a project, in this context, is thoroughly philosophical.[18]

4. CONCLUSION

Our attempt in this paper has been to isolate, diagnose, and gesture toward a way of alleviating the central cause of the lack of critical interaction in contemporary philosophy of technology. In part we have worked through this problem by endorsing a descriptivist approach to the field; in part we have tried to explicate this concern by critically interacting with one of the few theorists who has identified the need for improved descriptions in the field. Our criticism, however, should not be taken as a distaste for the field or its history. For its young age, philosophy of technology has suffered from an embarrassment of riches. The short history of the field is cluttered with thinkers of surpassing imagination and humanitarianism. Nevertheless, what has been the charm of the field is becoming a liability; namely, the field largely remains a collection of idiosyncratic thinkers united more in name than in philosophical orientation. Philosophy of technology has not taken on a life of its own, growing organically from the roots of shared descriptions and central questions; rather, it is rebuilt nearly from the ground up with each new account. This has and will continue to prevent the field from taking a respected place alongside others in mainstream applied philosophy.[19]

It is important to point out that we are not alone in our efforts here. Though few have made an explicit point of acknowledging it, there are several philosophers of technology who have embraced one form or another of descriptivism, however different their interpretation may be from ours. Certainly the work of Kristin Shrader-Frechette qualifies, though it is restricted to the narrow range of technologies associated with risk management.[20] Don Ihde's phenomenological approach is also clearly descriptivist, and the more recent work of Andrew Feenberg may also avoid the problem.[21] One last example worth mentioning can be found in Hugo Padilla's paper, "Technological Objects and Their Epistemological Base."[22] There Padilla outlines a formal account of the epistemic status of technological objects, their relation to theory construction in applied and pure science, and the underlying relation between applied and pure science in itself.

While the approach we have offered differs from those of the authors just mentioned, we hope that it is supportive in spirit. All of these approaches demonstrate a renewed focus on descriptions. Importantly, like most of these theorists, we do not believe that social-normative analysis should not occur, nor that descriptions must be normatively neutral to be legitimate. We simply think that a healthy philosophical discipline devoted to technology should comprise a shared space in which competing social-normative accounts can engage one another. We believe that the field will be healthier when each account within it is not vulnerable to the contestation of one or all of its social-normative claims. And we believe that a renewed focus on the art of philosophical description is a promising way to encourage the critical intradisciplinary conversation that is the hallmark of a robust philosophical community. We hope that our gesture toward a Wittgensteinian descriptivism contributes to that conversation.[23]

NOTES

1. Since 1978 a hard cover annual, *Research in Philosophy and Technology*, published by JAI Press, has served as the primary journal in the field. Another book series, *Philosophy and Technology*, was published in hard cover from, 1981 to, 1994 by Kluwer Academic Publishers. The Society for Philosophy and Technology (SPT) has met continuously since, 1977, holding international conferences every other year since, 1981 and meeting regularly at the divisional meetings of the American Philosophical Association. The Society now maintains a quarterly electronic journal at http://borg.lib.vt.edu/ejournals/SPT/spt.html. In the U. K. the SPT has been much less active. However, a supplemental volume of the Royal Institute of Philosophy was published in, 1995: *Philosophy and Technology*, edited by Roger Fellows. But as Fellows makes clear in the first line of his introduction, the essays in the volume do not represent philosophy of technology in the sense which "Don Ihde requires" (p. 1). We take it that this means that these essays are not in the field as it has been staked out by the SPT, and so are outside the scope of our essay. Whether our proposal for a more descriptivist philosophy of technology will help to bridge the gap between SPT style philosophy of technology and the work represented in the Fellows collection we will leave to another day.

2. Pitt, 1995, paragraph 9. (All citations from this journal are by paragraph because of the lack of page numbers in its world wide web format.) For an updated version of this argument, see also Pitt, 1999.

3. Throughout this chapter, when we refer to the "social-normative" implications or assessments of technology, we mean arguments about the social, political, moral, or otherwise axiological assessments of technology.

4. See for example Ellul, 1964 and Heidegger, 1977.

5. It is important to note that the descriptivist thesis does not imply the instrumental view of technology. Social-normative arguments are one way to distinguish technologies, for technologies certainly elicit social-normative effects; however, they should not be the only or even principle way to distinguish technologies.

6. See Borgmann, 1984. The device-thing distinction was later followed by a distinction between the real and the hyperreal. Essentially, the normative implications of the real are the same as are found in things, and the implications of the hyperreal are the same as are found with devices. See also Borgmann, 1992.

7. See Feenberg, 1996.

8. For Feenberg's other arguments on such issues see his *Critical Theory of Technology* (1992a), "Subversive Rationalization: Technology, Power and Democracy" (1992b), and *Alternative Modernity: The Technical Turn in Philosophy and Social Theory* (1995).

9. See Mitcham, 1994.

10. We certainly do not mean to imply that greater technical acuity and accuracy in philosophical accounts of technology are unworthy goals. Enriched understanding of engineering detail will give philosophical accounts greater credibility, not only in the eyes of engineers, but in the eyes of the public. But, we argue, such a change will not necessarily affect the philosophical orientation of those accounts.

11. It may be the case that the burden here is on what Mitcham means by "distinctions" in the quotation under discussion, but the meaning of this turn of phrase

is not made clear in the text. We assume that both the distinctions and the operations in question are technical/mechanical.

12. Wittgenstein 1958a, § 109. All further citations are given in the text by section number from *Philosophical Investigations* unless otherwise indicated.

13. Gert (1997). Our discussion of Wittgenstein, particularly the following few paragraphs, owes much to Gert's interpretation.

14. Many readers, particularly those accustomed to traditional debates in applied ethics, might view this discussion of language as beside the point. However, it is part of our Wittgensteinian account that natural language contains descriptions of much more complexity and breadth than those generally found in top-down accounts of technology derived from social-normative judgments. Thus, an examination of Wittgenstein's attitude toward language is necessary.

15. There are other descriptivist strategies less circuitous than Mitcham's. See, for example, the phenomenological account of Don Ihde in *Technology and the Life World* (1990), and the critical theory of Andrew Feenberg in his *Questioning Technology* (1999). Our reason for not focusing on either of these figures is that neither of them explicitly addresses the descriptivist problem in philosophy of technology. Certainly their phenomenological and neo-Marxist answers to coming up with philosophical descriptions of technology are legitimate alternatives to our Wittgensteinian approach. The next step will be to articulate criteria for preferring one or another of the approaches. Of course, such debates are what we think should be at the center of the field. For example, one could also claim that a virtue of Larry Hickman's pragmatist philosophy of technology is that it does not rely on an ontology of artifacts. But such a claim only carries weight if the question of forming an ontology of artifacts is placed at the center of our account of the philosophy of technology. See Larry Hickman, *John Dewey's Pragmatic Technology* (1990).

16. Importantly, many of the descriptions in natural language contradict, nullify, or at least fit uncomfortably with others; it is precisely here where compelling philosophical differences are manifest.

17. See Stefik, 1997.

18. In the interest of brevity, we have neglected to address one important issue that Wittgenstein scholars will no doubt raise: Are Wittgenstein's descriptive strategies separable from his larger philosophical project? Wittgenstein's descriptive clarity was not meant to serve as a starting point for further philosophical explication; in fact, he hoped that clarity would entirely obviate the need for philosophy: "For the clarity we are seeking is indeed complete clarity. But this simply means that the philosophical problems should completely disappear" (Wittgenstein, 1958a, §133). Philosophy, as Wittgenstein conceived it, is a misunderstanding, an aberration. It asks questions based on a skewed sensibility, questions which need to be un-asked, not answered (this argument is made at length by James C. Edwards in *Ethics Without Philosophy*, 1982). The question is, Can someone who does not share this view of philosophy still make use of Wittgenstein's descriptive techniques, which in his work are used to point out the fallacy of several traditional philosophical questions?

On this point, we are divided. One of us (A.L.) is inclined to answer that yes, they can be separated. He would have us simply assert that we are not orthodox Wittgensteinians; we simply want to borrow some of Wittgenstein's heuristic methods. As for Wittgenstein's antipathy toward philosophy, it constitutes a separable issue. The other of us (D.R.) is inclined to think that Wittgenstein's descriptive strategies are necessarily linked with his larger (anti)philosophical project, that they exist in service of and by virtue of that project. But if they are inseparable, how can they be put to use by the descriptivist philosopher? D. R. is inclined to say that the descriptivist who uses this strategy must avoid metaphysical or essentialist explanations, which are the targets of Wittgenstein's satire and ridicule. The Wittgensteinian philosopher of technology's task is to clarify and illuminate patterns of human engagement with the various things and practices referred to as "technology," not to justify or explain how any particular thing or practice is possible, nor to isolate the "essence" of technology. Despite our disagreement, we both believe that the descriptivist thesis is valid independent of this question.

19. This is in contrast to, for instance, environmental philosophy, a subdiscipline in which central questions were established fairly quickly.

20. See Shrader-Frechette, 1985.

21. See note 14.

22. See Padilla, 1993.

23. We are grateful for helpful comments on this manuscript by Albert Borgmann, Philip Brey, Scott Noam Cook, Andrew Feenberg, Heather Gert, and Larry Hickman.

REFERENCES

Borgmann, A. (1984). *Technology and the Character of Contemporary Life*. Chicago: University of Chicago Press.

Borgmann, A. (1992). *Crossing the Postmodern Divide*. Chicago: University of Chicago Press.

Edwards, J. C. (1982). *Ethics Without Philosophy*. Tampa, FL: University of South Florida Press.

Ellul, J. (1964). *The Technological Society*. Trans. J. Wilkinson. New York: Vintage.

Feenberg, A. (1992a). *Critical Theory of Technology*. New York: Oxford University Press.

Feenberg, A. (1992b). Subversive Rationalization: Technology, Power and Democracy. *Inquiry*, *35*(3/4), 301–322.

Feenberg, A. (1995). *Alternative Modernity: The Technical Turn in Philosophy and Social Theory*. Berkeley, CA: University of California Press.

Feenberg, A. (1996). *Four Lectures on Technique and Modernity*. Unpublished. Delivered at the Ecole des Hautes Etudes en Sciences Socialies, Paris, France.

Feenberg, A. (1999). *Questioning Technology*. New York: Routledge.

Fellows, R. (Ed.) (1995). *Philosophy and Technology*. New York: Cambridge University Press.

Gert, H. J. (1997). Wittgenstein on Descriptions. *Philosophical Studies*, *8*(3), 221–243.

Heidegger, M. (1977). *The Question Concerning Technology*. Trans. W. Lovitt. New York: Harper and Row. Originally published in German as 'Die Frage nach der Technik', in *Vorträge und Aufsätze* (1954).

Hickman, L. (1990). *John Dewey's Pragmatic Technology*. Bloomington, IN: Indiana University Press.

Ihde, D. (1990). *Technology and the Life World*. Indianapolis, IN: Indiana University Press.

Mitcham, C. (1994). *Thinking Through Technology: The Path Between Engineering and Philosophy*. Chicago: University of Chicago Press.

Mitcham, C. (1995a). Notes Toward a Philosophy of Meta-Technology. *Society for Philosophy and Technology Electronic Journal*, *1*(1–2). URL: http://borg.lib.vt.edu/ejournals/SPT/v1n1n2/mitcham.html.

Mitcham, C. (1995b). Philosophy of Technology. In: W. Reich (Ed.), *Encyclopedia of Bioethics*, Rev. ed., Vol. 5. (pp. 2477–2484). New York: Simon and Schuster.
Mitcham, C., & Leonard, W. (1995). Rethinking Technology: Conference Report. *Research in Philosophy and Technology*, *15*, 235–239.
Padilla, H. (1993). Technological Objects and Their Epistemological Base. In: C. Mitcham (Ed.), *Philosophy of Technology in Spanish Speaking Countries*. (pp. 121–131). Dordrecht, Netherlands: Kluwer.
Pitt, J. (1995). On the Philosophy of Technology, Past and Future. *Society for Philosophy and Technology Electronic Journal*, *1*(1–2). URL: http://borg.lib.vt.edu/ejournals/SPT/v1n1n2/pitt.html.
Pitt, J. (1999). *Thinking About Technology*. New York: Seven Bridges Press.
Shrader-Frechette, K. S. (1985). *Risk Analysis and Scientific Method*. Boston: D. Reidel.
Stefik, M. (1997). *Internet Dreams: Archetypes, Myths, and Metaphors*. Cambridge, MA: MIT Press.
Ströker, E. (1983). Philosophy of Technology: Problems of a Philosophical Discipline. In: P. Durbin & F. Rapp (Eds.), *Philosophy and Technology*. Boston Studies in the Philosophy of Science, vol. 80 (pp. 323–336). (Dordrecht, The Netherlands: D. Reidel).
Wittgenstein, L. (1958a). *Philosophical Investigations*. Trans. G. E. M. Anscombe. Englewood Cliffs, NJ: Prentice Hall.
Wittgenstein, L. (1958b). *The Blue and Brown Books*. New York: Harper and Row.
Wittgenstein, L. (1980). *Culture and Value*. Trans. Peter Winch. Chicago: University of Chicago Press.

TOWARD A PHENOMENOLOGY OF THE BODY IN VIRTUAL REALITY

Craig D. Murray

This chapter discusses the phenomenology of the body in virtual reality (VR). VR denotes the use of 3-D computer graphics technology to generate artificial environments that afford real-time interaction and exploration. The development of this, still relatively new, technology continues to evolve to further heights of psychological, sensorial, and bodily immersion. The distinction between sensorial and bodily immersion is mainly a practical one. 'Sensorial', as used here, refers to properties of the senses (to be able to see, hear, and touch in the virtual environment), while 'bodily immersion' refers to the (re)presentation of bodily parts in the virtual environment.[1]

Vision, the most privileged of the senses, continues to play a principal role in the habitation of this cyberspace, 'inputting' our gaze and transforming our intentionality.[2] We might wonder why VR has been, and largely remains, an optical technology. In many ways it could be argued that vision is (technically) the least problematic of the senses to 'input' into the computer environment. However, it is also true that vision has long been regarded as the finest of the senses (seeing is believing). Still another persuasive influence has been the insights (in-*sight*) that science has gained from optical technologies (Ihde, 1990), such as the microscope and the telescope, which reinforce the *view* that the acquisition of knowledge is primarily a visual enterprise. It is these elements which facilitate a dominant eye in VR.

It is a more recent recognition that vision by itself is incomplete[3] that has prompted the development of peripherals to capture and project the body in all its complexity into the 'terminal reality' (Bukatman, 1993) that is VR. Virtual

Metaphysics, Epistemology, and Technology, Volume, 19, pages 149–173.

ISBN: 0-7623-0681-5

reality is no longer solely characterized by a 'disembodied' gaze, a projection of our selves into an optic panorama. The technical principles of the familiar data glove (a lightweight Lycra glove fitted with optical fibers along the backs of the fingers) have been adapted to a fully instrumented body suit, a Lycra cat suit fitted with flexible sensors to capture and render joint positions and movement. Such apparatus enables the animation of a virtual body viewable via a head-mounted visual display (Ellis, 1995).[4] Indeed, a 'compelling' VR experience is created by 'blocking' sensory impressions from physical reality (Biocca and Levy, 1995). The eyes, possibly the ears, hand, and whole body, are 'covered' by VR peripherals. Reminiscent of procedures associated with sensory deprivation, it is, in fact, a substitution of sensory information.[5] From the data glove to the body suit, VR technologies are becoming all embodying, perhaps even re-embodying; they are becoming what Anne Balsamo (1995) calls "new technologies of corporeality" (p. 215).

The predominantly visual characteristic of VR (which is why it can be considered an optical technology) has facilitated neo-Cartesian readings of computer (dis)embodiment.[6] The body, the story goes, remains docked, immobile at the interface, while the mind wanders the pixelled delights of computer programmers' creations.[7] William Bogard (1996), for example, while recognizing the transparency of the virtual system (a feature which will be elaborated within this chapter), contends that the 'operator too' disappears, giving "way to the disembodied traveller, the astral projectionist, the 'interface, data cowboy' in cyberspace" (p. 37).

Even when the user is represented by a visible virtual body, Cartesian readings continue to abound of VR. Simon Penny (1993), for instance, argues that "Virtual reality reinforces the Cartesian duality, replacing the experiential body [that is, the phenomenal body] with a body image [the virtual body], a creation of mind . . ." (p. 20). All this, despite Sandy (Allucquere) Stone's (1992) caution to avoid the 'Cartesian trick': "No refigured virtual body," she warns,

> no matter how beautiful, will slow the death of a cyberpunk with AIDS. Even in the age of the technosocial subject, life is lived through bodies (p. 113).

Indeed. But as well as the recognition of a material body behind the VR experience, the body deserves recognition for its phenomenal primacy in the VR encounter (see Hayles, 1996). Therefore, it will be argued here that although the image presented in a virtual environment is 'a creation of mind', the experience of 'inhabiting' that 'body' is not prescribed by the VR developer, but has an existence and direction of its own.

PHENOMENOLOGY AND THE MALLEABILITY OF THE BODY IMAGE

Phenomenology, as a descriptive technique of subjective experience, and with a fundamental concern with perception and bodily activity (Ihde, 1990), enables us to explore phenomena as they are lived and experienced (Valle & King, 1978). Allied to the existentialism[8] of Jean-Paul Sartre and Maurice Merleau-Ponty, phenomenology has proved to be a method well suited to exploring the experience of being embodied (embodiment).

Framing the body, quite literally, in terms of its 'sensorial architecture' (Tomas, 1989), or its structures of corporeality – that is, the sensorial/perceptual experiences arising from the body – has been a useful method of understanding how an individual comes to 'know' the boundaries of their body. Therefore, discussions of how we experience the body focus on sensation and function (control of our bodies) (see Rom Harre, 1991, for example). Visual inspection of our bodies contributes to our experience of a bounded body. However, our bodies are also enveloped by skin, providing us with a sentient and dimensional bodyscape. Didier Anzieu (1989) extols all the virtues of the skin with a fervor that most reserve for sight,[9] noting that it comprises 'several sense organs' (touch, pressure, pain and heat) and is closely entwined with our other senses of hearing, sight, smell, and taste as well as an awareness of body movement and balance. In addition, Anzieu comments on how the skin also remains continually open to the world in a manner different to the other senses: the eyes can be closed, the ears and nose 'stopped up', but the skin "cannot reject any vibro-tactile or electro-tactile sign" (pp. 14–15).

We are able to move our limbs and twist our torsos, providing ourselves with a stream of proprioceptive feedback (tacit knowledge of the movements and placement of the body and its parts extended in space). In fact, the body occupies many phenomenal spaces. James Anthony (1968) discusses the various body spaces that arise out of different activities: there is "a mouth space, a visual space, an auditory space, a tactile space, [and] postural and kinaesthetic spaces" (p. 1108). Anthony's observations can be seen as characteristics of phenomenal embodiment. Our interactions of body and world (Merleau-Ponty's "being-in-the-world") animate these different spaces effortlessly on our part. At times these spaces shout out their existence: for example, John and Marcia Goin (1981) talk of how at the dentist's we "suddenly become all mouth" (p. 62).

However, these markers of corporeality are artificially divided. In reality they all form a synthesis of mutually complementary body experience. Only when the body breaks down do these corporeal threads unravel: consider the person with paralysis who has visual confirmation of his or her body, but no tactile or proprioceptive knowledge (Trieschmann, 1988). Such corporeal experience provides us with a 'body image', a phenomenological understanding of our bodies extended in space and time.[10] The term 'body image' has been much used by different theorists to denote different things. In psychological interpretations it has been used to denote 'knowledge' of one's own body, present at the borders of consciousness, or functioning unconsciously (Tiemersma, 1989). In employing the term in a phenomenological context I, much like Merleau-Ponty,[11] use it to refer to an immediate knowledge (impression) of my body as it juts into the fabric of experiential space and time. For this reason, we cannot speak of the phenomenological body image as a stable entity: the term is a schematic way of talking about the behavioral bonds between body and world (Moss, 1978).

The historical opposition of mind and body, Descartes' dualism, has an incomplete phenomenological basis – in Drew Leder's (1990) terms, a 'proto-phenomenology'. We neglect the body at the times when it recedes from awareness, which is most of the time. Only with various forms of corporeal breakdown do we pay attention to the body, and at these times the body appears 'other'. Cartesian dualism is built on this selected reflection of the body. A more disciplined phenomenology reveals that the body in the world is both foreground and background. It constitutes our locus, so that we are 'here' rather than 'there'. At the same time the body recedes from conscious reflection. At once an holistic sense organ, and yet an assemblage of sense apparatus, the body recedes from awareness in its perceptual activity. It is the very disclosure of the world that the sense organs provide which leads to their mindful demise. The body recedes in activity but is always implicitly present and known. For instance, "To see something as reachable and thereby open to my use is to implicitly experience my body's capacity of reach" (Leder, 1990, p. 22). Leder further notes,

> My body is always a field of immediately lived sensation .... Its presence is fleshed out by a ceaseless stream of kinaesthesias, cutaneous and visceral sensations, defining my body's space and extension and yielding information about position, [and] balance ... (ibid).

However, while the "rim of felt embodiment" is sketched out by how our bodies 'feel', it is refined by engagement in practical activities (Harre, 1991, pp. 20–21).

Through infantile exploration we become a body. Herbert Plugge (1970) suggests that to the young child the legs are "strange and distant things" which come to be mastered, incorporated "through feeling and kicking" (p. 304). Our early development includes a process of becoming embodied. We have a corporeal history: both ontological and phylogenetic. Along with our evolutionary corporeal history,[12] the "passage of bodily time" and its concomitant experiential activity molds our embodiment (Zaner, 1981). Marcel Maus (1992) recognized this when he argued that the body is our first and most natural technical object. Our clothes – think of how high heels shape "the gestalt of a walking body" (Falk, 1995, p. 96) – and techniques of the body[13] work upon the body-object but also upon the body-lived, producing our embodied experience. Susan Foster (1992) provides a vivid example of this in her discussion of the body of the dancer, where she argues that our corporeality can be manipulated through training. The dancer's body is perceived, argues Foster, via "visual, aural, haptic, olfactory, and, perhaps most important, kinaesthetic" sensory patterns (p. 482). Foster does not see the dancer's training as just a finer tuning of corporeal awareness, but contends that "over months and years of study, the training process repeatedly reconfigures the body" (p. 484). Such phenomena suggest the possibility of a manipulable body, a plastic corporeality, a malleable body image.

A PHENOMENOLOGY OF EMBODIMENT IN VIRTUAL REALITY

A sense of 'presence', of feeling 'there' in a virtual environment is, perhaps, the ultimate aim of VR research.[14] This calls for a dampening of awareness in 'reality' and a heightened 'acceptance' of the surrounding virtuality. The hardware of VR must recede and become transparent for this sense of presence (or 'telepresence' – the experience of being fully present within the computer-generated simulation) to occur. As Mark Lajoie (1996) notes,

> To the extent that the terminal, as an interface, acts as an object, it is a constant reminder to the user of their inability to become fully subject in the virtual space. In effect, it marks their lack of presence as subject within the virtual reality (p. 163).

One line of argument proposed is that for a sense of presence in the virtual environment of the computer, then the virtual body must closely resemble (both visually and sensorially) the body of the user. Some ask how the geometric mappings of the body within the virtual and physical environments, relative to each other, contribute to a sense of presence (Sheridan, 1992). The general consensus is that identification, and therefore telepresence, would be increased by a similarity in the visual appearance of the user and the virtual body[15] (which can be contrasted with the speculative discourse on the polymorphous potentiality of VR, discussed later).

As an example of how VR technology is progressing in its attempts to intentionally incorporate the user's body, we may consider the work of Mel Slater & Martin Usoh (1994). Slater & Usoh provided an experimental group with a (visually) entire virtual body representation.[16] The directly manipulable components of the virtual body consisted of a representation of their right hand attached to an arm that could be bent and twisted in response to similar movements of the real arm and wrist. When participants turned their real head around by more than 60 degrees, then the virtual body turned appropriately. Slater and Usoh argue that a high correspondence between proprioception of the physical body and other (namely visual) sensory data of the virtual body make it more likely that a VR user will identify with the virtual body, and experience a greater sense of presence in the virtual environment. But to illuminate our understanding of presence in virtual environments we need to return to the experience of embodiment. The perceptual effect that VR has on the experience of the body is little understood or explored.[17] In order to fully understand how a user experiences the body in VR, in order to identify the "phenomenological dimensions of the technologically mediated body" (Balsamo, 1993), we need to build up a rich phenomenology of user embodiment. This can come through a detailed, qualitative analysis of user's accounts of their immersion; through a thorough exploration of our own VR encounters; and through observation of cybernauts at 'play'.

At a recent conference on VR in the United Kingdom I tried perhaps one of the most ubiquitous forms of VR technology. My eyes 'entered' the virtual environment by means of a fairly light head mounted display (HMD), and movement was achieved by pressing my thumb on a hand-held navigator. My perception of 'travel' through this virtual environment vacillated between moving outward with some velocity, to pulling in the environment towards 'me'. However, where was my body in this experience? My legs, it might be argued, were not needed – my hand gripped the navigator with assuring firmness as my thumb toyed with the forward-backward power of the smooth button. However, it was my legs which gave me my height in this world.[18] My head movements directed my gaze. And the proprioceptive frame of my body, in which I was tacitly aware of the movement and position of my body and its parts, gave me a depth in this virtual world that a 'disembodied gaze' should not allow.[19]

In her essay, 'The Virtual body in Cyberspace', Anne Balsamo (1993), prompted by her own 'trip', considers embodiment in VR systems. The cyberspace experience, argues Balsamo, necessitates the wilful repression of the body. The 'disembodied gaze' of most VR applications mime the disembodied camera eye of film: in both the point of view (pov) approximates the height of an embodied eye. Movement of perspective is simulated by the camera with rarely any overt (visual) reference to the body – in this sense the body of the eye is repressed. In such applications " 'the body', as a sense apparatus is nothing more than excess baggage for the cyberspace traveller" (p. 126). This is not to say that Balsamo sees VR as a disembodying medium: she recognizes that the virtual experience 'makes sense' because of *an eye-level* perspective which changes with a *turn of the head.* In our everyday experience the body is the 'surpassed' but ever present (Sartre, 1970). In VR, when the body of the user is not (re)presented the body-as-surpassed takes on increased significance for embodied experience. For whereas in the physical environment the body 'withdraws' from (thematized) perceptual awareness, in VR its disappearance is manufactured in one mode of bodily knowledge – that is, visually. This is a poor method of eliminating the body; the body, in its kinaesthetic and proprioceptive modes of presence continues to surface in perceptual experience – the phantom body claims its territory and demands recognition.

Michael Heim (1995), inspired by his own VR encounter, describes the perceptual nausea following virtual immersion, a "forewarning of . . . an acute form of body amnesia" (p. 67). There is a lag, Heim contends, between the cyberbody (the body which is simulated as part of the virtual environment, the virtual body) and the actual physical body, as a result of sequential switching between their two domains. Heim gives anecdotal evidence from Stephen Ellis whose work with VR, apparently, has him

> unconsciously gesturing in the primary world in ways that function in the virtual world. He points his finger half expecting to fly (as his cyberbody does under convention of the virtual world). His body needs to recalibrate to the primary world (p. 68).[20]

Heim entreats us to observe someone emerging from a VR system:

> Watch their first hand movements. Invariably, the user stands in place a few moments . . ., takes in the surroundings, and then pats torso and buttocks with their hands – as if to secure a firm landing and return presence in the primary body (p. 68).[21]

Although it is argued here that the 'disembodied gaze' is experientially a misnomer in that the physical body is always (quasi)present, a case may be made for envisaging a dissolution between social and bodily space, or at least that such a distinction becomes ambiguous, problematic. Erwin Straus (1966) argues that this social space is an expansion of the 'body scheme' (an unconscious, pre-reflective organizing principle of bodily action).[22] It belongs to the body, but not completely: "it is not an indisputable property but a variable possession." This "intervening space is a medium between me and the world," and that is its social significance (p. 153). Ihde (1990) notes that what we take as our 'real' or 'naked' bodily space is transformed through optical technologies such as the microscope and eyeglasses.

> The transformation of vision through lenses changes, however slightly, our sense of bodily space. What was farther is now nearer; and when motility is involved, this calls for a new adjustment (p. 48).

VR as a predominantly optical technology has the same and additional properties as those of the traditional lens technologies. Along with the ability to move forever forward (magnify) and backward (miniaturist) in relation to an image in VR will also come a transformation of our 'naked' space – particularly when we do not have a virtual body (re)presentation.

The transparency of visual, kinaesthetic, aural, and other displays are considered a key determinant of feeling present in a virtual environment. However, Simon Penny (1994) argues that, even with smaller, less cumbersome peripherals, the fact of their attachment to the "outside of the body" creates a 'dislocation' between the virtual body and the corporeal body. I would like to turn my attention to this point. The continuity of the-body-as-I in VR calls for an assimilation of both technological peripherals and the virtual (re)presentation.[23] In the case of VR peripherals such as HMD, gloves, and body suits, we can illuminate this process by way of recourse to Martin Heidegger and Merleau-Ponty.

In Heidegger's (1962) analysis of tool use, he appeals to the example of a hammer. For a tool to be ready-to-hand it must, in Heidegger's terms, 'withdraw'. In so doing, the tool becomes the means rather than the object of the experience (Ihde, 1990). This has a parallel with the experience of the body itself: Sartre (1970) sees the body as the perpetually 'surpassed' (p. 233). The tool itself is also surpassed as it withdraws into the architecture of the body, forming what Ihde terms "an embodiment relation."

Merleau-Ponty (1970) invokes the example of the blind person's cane. The cane for the blind person is no longer an object, but an extension of the realm of the senses.[24] With the cane as a 'familiar instrument', touch is experienced at its end point ('its point has become an area of sensitivity', p. 143) rather than at the hand. This body familiarity (or 'habit') of the tool is the result of being "transplanted into them, or conversely to incorporate them into the bulk of our own body" (p. 143). The incorporation of the tool into the body gestalt is what Leder (1990, p. 34; see also Grosz, 1994) refers to as a "phenomenological osmosis," "the body allows instruments to melt into it" (Kujundzic & Buschert, 1994, pp. 207–208).

We can further illuminate our understanding of the above with Ihde's phenomenology of technics.[25] Ihde employs the example of eyeglasses, and expresses the optically mediated world when wearing them with his triptych separation of 'I-glasses-world'. However, in wearing eyeglasses, the glass is 'transparent'. The weight of the glasses on the ears and the bridge of the nose also 'withdraws'. Insofar as we take technologies into our experiencing by perceiving through them, the technology becomes embodied. 'I-glasses-world' becomes '(I-glasses)-world'. Ihde maintains that the same process occurs for the hearing aid (for hearing) and the blind person's cane (for tactile motility).

The above phenomenal examples of how the body incorporates tools into its structure have implications for how we experience peripheral devices of VR technology. The 'dislocation' between physical and virtual bodies that Penny talks of appears invalid – providing the virtual environment and virtual body facilitate these devices.[26] Light and small apparatus will always contribute to a transparency of the technology. A virtual body which incorporates the depth, expanse, and weight of VR input/output devices, as well as a 'sympathetic' virtual narrative, will facilitate the melding of the physical and virtual into one phenomenal body: (I-peripherals)-virtual world.[27]

ATYPICAL AND DISRUPTED BODIES: IMPLICATIONS FOR THE (RE)DESIGN OF THE BODY IN VIRTUAL REALITY

Instances of atypical embodiment, and the body images they provide, inform us of the importance of sensation, function, and morphology upon the self's corporeal moorings to the body. They tell us about a process of re-embodiment – a radical metamorphosis in the architecture of the body. Amputees and people with congenital limb absence who use prostheses can be considered as a case of "the extended body," while people's experience of paralysis can be characterized as "the receding body." Plastic surgery provides us with an example of a "technology of embodiment" for the changes it exerts upon a person's corporeality (visual, tactile, proprioceptive). These experiences of embodiment have important implications for the (re)design of the body in VR.

Limb Absence and Prosthesis Use

With amputation the person loses one of his or her "conceptual links to the world, an anchor of his very existence" (Murphy, 1987, p. 85). In objective terms, the amputee loses all modes of bodily knowledge about a lost limb. In the amputee's 'reality' this is not the case, since phantom phenomena permeates his or her experience. Typically amputees report feeling as if their amputated limb is 'still there'. The phantom may be experienced as painful, intermittent, and locked in the same position (which may match the limb's preamputation position, particularly in the case of trauma). Conversely, the phantom may not be painful, enduring, and capable of 'movement' (sometimes under the amputee's conscious control). Any combination of these phenomena might characterize a particular amputee's experience.

My recent research has centered upon the phenomenological boundaries of prosthesis users' bodies. Both people with acquired amputations and people with congenital limb absence will be encouraged to use prostheses. In addition to a frequently expressed emotional attachment to an artificial limb, users often report that the limb feels 'part of' them. As an example of this consider the following excerpt from an interview with a female with a lower-limb amputation: "[M]any amputees feel that their artificial limb is somehow part of them, a simple example of this is that I wouldn't like just anyone putting their hand on my knee, even though it is not actually part of my body's flesh, it is still mine even though it's a piece of plastic and metal." Another interviewee, a female with congenital absence of her right forearm, told me

> [I]t's [the prosthesis] a part of me now, that's the only way I can describe it. To me it's as if, though I've not got my lower arm, it's as though I've got it and it's a part of me now. It's as though I've got two hands, two arms.

As well as this direct assertion, amputees and people with congenital limb absence provide rich descriptions of prosthesis use which are reminiscent of the descriptions given of embodied technologies by Heidegger, Merleau-Ponty, and Ihde. The body, in Sartre's terminology, 'extends across' the prosthesis to incorporate it into the phenomenological body image. In the context of VR, this phenomenon is informative about the possibility of 'providing' VR users with additional limbs. Prosthesis use is, for those with congenital limb absence, a redesign of the user. The incorporation into the body (Ihde's 'embodiment relation') of the prosthesis is indicative of how VR users might adjust to altered bodies in VR applications.[28]

Paralysis

Paralysis is a condition that involves the loss of multiple modes of knowledge about the body (e.g. tactile and proprioceptive). Robert Murphy (1987, p. 85) wrote about his own experience of quadriplegia in *The Body Silent*:

> I no longer know where my feet are, and without the low-level pain I still feel, I would hardly know I had legs. . . . I have . . . become emotionally detached from my body, often referring to one of my limbs as the leg or the arm . . . . My solution to this dilemma is radical dissociation from the body, a kind of etherialization of identity.

Trieschmann (1988) relates how some spinal cord injury patients (particularly those with quadriplegia) describe their embodiment as a sensation of "feeling like a balloon floating in space" (p. 61); they have no 'perceptible ties' to the ground. When one considers that these patients have sensation from the neck and head only, the metaphor speaks a phenomenological truth. The rest of the insentient body has only a visual life for such patients.

Murphy (1987) notes that "Many paralytics say that they no longer feel attached to their bodies, which is another way of expressing the shattering of Merleau-Ponty's mind-body system" (p. 87). Another person with cerebral palsy, who had involuntary movements of his limbs, referred to his arms as 'the bandits' and did not experience them as part of his body (Stensman, 1989). What accounts such as these demonstrate is the importance of 'authorship of action' (Harre, 1991), to be able to control the body, in contributing to a 'ownership' or identification of the self with the body.

Plastic Surgery

Plastic surgery is generally perceived as a practice or discipline of aesthetics. In the context of this paper I would like to discuss plastic surgery (cosmetic and corrective) as a technology of embodiment, for the changes it exerts upon a person's felt corporeality. The most obvious feature of plastic surgery (to the gaze of the other) is a change in appearance. However, for the embodied patients/customers there are also sensory changes (of which their own visual inspection will be one). In contrast to the body observed in the mirror, the observable result of surgery, there are more subjective changes of embodiment (such as changes in tactility and proprioception) which are experienced-as-being, rather than just observed. In the examples of breast augmentation and reduction, there may be reduced tactile sensations ("crude touch, and light pressure sensation" (Courtiss & Goldwyn, 1976)) diminished pain perception (Pruzinsky & Edgerton, 1990), and subjective changes in the position, weight, and form of the breasts (Goin et al., 1977). One woman, who had a breast reduction, is reported to have found difficulty in locating itching areas of her breasts: "what used to be in one place is now in another" (Goin & Goin, 1981, p. 206).

It has been noted that sensory changes accompany "virtually all types of cosmetic surgery" (Pruzinsky & Edgerton, 1990, p. 221). In eyelid surgery (a technique employed by some to 'remove' ethnicity) there may be an increase in the scope of sight; operations on the nose may result in permanent changes of respiratory sensations (Pruzinsky & Edgerton, 1990). The ubiquitous face-lift operation results in a hyperthesia of the skin on the face and neck which blurs the 'self-non-self boundary' where patients may experience a "loss of 'boundary sensations' when sleeping with their partially numb faces against the pillow" (Goin & Goin, 1981, p. 146).

These changes in embodiment can lead to a dissociation from the body. A 'loss of identity' (Pruzinsky & Edgerton, 1990, p. 232), which may be short or long term, has been associated with both cosmetic and corrective plastic surgery. The sensory changes that accompany a change in the topography of the body take time to weave their way into the body image of the patient. We may consider the examples provided by Richard Druss (1973), who underwent breast augmentations. Typically, these patients examined their breasts visually and tactually in a "exploration of the[ir] newly altered body contour" (Goin & Goin, 1981). We can look upon this as a sensory convergence, a unification and continued integration of the senses in the phenomenological body image.

In direct relation to VR, we can see that the explosion of sensory information that comes with plastic surgery (and with amputation too) takes time to crystallize into a concise body psyche, whereupon a sense of completeness allows a reliable body image once more. The dizzying sensory changes that accompany VR experiences will likewise disorient our 'sense of body' before, with exploration and time, a coherent body is experienced. In the later sections of this paper we will see how the key phenomena of phantom limb (as experienced by the amputee) and the dissociation with or objectification of the body (as experienced by the paralytic) are manifested in current VR applications.

THE BODY IMAGE, THE PROSTHETIC, AND THE PHANTOM

Merleau-Ponty (1970) understands the phantom limb phenomena as a result of 'being-in-the-world'. For an amputee to have phantom phenomena is to 'remain open' to a customary world, and to "retain the practical field which one enjoyed before mutilation" (pp. 81–82). The customary world continues to elicit habitual intentions (lost capabilities). The amputee is torn between two bodies – the customary or habitual body, and the "body at this moment," body reality. For Merleau-Ponty, "the problem how I [the amputee] can have the sensation of still possessing a limb which I no longer have amounts to finding out how the habitual body can act as guarantee for the body at this moment" (p. 82). In the case of the amputee who employs a prosthesis, the habitual body is buttressed by an effective combination of the artificial and phantastic: the prosthetically enhanced body-at-this-moment acts as guarantee for the habitual body. We may consider this phenomena with regard to one amputee who told me:

> I have always worn a prosthetic for all my waking hours unless I was injured. I feel connected to the ground much like I do with my real foot, maybe this is because of phantom limb experience.

Here we see the prosthetic and phantom limb phenomena entwining their different properties (the 'legness' of the phantom, the solidity and 'reality' of the prosthetic) which anchors (a 'connectedness-to-the ground') the amputee to their habitual world.

The visual aesthetic, the change in physical appearance, is just one aspect of the improvement that breast reconstruction with prosthetic implants have for some women. Changes in proprioception, and tactility are intimately experienced-as-being, possibly more compelling than the optic sense, and ever-present. The contribution a phantom may make to the acceptance of prosthetic breast implants is evidenced by one women who, one and a half years after her mastectomy, was still experiencing phantom breast sensations and "immediately experienced the implants as being her own breasts" (Goin & Goin, 1981, p. 185).

The prevailing view is that an external breast prosthesis is never experienced as an integral part of the body, that it is psychologically always perceived as an external device (Goin & Goin, 1981). Wendy Schain et al. (1985) report that "Approximately 75% of our sample indicated a wish . . . to be rid of their external breast prosthesis as [one of] the major reasons for wanting a breast reconstruction," and conclude that "an external breast device is never incorporated into a women's body image" (p. 45). Such experiences as these contest arguments that "Anything that comes into contact with the surface of the body and remains there long enough will be incorporated into the body image" (Grosz, 1994, p. 80).

THE PHANTOM AND OBJECTIFIED BODY IN VR

In ancient Egypt the word 'ka' referred to an "ethereal and less dense" copy of the human body. This "invisible but still material analogue" of the soul "inhabited and animated" the physical body (Grosz, 1989, pp. 62–63). Such a cultural myth echoes the state of the body in current virtual reality systems: we might consider that the dense human body inhabits an ethereal virtual one,[29] or that the physical body becomes more ethereal (less tangible) in the virtual experience while perceptually and experientially the virtual body becomes more dense. Heim (1995) implies much the same thing when he says that

> HMD [head-mounted display] immersion results from the primary body giving away priority to the cyberbody . . . . The user undergoes a high-powered interiorization of a virtual environment but in the process loses self-awareness (p. 72).[30]

This is a question of phenomenal embodiment in VR systems: our technological embodiment may vacillate between the two: but two there are. This is what Penny (1994, p. 242) calls the 'split body condition' or the 'double body'. In VR, part of the sensorial architecture of the body remains in the physical world, while another is projected into the virtual one. The corporeal body in the physical environment remains ever present to mind, while an "electronic 'body image'" weakly echoes and competes with it. When only part(s) of the body are absorbed by VR technology, phantom phenomena will occur. The degree to which our visual corporeality dominates our embodied experience will influence the tangibility of our remaining body outside the VR experience.

I want to explore some of the phenomena of amputees' and paralytics' experiences of body in relation to embodiment in VR technologies. As an example of the informative value of atypical embodiment when applied to VR, it is useful to consider some of the findings of Slater & Usoh (1994). The range of movement offered by the virtual body in the studies reported here was limited. Following a pilot study (in which only one arm and gross torso movements were represented) some participants commented that their virtual body was "a dead weight," "a useless thing," and "nothing to do with me" (p. 134). Slater & Usoh provide their own analogy of this phenomena. They use an example of atypical embodiment, namely the loss of proprioception, to inform us about the participants' behaviors. The example is drawn from Oliver Sacks' (1985) essay 'The Disembodied Lady'. Sacks writes of Christina who lost all sense of her muscle, tendon and joint positions. By her own account, Christina could not 'feel' her body, she felt 'disembodied'. Only by careful (visual) observation of her movements could Christina accomplish motor tasks.[31] Without this close scrutiny her body was 'blind' to itself: she would 'lose' her arms for instance, "I think they're one place, and I find they're another" (p. 46).

However, the responses from the participants in Slater and Usoh's studies are remarkably similar to the comments made by people with paralysis of various parts of their bodies. In both cases a dissociation with or an objectification of the body (or its parts) is articulated. The dissociation that Murphy (discussed earlier) experienced with regard to his body is similar to that experienced by some participants in Slater and Usoh's studies. We can relate the objectification of these bodies[32] to their lack of movement and direct sensation. However, this is only part of the story. While Murphy demonstrated a lack of movement and sensation from his body, 'inhabitants' of current virtual bodies do have their physicality to fall back on. For instance, Slater and Usoh comment that "one subject on noticing the fixed virtual left arm began to move her real left arm very rapidly, in a manner indicating panic," while another related that "I thought there was really something wrong with my [left] arm" (p. 133). One reading of these comments is that participants experienced the contradiction between their virtual and physical bodies as phantom phenomena (a feature of the amputee's experience). While they experienced a kind of paralysis in the virtual environment, they also experienced a moveable body in the 'primary world' (Heim, 1995). Such an interpretation would also appear to explain another comment made by a participant in relation to 'walking' in the virtual environment: "Sometimes [I had] a desperate need to actually walk when virtually walking, there does seem to be a conflict between what the eyes see and the body feels – e.g. my feet appear to be floating but I can feel my feet on the ground" (Slater & Usoh, 1994, p. 135).

When our familiar embodiment is disrupted through illness, paralysis, amputation, or plastic surgery, we are forced to thematize our bodies. This thematization may lead to a dissociation from the body that slowly subsides as a mastery and familiarity with the body is regained. These examples shed light on what a (re)design of the body in VR might entail.

(RE)DESIGN AND (RE)PRESENTATION OF THE BODY: SPECULATIVE DISCOURSE ON POLYMORPHOUS VIRTUAL REALITY

There is a recurring line of speculative virtual reality discourse that argues future developments of these technologies will fulfil what Balsamo (1993) calls "a utopian desire for control over the form of personal embodiment." Lajoie (1996) tells us that "The matrix . . . promises to do two things, one of which is to completely effect our body and the space in which it is located, and second, to inscribe subjectivity within a computer-generated phenomenal space, a space in which all desires can be fulfilled" (pp. 162–163). There is a sense in which the computer programmer or VR designer can be envisaged as wielding a scalpel on the virtual body of the user. As with plastic surgery, VR is a technology of embodiment, manipulating the corporeal structures of experience (and the experience of corporeal structures).

Scott Fisher (1991) talks of the ability of VR to allow users to be represented by any chosen form ("a kind of electronic persona"). Bukatman (1993) recognizes the simulated, but powerful, polymorphism that is at work in such applications. "To be installed into such an apparatus would be to exist on two planes at once: while one's objective body would remain in the real world, one's phenomenal body would be projected into the terminal reality" (Bukatman, 1993).

Such polymorphic experiments have been conducted by Jaron Lanier and colleagues, where the relationship between bodily movement and sensory feedback were reorganized, and have been reported by Frank Biocca & Mark Levy (1995) and Penny (1994). Here, body parts were presented in different proportions in the virtual world for perceptual effect. Lanier himself comments on the 'slippery' distinction between the body and the world in VR. He asks

> How are you connected to the world? What if your eyes were on your fingers? . . . What if you took all the measurements and the movements of your physical body and somehow put them through a mathematical function that allowed you to control six arms at once with practice' (Biocca & Lanier, 1992, p. 158).

This transfiguration of corporeality discourse is further applied to people with disabilities. So, for instance, someone with paralysis, it is argued, would be able to employ 'sip and puff' control strategies for their virtual limbs. The premise is that since (re)presentation in virtual environments is digital, we will be able to 'define ourselves' as anyone or anything. Quadriplegics will walk – or better still, be Olympic athletes (Fodor, 1994). In fact the peripheral input/output devices so characteristic of VR become, in this speculative discourse, the means of abolishing or creating disability. The distinction between able and disabled 'disappears': "In Virtual Reality everyone is disabled to the extent that we cannot participate without 'assistive devices', and on the other hand, there need be no limit to what we can accomplish with them" (Lasko-Harvill, 1993, p. 61).

The issue that statements such as these raise is whether such control strategies will allow users to experience these virtual limbs as part of their phenomenal bodies. Penny (1994) talks of designer reality where we will be able to choose the morphology and style of our bodies. Here, the morphology of the body may be manipulated as additional sensors are employed to animate extra limbs. Penny tells us that "The mind maps to this new body almost effortlessly." Lanier's report that it only takes a matter of minutes to adjust to arbitrarily placed sensors as controllers for extra limbs are seized upon by Penny to suggest that the mind can quickly draw a new internal body representation to allow control of the new body. Ease of control here is (erroneously I believe) being held as evidence for 'habitation' of a new body. Many amputees whom I have interviewed are adept users of their prosthetic technology, but still some of them insist on the 'otherness' of these limbs. From the experience of amputees, paralytics, and recipients of plastic surgery, I would argue that sensory changes in the architecture of the body take time to weave into an altered body image; VR applications do not appear to be an exception to this. A literary example helps provides us with an idea of the phenomenology of a radically refigured body in VR:

> As Gregor Samsa awoke one morning from uneasy dreams he found himself transformed in his bed into a gigantic insect. He was lying on his hard, as it were armour-plated, back and when he lifted his head a little could see his dome-like brown belly divided into stiff arched segments on top of which the bed-quilt could hardly keep in position and was about to slide off completely. His numerous legs, which were pitifully thin compared to the rest of his bulk, waved helplessly before his eyes (Kafka, 1952, p. 19).

For Gregor Samsa, the legs had become, in Plugge's (1970) characterization of the infant's experience, "strange and distant things" once more. To be 'installed' in a radically reconfigured body in VR would, at least initially, be a similar experience.

However, for a compelling transfiguration in the perception of the body using VR technology, I would argue that the physical body itself must be manipulated. The visual confirmation of a reconfigured body is not sufficient. A good example of a manipulation of the body which may be amenable to polymorphic VR applications is provided by Catherine Richards.[33] Richards used a vibrator to stimulate the surface of the arms and hand on blindfolded participants. One participant felt her neck shrink and thicken, until finally it sunk into her chest; another women felt her arms grow longer, stretching more than six feet away from her body. Hayles (1996) notes how such a simple device is capable of destabilizing and reconfiguring our perceived body. How much more compelling, she asks, would this experience be if accompanied by simulated visual information via a head-mounted display?

SUMMARY

In this chapter I have attempted to address and elaborate the phenomenology of the body in immersive virtual reality systems. This has involved a discussion of VR's predominantly optical nature and a critique of the dualism that pervades current accounts of VR. Throughout I have attempted to explore the core aspects of embodiment and how they apply in VR applications. Various embodiment theorists have been drawn upon to demonstrate our lived, sentient knowledge of our bodies extended in space and time (the "phenomenological body image"). Along with a consideration of atypical or disrupted bodies (limb absence, paralysis, plastic surgery), this has enabled a consideration of the phenomenological experience of embodiment in current and (possible) future VR applications. Informative comparisons of phenomena associated with such disrupted bodies (phantom limb, dissociation of the 'self' from the body) to the experience of the body in VR (the phantom and objectified bodies in current VR applications) have been made.

While immersive virtual environments can be created and perceived using specialized peripheral devices (such as a head-mounted display and instrumented clothing), the extent to which such devices may be used to transform the phenomenal properties of the body, to (re)define experiential human morphology, remains an open question. In order to avoid an over riding emphasis on the technology we need to explore the experience(s) of people who are installed in VR apparatus. Therefore, this paper is intended to serve as a building block for further research, in which a phenomenological approach to technological embodiment is adopted.[34]

NOTES

1. The evolution of this development has progressed from no representation, to an arrow (as hand), to the hand, and finally 'full body' representations. However, all of these forms exist and continue to be developed. A decision on whether to include a body representation will often be made on its perceived desirability for the particular application in question.

2. 'Intentionality' derives from the Latin 'intendere' meaning 'to stretch forth'. As employed in phenomenology, 'intentionality' refers to the reaching out of the mind to phenomenal reality, whereby we experience an intentionally interpreted world (Spinelli, 1995).

3. Hans Jonas (1970) presents a phenomenology of the senses in his essay in which he finds "the ancient claims for sight substantiated and at the same time qualified" (p. 313). He recognizes that though a very special sense, sight is incomplete by itself. Jonas highlights hearing and touch as two senses that complement and help sight achieve its full potential. In fact Jonas contends that "in his hand man possesses a tactile organ which can take over some of the distinctive achievements of his eye" (p. 318). For Jonas, sight and touch are two organs of a more general faculty of 'vision'. Kurt Kaczmarek & Paul Bach-Y-Rita (1995) also note that "The skin shows a number of functional similarities to the retina of the eye in its capacity to mediate information" (p. 349).

4. The history of VR has been to 'input' the eyes, then the hand, while the rest of the body sensorium and motorium has been neglected or considered peripheral to the immediate aims of VR. Our evolutionary history includes the development of an upright posture, along with which 'the hand' has become "an organ of active gnostic touching – the epicritic, discriminative instrument par excellence" (Straus, 1966, p. 150). In some forms of 'bodyless' VR, the properties of the eye (via a head-mounted display) and hand (via a disembodied representation of the hand) are combined to form one synergistic tool for activity in VRs. Given that we can characterize, as Jonas does, sight and touch as two organs of a general faculty of 'vision', this development might not be considered surprising. A more erotic explanation is provided by Marcus Novak (1992), who argues that "[t]he course of invention has been to follow desire, with its access to the parts of our bodies that have most nerve ending. When we enter cyberspace we will expect to feel the mass of our bodies, the reluctance of our skeleton; but we will choose to *control* with our eyes, fingertips, lip, and tongues, even genitals" (p. 241).

5. It is interesting to compare embodiment under conditions of sensory deprivation with immersion in VR. Seymour Fisher (1973) gives the example of a sensory deprivation study carried out by the psychiatrist, John C. Lilly. Lilly submerged himself in a tank of water, the temperature of which matched his body. As he floated in the tank with all light and sound excluded ('blocked'), Lilly began to feel "merged and indistinguishable from all that surrounded him," "he could not distinguish where his body left off and the water began . . ." (p. 22). With no sensory detail (including diminished proprioceptive and kinaesthetic frames of reference), Lilly's body boundaries became ambiguous. Even the temperature of his body no longer 'framed' his body against the surrounding environmental temperature. The condition in VR is (partial) substitution of sensory information, though a deprivation (of physical reality), as articulated by Biocca & Levy, is an integral part of a 'compelling' VR experience (particularly when the body is to be 're-morphed').

6. This is a characteristic of optical technologies, a proto-phenomenology or selective reflection on embodiment facilitates such Cartesian readings. For instance, Robert Romanyshyn (1994) reflects this when he talks about the telescope: "In 1609 Galileo's telescopic eye danced on the moon. Aided by this newly invented instrument of the telescope, Galileo took another step in the development of a kind of consciousness that required him to organise his experience and to make sense of the world by leaving the body behind." In a richer phenomenology, Sartre (1970) identifies the body as a whole, as a locus of being, as that which "extends across the tool it utilises: . . . *it is at the end of the telescope which shows me the stars* . . ." (p. 232, my emphasis).

7. In another dualistic twist, 'we' are the mind to our cyberbody which will perish while the cyberbody is 'immortal' (Novak, 1992).

8. The etymology of 'existentialism' (or 'existence') derives from the Latin verb 'existere', "meaning to stand out, or to become, to emerge" (Misiak & Sexton, 1973, p. 72). Applied specifically to the phenomenal experience of the body, 'to stand out' has increased significance; though world and self are intimately related our bodies (as the locus of self) are the foreground of our existence.

9. "Of all the sense organs," argues Anzieu, "[the skin] is the most vital: one can live without sight, hearing, taste or smell, but it is impossible to survive if the greater part of one's skin is not intact. The skin . . . occupies a greater surface (. . . 18,000 [square centimeters] in the adult) than any other sense organ . . ." (p. 14).

10. Merleau-Ponty (1970), stressing the experiential nature of space and time, cautions that "We must avoid saying that our body is in space, or in time. It inhabits space and time" (p. 139).

11. Merleau-Ponty (1970) tells us that "my body for me is not an assemblage of organs juxtaposed in space. I am in undivided possession of it and know where each of my limbs is through a *body image* in which all are included" (p. 98). As an example of the immediacy of the body image extended in space ("a spatiality of situation"), he continues, "If I stand in front of my desk and lean on it with both hands, only my hands are stressed and the whole of my body trails behind them like the tail of a comet. It is not that I am unaware of my shoulders or back, but these are simply swallowed up in the position of my hands, and my whole posture can be read so to speak in the pressure they exert on the table" (p. 100).

12. For instance, Straus (1966) notes that the development of bipedality in humans brought with it a "transition of sensorium and motorium, of periphery and centre, of

form and function" (p. 164). As well as the implications for perception, there are accompanying changes in the structure of thought itself. Maxine Sheets-Johnstone (1988) contends that our bodies have served as semantic templates throughout our evolutionary history. Our concepts of hardness and number, for example, are said to have arisen through the lived body (hardness of teeth, the binary rhythm of walking).

13. This is Maus' term and is most appropriately, for the theme of this paper, demonstrated in "techniques of activity, of movement" (such as walking, running, jumping and climbing) (Maus, 1992, p. 469) that are taught and culturally transmitted.

14. 'Photo-realism' (intimately related to the quest for 'presence') may arguably be another version of the holy grail for some computer scientists.

15. Richard Held & Nat Durlach (1992) were primarily talking about teleoperator systems (controlling remote devices/robots), although they apply their analysis to virtual environments. Jack Loomis (1992), too, notes that when highly anthropomorphic designs are employed in teleoperator systems (torso, limbs, video and auditory displays) a compelling impression of telepresence is often reported by the users.

16. "The VB [virtual body] used in these experiments was a crude polygonal representation of a real human body" (Barfield et al., 1995, p. 506).

17. A notable exception is a study reported by Giuseppe Riva (1998) who attempted to look at the effects of an immersive virtual environment on body experience, and claimed to have found that even a short-term immersion in the virtual environment is able to modify the perceptual/cognitive contents of the body image. The representation of participants in this environment was a solitary virtual hand. The methodology employed in this study involved questionnaires to 'measure' changes in participant's body images, but the participants verbal accounts of their experiences of corporeality in this environment was not explored. Riva also cites a study by G. Cioffi (1993) which reported that subjects felt as if they had dematerialized or were "in the absence of gravity." In addition high numbers of participants claimed not to feel their bodies.

18. Dennis Proffitt explains that the "point of projection" in VR is at standing height. The perspective offered to viewers mimics their experience in the world, and they measure objects in the VE as they do in 'reality', that is against their own bodies. "You turn your head and see a stool in the corner, it appears below your line of vision, making it appear shorter than you are" (Azar, 1996a, pp. 1 and 25).

19. The supporting role of the whole body in any perceptual activity is elaborated on by Leder (1990): "When I gaze at a landscape I dwell most fully in the eyes. Yet this is only possible because my back muscles hold my spine erect, my neck muscles adjust my head into the proper position for viewing .... My whole body provides the background that supports and enables the point of corporeal focus. As such, the body itself is not a point but an organised field in which certain organs and abilities come to prominence while others recede."

20. Apart from 'motion sickness', other "after-effects include flashbacks, illusory sensations of climbing and turning, and reduced motor control" have "been documented as lasting up to two weeks" (Azar, 1996b, p. 25).

21. The assuring pats on the buttocks and torso that Heim observes in 'emerging' VR users is reminiscent of behavior observed in some schizophrenics and psychotics who express their reasons for caressing, cutting, or banging various parts of the body as a wish to "regain a clear picture of the dimensionality of their bodies, which had become vague or 'deadened' " (Fisher, 1973, pp. 23–24).

22. Henry Head (1920) posited the body scheme, a plastic neural representation of the body, to explain how postural adjustments and the localization of bodily sensations are achieved. The body schema is dynamic, being constantly restructured by the acquisition of new skills and movements. It is presupposed in the awareness and movement of the body (Tiemersma, 1989), enabling the functional distinction between corporeal and extracorporeal space. As such, the body scheme provides the basis for meaningful action within the environment. Head famously stated that the body scheme extended to the tip of the feather in a woman's hat (Head & Holmes, 1911). The body scheme, then, enables the projection of "our recognition of posture, movement, and locality beyond the limits of our bodies to the end of some instrument held in the hand" (ibid). Without it we would be unable to carry out even seemingly mundane tasks, such as feeding ourselves with a knife and fork.

23. There is also a sense whereby 'donning' HMDs, gloves and body suits becomes a ritual. In order to 'enter' the virtual world these devices have to be worn, both literally and symbolically. For example, consider a study by Mel Slater, Martin Usoh & Anthony Steed (1994) where a greater sense of 'presence' was induced by having participants *simulate the process of entering the virtual environment* while already immersed in a VR. This simulation included repetition of 'donning' a virtual HMD to enter different virtual environments.

24. Indeed, Merleau-Ponty (1970) speaks of the blind man's cane as "an instrument *with* which he perceives. It is a bodily auxiliary, an extension of the bodily synthesis" (p. 152). Similarly, Leder (1990) contends that "In its use of tools and machines the body supplements itself through annexing artificial organs" (p. 30).

25. Ihde defines 'technics' as "the symbiosis of artefact and user within a human action" (p. 73).

26. For instance, Rupert England (1995) makes the point that protective clothing impinges on certain tasks (e.g. thick gloves prevent, or hinder, fine manipulations) in the real environment. Therefore, VR peripherals should be built into the task, environment, and virtual body of the user. A 3-D mouse for instance (a piece of wired apparatus equipped with buttons which facilitate movement within virtual environments) might take the form of a hammer in the virtual world.

27. This is not to say that we cannot become aware of them. Perception takes place 'through' the peripherals, but as a 'fringe phenomenon' we can become aware of, for instance, the light pressure of eyeglasses on the bridge of the nose – but the 'focal' phenomenon is achieved by the perceptual transparency of the peripherals. This is what Ihde (1990) refers to as a 'ratio' between the objectness of the technology and its transparency in use. At the extreme height of embodiment, background presence of the technology may be detected. However, this does not imply that a 'dislocation' will inevitably be experienced between the corporeal and the virtual body.

28. This statement has at least one caveat: the prosthesis user may reach this type of incorporation after only many years of use, if at all. I am currently researching the circumstances surrounding this incorporation of the prosthesis into the phenomenal boundaries of a user's body. However, there is some indication, as discussed later, that it is possible to manipulate the physical body and (re)present these manipulations in a virtual environment.

29. Romanyshyn (1994) argues that it is closer to the phenomenology of cyberspace experience to see the virtual or cyberbody as 'haunting' the virtual world. The virtual body has a 'ghostly presence', "its interactions with its virtual environment leave no

tangible marks upon its digital flesh." Romanyshyn"s description, I would argue, is more suited to a distanced 'observer' than to the 'inhabitant' of the virtual body.

30. A fictional example of this is provided by the British television sci-fi comedy, *Red Dwarf*. In one episode one of the main characters, Cat, attempts to leave an 'artificial reality console' by taking off the peripheral devices. After taking off the left glove and boot, Cat's (virtual) body is rendered paralysed down his left-hand side. Although Cat's physical body is ostensibly not paralysed, it is the virtual reality that remains his dominant bodily perception.

31. This is actually very true of tasks in VR when manipulation is required and where force-feedback is absent.

32. As a case of extreme objectification of the virtual body we may consider a participant in a study reported by Mark Draper (1995): "one pilot subject attempted to separate the free floating right arm from the body for use as a detached virtual ruler!"

33. For a number of other ingenious and simple methods of destabilizing the body image see Vilayanur Ramachandran & Sandra Blakeslee (1998), pp. 59–60.

34. This paper was written partly during my employment on the 'eSCAPE' project with the financial support of the European Strategic Programme for Research and Development in Information Technologies. I would also like to thank Dr. Paul Arnold and Joanne Wynne for their unwavering support during this work.

REFERENCES

Anthony, E. J. (1968). The child's discovery of his body. *Physical Therapy Practice, 48*, 1103–1114.

Anzieu, D. (1989). *The Skin Ego*. Trans. C. Turner. London: New Haven Press.

Azar, B. (1996a). Psychologists dive into Virtual Reality. *American Psychological Association Monitor, 27*(3), 1, 25–26.

Azar, B. (1996b). The VR pricetag: nausea, dizziness and disorientation. *American Psychological Association Monitor, 27*(3), 25.

Balsamo, A. (1993). The virtual body in cyberspace. *Research in Philosophy and Technology, 13*, 119–139.

Balsamo, A. (1995). Forms of technological embodiment. *Body & Society, 1*(3–4), 215–237.

Barfield, W., Zeltzer, D., Sheridan, T., & Slater, M. (1995). Presence and performance within virtual environments. In: W. Barfield & T. A. Furness (Eds), *Virtual Environments and Advanced Interface Design*. (pp. 473–513). New York: Oxford University Press.

Biocca, F., & Lanier, J. (1992). An insider's view of the future of virtual reality. *Journal of Communication, 42*(4), 150–172.

Biocca, F., & Levy, M. R. (1995). Communication applications of Virtual Reality. In: F. Biocca & M. R. Levy (Eds), *Communication in the Age of Virtual Reality*. (pp. 127–157). Hillsdale, NJ: Lawrence Erlbaum Associates.

Bukatman, S. (1993). *Terminal Identity: The Virtual Subject in Post-Modern Science Fiction*. Durham, NC: Duke University Press.

Bogard, W. (1996). *The Simulation of Surveillance: Hypercontrol in Telematic Societies*. New York: Cambridge University Press.

Cioffi, G. (1993). Le variabili psicologiche implicate in un 'esperienza virtuale'" [Psychological variables that influence a virtual experience]. In: G. Belotti (Ed.), *Del Virtuale*. Milan: Rostro.

Courtiss, E. H., & Goldwyn, R. M. (1976). Breast sensation before and after plastic surgery. *Plastic and Reconstructive Surgery, 58*, 1–13.

Draper, M. (1993). Exploring the influence of a virtual body on spatial awareness. *Master's Thesis*. Human Interface Technology Laboratory, University of Washington. Available from: http://www.hitl.washington.edu/publications/draper/3.html

Druss, R. G. (1973). Changes in body image following augmentation breast surgery. *International Journal of Psychoanalytic Psychotherapy, 2*, 248–256.

Ellis, S. R. (1995). Origin and elements of virtual environments. In: W. Barfield & T. A. Furness (Eds), *Virtual Environments and Advanced Interface Design*. (pp. 15–57). New York: Oxford University Press.

England, R. (1995). Sensory-motor systems in virtual manipulation. In: K. Carr & R. England (Eds), *Simulated and Virtual Realities: Elements of Perception*. (pp. 131–177). London: Taylor and Francis.

Falk, P. (1995). Written in the flesh. *Body & Society, 1*(1), 95–105.

Fisher, S. (1973). *Body Consciousness*. London: Calder and Boyars.

Fisher, S. (1991). Virtual environments: Personal simulations and telepresence. In: S. K. Helsel & J. P. Roth (Eds), *Virtual Reality: Theory, Practice and Promise*. (pp. 101–110). Westport, CT: Meckler.

Fodor, J. (1994). VR and disabilities: Some ethical considerations. In: *Proceedings: Technology and Persons with Disabilities, Ninth Annual Conference*. (pp. 143–147). California State University: Northridge.

Foster, S. L. (1993). Dancing bodies. In: J. Crary & S. Kwinter (Eds), *Incorporations* (pp. 480–495). (New York: Zone).

Goin, J. M., & Goin, M. K. (1981). *Changing the Body: Psychological Effects of Plastic Surgery*. London: Williams and Wilkins.

Goin, M. K., Goin, J. M., & Gianni, M. H. (1977). The psychic consequences of a reduction mammaplasty. *Plastic Reconstructive Surgery, 59*, 530–534.

Grosz, E. (1994). *Volatile Bodies: Toward a Corporeal Feminism*. Bloomington and Indianapolis: Indiana University Press.

Harre, R. (1991). *Physical Being*. Oxford, UK: Blackwell.

Hayles, N. K. (1996). Embodied virtuality: Or how to put bodies back into the picture. In: M. A. Moser (Ed.), *Immersed in Technology: Art and Virtual Environments*. (pp. 1–28). Cambridge, MA: MIT Press.

Head, H. (1920). *Studies in Neurology*, Vol. 2. London: Oxford University Press.

Head, H., & Holmes, G. (1912). Sensory disturbances from cerebral lessions. *Lancet, 1*, 144–152.

Heidegger, M. (1962). *Being and Time*. Trans. J. Macquarrie and E. Robinson. New York: Harper and Row.

Heim, M. (1995). The design of virtual reality. *Body & Society, 1*(3–4), 65–77.

Held, R., & Durlach, N. (1992). Telepresence. *Presence: Teleoperators and Virtual Environments, 1*(1), 109–112.

Ihde, D. (1990). *Technology and the Lifeworld: From Garden to Earth*. Bloomington and Indianapolis: Indiana University.

Jonas, H. (1970). The nobility of sight: A study in the phenomenology of the senses. In: S. F. Spicker (Ed.), *The Philosophy of the Body*. (pp. 312–333). New York: Quadrangle.

Kaczmarek, K. A., & Bach-Y-Rita, P. (1995). Tactile displays. In: W. Barfield & T. A. Furness (Eds), *Virtual Environments and Advanced Interface Design*. (pp. 349–414). New York: Oxford University Press.

Kafka, F. (1952). *Selected Stories of Franz Kafka*. Trans. W. Muir & E. Muir. New York: Modern Library.

Kujundzic, N., & Buschert, W. (1994). Instruments and the body: Sartre and Merleau-Ponty. *Research in Phenomenology, 24*, 206–215.

Lajoie, M. (1996). Psychoanalysis and cyberspace. In: R. Shields (Ed.), *Cultures of Internet*. (pp. 153–169). London: Sage.

Lasko-Harvill, A. (1993). User interface devices for virtual reality as technology for people with disabilities. In: *Proceedings: First International Conference, Virtual Reality And Persons With Disabilities*. (pp. 59–65). California State University: Northridge.

Leder, D. (1990). *The Absent Body*. Chicago: University of Chicago Press.

Loomis, J. (1992). Distal attribution and presence. *Presence: Teleoperators and Virtual Environments, 1*(1), 113–119.

Maus, M. (1992). Techniques of the body. In: J. Crary & S. Kwinter (Eds), *Incorporations*. (pp. 455–477). New York: Zone.

Merleau-Ponty, M. (1970). *The Phenomenology of Perception*. Trans. C. Smith. London: Routledge and Kegan Paul.

Misiak, H., & Sexton, V. S. (1973). *Phenomenological, Existential, and Humanistic Psychologies: A Historical Survey*. New York: Grune and Stratton.

Moss, D. (1978). Brain, body, and world: Perspectives on body image. In: R. S. Valle & M. King (Eds), *Existential-Phenomenological Alternatives for Psychology*. (pp. 73–93). New York: Oxford University Press.

Murphy, R. (1987). *The Body Silent*. London: J. M. Dent.

Novak, M. (1992). Liquid architecture in cyberspace. In: M. Benedikt (Ed.), *Cyberspace: First Steps* (pp. 225–254). London: MIT Press.

Penny, S. (1993). Virtual body building. *Media Information Australia, 69*, 17–22.

Penny, S. (1994). Virtual reality as the completion of the enlightenment project. In: G. Bender & T. Druckery (Eds), *Culture on the Brink: Ideologies and Technology*. (pp. 231–248). Seattle: Bay Press.

Plugge, H. (1970). Man and his body. In: S. F. Spicker (Ed.), *The Philosophy of the Body*. (pp. 293–311). New York: Quadrangle.

Pruzinsky, T., & Edgerton, M. T. (1990). Body-image change in cosmetic plastic surgery. In: T. F. Cash & T. Pruzinsky (Eds), *Body Images: Development, Deviance and Change* (pp. 217–236). London: Guilford Press.

Ramachandran, V. S., & Blakeslee, S. (1998). *Phantoms in the Brain*. London: Fourth Estate.

Riva, G. (1998). Modifications of body-image induced by virtual reality. *Perceptual and Motor Skills, 86*, 163–170.

Romanyshyn, R. (1994). The dream-body in cyberspace. *Psychological Perspectives, 29*, 90–103.

Sacks, O. (1985). *The Man Who Mistook His Wife For a Hat*. London: Duckworth.

Sartre, J.-P. (1970). The body. In: S. F. Spicker (Ed.), *The Philosophy of the Body*. (pp. 218–240). New York: Quadrangle.

Schain, W. S., Wellisch, D. K., Pasnau, R. O., & Landsverk, J. (1985). The sooner the better: A study of psychological factors in women undergoing immediate versus delayed breast reconstruction. *American Journal of Psychiatry, 142*(1), 40–46.

Sheets-Johnstone, M. (1988). *The Roots of Thinking*. Philadelphia: Temple University Press.

Sheridan, T. B. (1992). Musings on telepresence and virtual presence. *Presence: Teleoperators and Virtual Environments, 1*(1), 120–125.

Slater, M., & Usoh, M. (1994). Body centered interaction in immersive virtual environments. In: N. M. Thalmann & D. Thalmann (Eds), *In Artificial Life and Virtual Reality*. (pp. 125–147). Chichester: John Wiley and Sons.

Slater, M., Usoh, M, & Steed, A. (1994). Depth of presence in virtual environments', *Presence: Teleoperators and Virtual Environments*, *3*(2), 130–144.

Spinelli, E. (1995). *The Interpreted World: An introduction to Phenomenological Psychology.* London: Sage.

Stensman, R. (1989). Body image among 22 persons with acquired and congenital severe mobility impairment. *Paraplegia*, *27*, 27–35.

Stone, A. R. (1992). Will the real body please stand up? Boundary stories about virtual cultures. In: M. Benedikt (Ed.), *Cyberspace: First Steps*. (pp. 81–118). Cambridge, MA: MIT Press.

Straus, E. W. (1966). *Phenomenological Psychology*. Trans. in part, E. Eng. London: Tavistock Publications.

Tiemersma, D. (1989). *Body Schema and Body Image: An Interdisciplinary and Philosophical Study*. Amsterdam: Swets and Zeitlinger.

Tomas, D. (1989). The technophilic body: On technicity in William Gibson's cyborg culture. *New Formations*, *8*, 113–129.

Trieschmann, R. B. (1988). *Spinal Cord Injuries: Psychological, Social, and Vocational Rehabilitation*. (2nd ed.). Arizona: Demos.

Valle, R. S., & King, M. (1978). An introduction to existential-phenomenological thought in psychology. In: R. S. Valle & M. King (Eds), *Existential-Phenomenological Alternatives for Psychology*. (pp. 3–17). New York: Oxford University Press.

Zaner, R. M. (1981). *The Context of Self: A Phenomenological Inquiry Using Medicine as a Clue*. Athens, OH: Ohio University Press.

ON THE COGNITIVE DEMARCATION BETWEEN TECHNOLOGY AND SCIENCE: AN ATTEMPT AT EPISTEMOLOGICAL CLARIFICATION

Yannis A. Niadas

This chapter aspires to present a demarcation scheme between technology and science, based on the assumption that technology should be considered as an autonomous form of knowledge. The main objectives toward this end are: (a) the clarification of the distinct cognitive characteristics of technology and engineering, (b) the assessment of the concept of 'efficiency' and its role in the technological context and, (c) in view of the above, an attempt to lay out more precisely the epistemological relationship between science and technology, taking into account the unique characteristics of engineering. This demarcation, however, seeks to do justice to the complex intertwining of science and technology experienced in our time, and also to serve as a tool for future philosophical evaluations of the technological activity.

1. INTRODUCTION: WHY DEMARCATING BETWEEN TECHNOLOGY AND SCIENCE?

In today's fast-paced world, modern science and technology have intertwined to such a degree that it is getting more and more difficult to distinguish clearly the contribution of each one party in any given human endeavor aiming toward some practical end. In fact, it is often said that it is exactly this close interaction

Metaphysics, Epistemology, and Technology, Volume, 19, pages 175–213.
2000 by Elsevier Science Inc.
ISBN: 0-7623-0681-5

of modern science and technology that is responsible for the cataclysm of technological achievements we perceive in the modern world. This intricate intertwining has led many thinkers to regard the present day techno-scientific complex as a concrete body of knowledge and activity which is constantly transforming our world – for better or worse. And the growing importance of this constant transformation seems to overshadow, perhaps rightfully so, other aspects of the subject. Cornelius Castoriadis, for instance, uses the term 'technoscience' to describe the techno-scientific complex, asserting that the present state of it constitutes a fundamental political problem because 'the problem of technoscience is the problem of the relationship between humanity, human knowledge, and its own might and power' (Castoriadis, 1990, p. 133). In view of the continuously pressing environmental crisis and the seemingly ongoing deterioration of this planet's ability to sustain life as we know it, the question 'How far do we go with science and technology?' seems to acquire an overwhelming importance. What is the need, then, in the present context, for a cognitive demarcation between them?

We shall argue later that there indeed exists a need to demarcate between the cognitive aspects of science and technology, both for purely philosophical as well as for practical reasons. Apart from the serious questions involved in assessing the power and the influential range of 'technoscience', another matter becoming philosophically important is that the blurring of boundaries between technology and science sometimes gives a fuzzy character to reflection on the status of the techno-scientific enterprise. Specialists themselves, much less philosophers, tend to be unable to discern whether any given accomplishment within the techno-scientific sphere belongs to a technological or scientific discipline, even after close examination. One can witness this confusion in an amusing way: Arageorgis & Baltas (1989) report an entertaining example of this kind of fuzziness furnished by B. R. Gaines; his illustration, considering a given paper and its content, is as follows:

> [C]onsider a study in control engineering; it clearly remains one if we replace the actual plant controlled with a computer model of that plant. It clearly remains so if we consider the plant model as a set of numeric equations. It continues to remain so if we consider the general algebraic form of these equations. And so on each step in itself is a small enough change that we agree that the content of the paper cannot have crossed a borderline between 'control engineering' and 'not control engineering'. Yet when the final paper appears (called 'Residues of Contraction Mappings in Banach Spaces'!), few control engineers will recognize it as belonging to their discipline (Gaines, 1976, p. 624).[1]

Let us first outline in brief the basic assumption that justifies any attempt to cognitively demarcate science and technology. This assumption is none other than the position that technology constitutes an autonomous form of knowledge, distinct of scientific knowledge. In the course of the last 30 years, many writers from different standpoints – philosophical, historical, engineering-oriented, etc. – have explored the singular cognitive status of technology. Many (but not all) question the popular belief that technology is simply 'applied science', a view still widely shared among engineers and technologists themselves. These accounts of the cognitive status of technology cover a wide range, beginning with rather narrow analytic views of technical[2] knowledge as 'sensorimotor skills' (Gilles, 1986), as a collection of technical maxims or recipes and descriptive laws (Carpenter, 1974; Gilles, 1986), or technological rules 'grounded' on scientific knowledge and integrated into technological theories (Bunge, 1972, 1985).

Other definitions try to capture the broader picture of technological cognition rather than insisting on providing a typology of particular elements. Thus, Robert E. McGinn identifies three components of technical knowledge, the first being 'knowledge of how to do certain things by making or utilizing certain material products or by transforming certain material objects'; the second being 'knowledge of the resources, especially energy and materials, used in technological activity'; and the third consisting of 'knowledge of the methods used in reaching the desired result of the activity' (McGinn, 1990, pp. 15–16). Hans Jonas speaks of technical knowledge as "a knowledge concerning the planned changing of the changeable," which 'springs from experience, not from theory or speculative reason' (Jonas, 1972, p. 335). I. C. Jarvie contends that technical knowledge is "knowledge of sorts" which "tells us about what works in this world": "It is, so to speak, true knowledge of what is effective" (Jarvie, 1972, p. 55). And Bohdan Walentynowicz asserts that technology is (among other things) "knowledge concerning technical objects and their functioning" as well as knowledge concerning the activity towards the production of these artifacts (Walentynowicz, 1968, p. 586). The two latter statements form a generalized concept of technical knowledge, which is widely accepted and shared among technologists and engineers.

However, in social science usage the term 'technology' has acquired a quite different content. Carl Mitcham (1994, pp. 143–151) has properly identified the different understandings of technology among the social groups of engineers and technologists, on the one hand, and social scientists, on the other. This difference in understanding is revealed in definitions of technology and technical knowledge given by sociologists or anthropologists, which explicitly stress the social and cultural effects of technology (and its artifacts) as opposed to the more narrow engineering view of technical knowledge that Walentynowicz offers. David J. Hess, for example, begins his book stating that he will understand technology as "materially embedded knowledge about how to create effects via artifacts (that is, via 'things' such as machines, tools, and architecture)" (Hess, 1995, p. 1).

In recent years, a growing body of literature has appeared, furnished by engineers-philosophers/historians who stress even further the arguments against technology and engineering as 'applied science', and focus on the unique cognitive aspects of engineering activity as they are manifested through engineering design and the optimization procedures followed in the engineering elaboration of any technical project. These writers also bring to light aspects of the engineering activity that are non-reducible to any other application of knowledge, such as the ability of engineering to handle problematic data and ill-defined situations, the 'visual' rather than verbal way of thinking and communicating in engineering, etc. (Petroski, 1985, 1986 and 1990; Ferguson, 1992; Vincenti, 1990; Layton, 1974, 1976, 1984 and 1991).

This plethora of definitions and arguments making the case for the cognitive autonomy of technology as distinct from its scientific counterpart is indeed lucid and quite convincing, but presents one problem: each argument tends to stress some particular aspect of technical knowledge over others as the cornerstone of its definition given. 'Efficiency' and 'artifacts' are the most common among those aspects that stand out as primary objects of technical knowledge. Of all the above definitions, only Jonas's definition attempts to express a broader concept of technical knowledge as related to human technological action within the world in which we live. As Mitcham has pointed out, technology itself "is not a univocal term; it does not mean exactly the same thing in all contexts. It is often, and in significant ways, context-dependent both in speech and in the world" (Mitcham, 1994, p. 152). Accordingly, we feel that any attempt to define technical knowledge should at least make clear the specific context in which the definition is drawn. For the same reasons, any definition should try to address the broader together with the narrower aspects of technical knowledge, as well as trying to connect this knowledge to the way technology is presented and manifested in the world we experience and create.

In any case, the above definitions enable us to sketch an outline of what technical knowledge is and is not considered to be in current philosophical thinking on technology:

(1) Technical knowledge *is not* reducible to plain 'sensorimotor' or 'neuromuscular' skills acquired through repetition. In other words, technical knowledge comprises much more than plain empirical acquaintance with materials and the handling of tools used in any technological process.

(2) Technical knowledge refers both to artifacts and to the process of their production. Knowledge of the production process implies that technical knowledge comprises knowledge of manipulating materials and resources as well as knowledge of methods and courses of action toward the intended purpose of the technological activity.
(3) Technical knowledge aims primarily to assess the most efficient way of bringing about the desired effect. Just as scientific knowledge aims to access the truth about the way the world is structured, technical knowledge concerns itself with discovering the most efficient solution to any technological problem it is called upon to face.
(4) Finally, technical knowledge involves unique cognitive procedures, specifically pertaining to technology and engineering, as manifested in the iterative procedures of engineering design[3] and optimization, in the engineering ability to confront ill-defined problem situations and in the technological use of *theoretical tools* in solving technological problems.

We come now to the reasons for which this attempt at a demarcation is undertaken: firstly, despite the growing number of philosophical studies on technology, philosophy of technology has not yet established an autonomous philosophical domain within which the philosophical questions concerning technology are discussed entirely from the standpoint of *technology as a philosophical subject*. It is fairly easy to recognize this by looking at the following chronology of statements of some of the participants in this field: in 1966, Henryk Skolimowski was exploring possibilities in order "to end the stagnation in a field which as yet has only a name – the philosophy of technology" (Skolimowski, 1972, p. 49). Eighteen years later, Albert Borgmann notes that "[t]he philosophy of technology is just beginning to develop as a discipline" and that "theories of technology in the wider sense have come [mostly] from the social sciences" (Borgmann, 1984, p. 8). A decade later still, Mitcham begins his extensive survey of the field by stating that "development of the philosophy of technology remains in its early stages; until quite recently there was little discussion that consciously saw itself as part of such a cooperative, reflective endeavor" (Mitcham, 1994, p. 17). It is obvious from such statements that the gap between the augmenting importance of the technical phenomenon in the modern world and its philosophical elaboration is still a very wide one. Although thinking about technology has a lot to benefit from a wide interdisciplinary approach, combining the expertise in fields such as cultural and social anthropology, sociology, etc., this nevertheless does not eliminate the need for establishing a philosophical domain pertaining to technology *per se* which does not overlap with philosophical traditions dealing primarily with other cognitive fields and, more specifically, philosophy of science.

A possible exploration of the full spectrum of reasons responsible for the underdevelopment of philosophical reflection on technology is outside of the scope of this paper and cannot be undertaken here. However, we may note that the overgrown (with respect to philosophy of technology) philosophy of science field is indirectly part of the problem. The intense philosophical elaboration of science has created a powerful array of patterns of thought, which, given the fuzziness of distinctions between science and technology, often implies and/or unconsciously dictates to thinkers on technology, certain conceptual manipulations and frameworks of analysis that are not always relevant to the philosophical questioning of technology. Indeed, patterns of thinking on science can be totally irrelevant to thinking on technology, since what is at stake in each case is often radically different.[4]

As already noted, one of the main points for classifying technology as an important philosophical subject *per se* is the assertion that technology constitutes an autonomous form of knowledge distinct from scientific knowledge. It follows that one primary concern in establishing such a philosophical domain should be the epistemological analysis of this distinct form of knowledge, a concern which is necessarily linked to the demarcation between science and technology.[5] In this respect, following Skolimowski (1972, p. 43), the epistemological clarification in question lies at the core of the philosophy of technology and therefore it is (still) a reasonable point of departure.

Secondly, the efforts toward such a demarcation so far suffer from two severe flaws: on the one hand, despite continuous criticisms of the notion of technology as 'applied science', technology is often readily reduced to a mere *modus operandi* stage, the technical statements and ensuing actions of which still need to receive some corroboration through discourse, while the serious cognitive aspects of the techno-scientific enterprise remain constantly with science. Also, the rich scientific overtones embedded in engineering lead many writers to assume some kind of scientific attestation of engineering practice (Radnitzky, 1978; Arageorgis & Baltas, 1989). On a cognitive basis, however, we maintain here that the two fields are entirely separated, though of equal importance. At the same time, all the above efforts leave far too many things to be desired as far as the role and nature of engineering is concerned. We do not have, as yet, a clear picture of where exactly engineering stands and how it fits into the overall scheme concerning the science-technology relationship. The writings of engineer-philosophers/historians such as Henry Petroski, Eugene Ferguson, Walter Vincenti, and Edwin Layton are quite lucid and correct in distinguishing engineering from scientific activity in general; however, since they are not strictly concerned with the science-technology relationship, they do not give an account that would satisfy both ends of the spectrum. An attempt is made here to clarify the cognitive aspects of the relationship.

It should be noted that the idea of technology and especially engineering being merely 'applied science' is dominant among engineers themselves and is the most likely answer an engineer would give for a description of his or her trade. There are two main reasons for this. First, the strength of the 'applied science' position among engineers is undoubtedly due in large part to the kind of education that engineers have been receiving in the post-war period. As Ferguson (1992) and also Stephen Johnston, et al. (1996) have shown, engineering education in recent times took an overly-scientific turn, gradually abandoning practical and field courses which in earlier times gave a more 'hands-on' experience to future engineers. Engineering education is presently stressing the elements of scientific theory and method, which are indispensably embedded in engineering, and at the same time underplaying the importance of practical considerations and empirical assessments, which constitute much of the engineer's work. As a result, the preeminence of the 'applied science' thesis among engineers seems almost inevitable. Second, as Layton (1991) has shown, there is a strong ideological basis, manifested historically, which is also responsible for the endorsement of the 'applied science' thesis by the engineering profession. Engineers historically pursued the linkage of engineering to science, according to Layton, in their effort to claim for themselves the role of the foremost agents of progress. As progress was (and still is) readily identified in this culture by the advancement of science, the linkage between engineering and science, with the former 'applying' the objective results of the latter in practical problems, provided the necessary foundation to give scientific credit to the accomplishments of engineering.

This situation adds to the need for a sufficient demarcation between science and technology. Apart from the blurring of boundaries between them, something which only makes the relevant analyses more difficult to perform with accuracy to the facts, the ideological framework working behind the scenes of 'technoscience' is furthermore confusing the picture by introducing patterns of mutual 'credibility-loans' to confront criticism of the feats produced by the techno-scientific enterprise. If one looks, for example, at current environmental disputes over large technological and engineering schemes, a constant claim on the part of those defending their projects is that the projects have been conceptualized, designed, and promoted according to the latest scientific research results on the disputed issues. On the other side of this ideological coin, scientists engaged in pure or basic research can claim more funding for their own projects since their findings are considered to expand the technical potential of engineering. A discussion of ethics, or of the politics of research and development strategies, which is necessarily conducted against this confusing background of interrelations, can only benefit from having at its disposal some concept for discerning the limits of the interaction between science and technology.

2. TECHNOLOGY AS KNOWLEDGE

In what follows, we will attempt to provide a sketch of an epistemology of technology, which will be later used in order to support the proposed demarcation scheme between technology-engineering, on the one side, and science, on the other. The epistemological account presented herein is greatly indebted to the writings of the aforementioned thinkers on the subject of technology as knowledge, as well as to other sources that will also be mentioned. The aim of this section is twofold: (a) to describe the nature of technical knowledge and to provide a rough examination of its cognitive characteristics, and (b) to provide an account of the guiding principles of the technological activity.[6]

Technical knowledge has been many times defined in the past as an expression of the human quality of learning referred to as *knowing-how* as opposed to *knowing-that*, which pertains to scientific knowledge.[7] *Knowing-how* can be defined as the human way of acquiring knowledge through action or apprenticeship, that is by way of doing things or acquiring skills by observation and through trial-and-error procedures. A simple example of this mode of knowing is the way we learn to ride a bicycle. In order to learn how to ride on a bike, it is inevitable that one engages in a trial-and-error procedure, falling off balance and trying again, until, at some point, the required positioning on the bike in order to balance gravitational forces in this specific task is found. Now, the learner may benefit from comments and directions given by others who have already mastered the art of balancing on a bicycle, but there is no way of finally getting to the desired effect other than to keep falling off balance until one finds out how not to. We could provide the learner with a multi-volume work beginning with the explanation of gravitational forces and the laws governing them and ending with systems analysis theory, but it would do little if any good. In order to learn how to ride a bicycle, one just has to learn by doing it.

Despite the fact that much of our common everyday knowledge is of this type (ranging from simple household techniques such as cooking and small scale repairing, to important inter-personal ways of communicating and socializing, including humor and love-making), *knowing-how* has always been regarded in the scientific age as knowledge of a secondary quality. This should come as no surprise considering the power and influence the Cartesian regime of thought has had over philosophical considerations of human knowledge. Gilbert Ryle, in *The Concept of Mind*, originally published in, 1949, provides an account of *knowing-how* in positive terms, useful for our purposes here. Ryle shows that the exercise of intelligence in performing certain tasks does not entail that a prior intellectual operation of considering reasons and criteria for so acting is taking place. Examining human performances in which intelligence is displayed, like appreciation of humor and the making of jokes as well as chess playing, Ryle shows that 'Efficient practice precedes the theory of it; methodologies presuppose the application of the methods, of the critical investigation of which they are the products' (1990, p. 31). Ryle attacks the intellectualist legend that for someone to act reasonably, a prior internal process of self-avowing to rules or maxims relevant in each case, is necessary:

> The crucial objection to the intellectualist legend is this. The consideration of propositions is itself an operation the execution of which can be more or less intelligent, less or more stupid. But if, for any operation to be intelligently executed, a prior theoretical operation had first to be performed and performed intelligently, it would be a logical impossibility for anyone ever to break into the circle .... Knowing how to apply maxims cannot be reduced to, or derived from, the acceptance of those or any other maxims (Ryle, 1990, pp. 31–32).

In this respect, 'to perform intelligently is to do one and not two things', i.e. when exercising a *knowing-how* ability, one is not engaged in 'a tandem operation of first considering prescriptions and then executing them'. Accordingly, any intelligent performance involves the application of some criteria in the conduct of the performance itself, but it is not imperative that the criteria or the rules governing the performance should be learned through instruction: 'We learn *how* by practice, schooled indeed by criticism and example, but often quite unaided by any lessons in the theory'. Just as we say for someone that s/he knows how to play chess when s/he exhibits the ability to make the right moves, while we do not assert the same when this person just knows to cite the rules of the game, *knowing-how* is manifested in the way the performance is actually taking place and not in the formulation of the rules behind it. Ryle also shows that practical skills and competencies are the product of practice and not sheer habit. Habits are acquired dispositions or second natures that are built up by drill, while practical skills are built up by training. Drill, being merely the imposition of repetitions, 'dispenses with intelligence; training develops it'. Training 'involves the stimulation by criticism and example of the pupil's own judgment' (Ryle, 1990, pp. 41–42).

> Knowing *how*, then, is a disposition, but not a single-track disposition like a reflex or a habit. Its exercises are observances of rules or canons or the applications of criteria, but they are not tandem operations of theoretically avowing maxims and then putting them into practice. Further, its exercises can be overt or covert, deeds performed or deeds imagined, words spoken aloud or words heard in one's head, pictures painted on canvas or pictures in the mind's eye. Or they can be amalgamations of the two (Ryle, 1990, p. 46).

Now, to contend that technical knowledge is an expression of the human quality of knowing through action or apprenticeship is certainly not enough. Technical knowledge is relying on the *knowing-how* capacity of human nature, but aims at a much more detailed level of information about the world. Most accounts on the nature of technical knowledge fail to pinpoint the exact relationship between the faculty of *knowing-how* and the proceedings of any technological activity. In other words, they do not give a concrete picture of where and when exactly in the process of a technological endeavor, actions, procedures or decisions relying on *knowing-how* cognitive dispositions are indeed manifested. The reader must bear with us on this point until we come to examine the practices of engineering design and optimization. For now, in order to assess clearly the level of technical knowledge, we must bear in mind what is the ultimate goal in any given technological process: technology and engineering aim at bringing-about a certain state of affairs and/or at creating a specific artifact or device. This goal is predetermined to the extent of the possible forecasting of its characteristics at the time of its conception. In what follows, we shall refer to this predetermined goal of technology, together with the specific technological context in which it is conceived and which is guiding its realization, as the *application* pursued by technological means. Strictly speaking then, technical knowledge is about *what functions (and how it functions) in the given world with respect to the intended application.* Technical knowledge does not care to access the truth about the world (the way the world is structured); it does not care why what works in the world indeed does so. It only cares about whether this or the other action works better or worse with respect to the intended application. Still, this knowledge is not entirely deprived of *knowing-that* cognitive aspects, and this is especially true as far as engineering is concerned. Nevertheless, the main cognitive qualities expressed in technical knowledge are always linked to the *knowing-how* capacity of the human agents who are engaged in the technological process, as we are going to see in the course of this work.

At this point, we would like to address two frequent counter-arguments against this account of technical knowledge. The first argument is often expressed by questioning the essence of the so-defined technical knowledge: What do we know exactly when we say we know how something works or not with respect to any predetermined goal we are pursuing? This questioning challenges the concreteness of technical knowledge as opposed to scientific knowledge (which is taken to be the model for all knowledge): when, in a scientific context, we affirm our belief in a scientific theory and the explanation of natural phenomena that the theory provides, we mean that we accept a logically formulated hypothesis to the extent that it has proven itself through experimental evidence and the way it adheres to our present conception of the world. A scientific theory, that is, represents a well-founded structural conception of a facet of reality. And this, to the extent of our present understanding of the world structure, is as concrete as knowledge can get. How does technical knowledge, being only a mere knowledge of some things working this way and of some others not, compare to that? What does it really tell us about the world we live in?

The argument would seem plausible if only it did not miss the point: what this argument really asks is what universal features does technical knowledge comprise. Well, it does not comprise any (at least in the scientific sense); and quite naturally, so we should say. Technical knowledge is by nature extremely *specific*: it applies to specific given applications under narrowly defined given conditions, and it has no intent to produce general truths. This is something simply outside of its scope. It does not however follow that technical knowledge is not as concrete as any human knowledge can be. It is just that technical knowledge is about specific things and specific arrangements of things.

We could phrase this in more philosophical terms by saying that, in the domain of *knowing-how*, all the cognitive faculties acquire a distinct *specificity*, explicitly related to the *application* that is at any given time the object of the technological process. For example, when engineers solve the problem of irrigation water needs in a rural area, we end up *knowing* how the particular arrangement of reservoirs, ponds, irrigation canals, flow regulation structures, storage tanks, and pressure pipe networks *does* provide an answer to the technical problem of providing irrigation water to cover the needs of the given area. There is no reason to deny the fact that we now *do know* something we did not before. And although this newly furnished knowledge is directly applicable only in the very narrow context specified by the rural area in question (its characteristics and its specific needs for irrigation water), it nevertheless constitutes a concrete knowledge of a certain arrangement of things working efficiently under given conditions and prerequisites. In short, technical knowledge is extremely specific by nature, just as scientific knowledge by its own nature aims to be as universal as possible.

This brings us to another frequently presented objection to accounts of technical knowledge based on the notion of *knowing-how*. The objection states that all *knowing-how* may be ultimately reduced to *knowing-that*, since, in each case, we end up knowing *that* an arrangement of materials and devices and relevant actions resulted in the solution of a particular technological problem. In the previous example, for instance, the objection would be that we now do know that the arrangement of waterworks in question did provide an answer to the technical problem we were facing.

But this is merely playing with words: to say that we know *that* a particular arrangement of waterworks resulted in the resolution of the problem of irrigation water supply in a given area, is equivalent with saying "we know *that* human technology furnishes solutions to human technical problems in general." Moreover, stating *that* one knows *that* a technological process A furnished result B in a particular case, is just a specific form of the general proposition 'one knows *that* technological processes furnish acceptable results in a multitude of technological problems'. When we contend that we know *how* a particular solution to a technological problem was brought-about, we do not refer to an external recognition of an attained stage of a train of events (consisting of technical decisions and actions), but to the inner workings of a (technological) process which precedes, deploys with, and concludes in the aftermath of that train of events, resulting in a desired effect which, as previously stated, was predetermined to the extent of the possible forecasting of its characteristics at the time of its conception. Yet we may never know precisely *how* the particular interplay within that process results in bringing-about the desired effect, although we always know *that* processes of this kind furnish more or less successful results. We know *how* to set up a process of this kind; and we know *that* processes of this kind *can* indeed be set up. So, the proposed objection is simply unfounded because it confuses knowledge concerning the setting-up of a technological process (its inner workings) with knowledge concerning the reality of setting-up technological processes (the fact of their deployment and outcome).

This discussion reveals one crucial point about technical knowledge: despite the specificity that characterizes it, subsequent technological processes inevitably create classes of similar applications. Across these classes, some elements of technical knowledge are common for all applications in any given class. A body of technical knowledge, while indispensably linked to the specific application realized by the technological process which furnished that body of knowledge, is always extremely valuable as practical experience – though not directly applicable – in other problems of the same class of applications. This property enables design operations to begin with a delimited set of alternatives, rather than 'getting lost' every time exploring all imaginable possibilities in resolving a particular technological problem. The range of delimited alternatives varies with the degree of existing previous experience in the particular class to which the problem at hand belongs. Technological innovation in this sense, can be seen as building upon and expanding the technical cognitive basis of a given class of applications, by introducing solutions that can later be used as a point of departure in dealing with subsequent technological problems of the same class.

We can now, in view of the above, provide a more exact definition of technical knowledge. What technology essentially does is a double alteration of the way the world is structured: it performs a *re-arrangement* of given world objects and of their existing arrangements (materials, forces, etc.); and/or, it *introduces* altogether new physical forms, objects, and arrangements of them (artifacts, systems, etc.) into the given world. Both actions manipulate reality aiming to achieve a predetermined goal, using natural and artificial means selected for each specific process by a system of criteria formulated within the context of this process and with respect to the sought-after application. In a nutshell, technology performs small or large-scale *transformations*[8] toward the realization of a predetermined goal. These transformations are of course not random; indeed, they are the direct product of a detailed design which varies in detail according to the level of the technological process to which it pertains. Technical knowledge is the knowledge used in and acquired through these transformations. Thus, technical knowledge can be defined as *the knowledge concerning the designed transformation of the world.*

It seems to us that the above definition manages to be specific enough (since every technological transforamtion of the real is specific in its nature, according to what has been said before) while maintaining a degree of abstraction necessary in order to include all aspects of technical knowledge. This is possible because 'designed' can be interpreted here both in the narrow context of specific applications and in the broader context of Cornelius Castoriadis' definition of *technique* as humanity's conscious, planned advancement into the world (Castoriadis, 1991), a notion that we think lies at the core of the technical phenomenon in general. Another advantage of this definition of technical knowledge is that it can be matched with both the narrow sense of technology in technological and engineering usage (technology as the production of material artifacts and/or processes and the knowledge concerning them) and the broader sense of technology as understood by social scientists (technology as relationship, cultural form, etc.)

Let us now examine in short the guiding principles of the technological activity. To begin with, we must specify what is lying at the core of this activity;

that is, we must ask ourselves what is the *norm* that underlies and unifies all singular technological actions within a specific technological process and even across different processes. As has been noted many times, technology seeks the most *efficient* way to accomplish its given task. The engineer, or the technologist in general, is trying to find out those methods, tools, resource management options, and design strategies that work better within the immediate and pragmatic context of the application at hand, as well as within the available resources. The concept of efficiency has been loosely defined as a ratio of inputs to outputs.

It has many times been affirmed (e.g. Carpenter, 1983) that the concept of efficiency, like technology itself, is context-dependent. Mitcham remarks on this that "Context dependency, however, need not deprive a concept of all formal characteristics; even variables can have boundaries" (1991, p. 100). Therefore, one should not attribute to the concept of efficiency the same degree of freedom that is manifested in the various uses of the term 'technology'. Under the engineering perspective, the seeking of efficiency in the process of bringing-about the intended application is exercised through the process of engineering design. Within the design process however, the concept of efficiency is manipulated in ways that are specific to the application being designed. It is impossible to explore here in full detail the inner workings of a design process. For now, it is sufficient to say that design is the process by which the variable 'efficiency' is confined to certain boundaries. And these boundaries are specific to the application that is the object of design, in that any solution to the technological problems dealt with in the course of the design process is an 'efficient' solution only within the scope of the application being designed.

Since we are dealing here with the task of demarcating between the cognitive aspects of science and technology as they are manifested in the techno-scientific enterprise predominant in our times, a good point to be made is to show the radical difference between the two guiding concepts of the technological and the scientific activity (*efficiency* and *truth*, respectively). Let us examine this difference with respect to 'normal' practice in technology and science. Efficiency cannot in any case be attributed a role in the technological context derived symmetrically from the role attributed to truth in the scientific context and vice versa. Efficiency is by definition a *dynamic* concept, whereas truth in the everyday scientific activity (what T. S. Kuhn describes as "normal science") acquires, at least for a certain time frame, a *static* character.

Let us examine this statement for a moment in light of T. S. Kuhn's writings: in Kuhn's depiction of the scientific activity (Kuhn, 1983), scientists work under the influence of a concrete set of maxims, laws, methods, and already approved scientific theories and hypotheses called a paradigm. The whole of the 'normal science' activity is carried out within the intellectual confines of this underlying set of beliefs, and consists in the 'solving of puzzles' that still do not fit in with what is predicted by the paradigm's theories, with the primary purpose of verifying the competence and the accuracy of these theories. Now, the 'scientific truth' in this line of thinking, is possible to change (and change dramatically) when the dominant paradigm falls into a state of crisis as a result of the inability of normal science to produce plausible answers to still unresolved puzzles, that is, answers valid within the underlying theoretical context of the paradigm.

However, for as long as the dominant paradigm has not fallen into a crisis, the set of underlying beliefs and assumptions which constitute the intellectual hard core of the paradigm, *are not* and *should not* in any case be questioned. These unquestioned theoretical beliefs and assumptions of course constitute propositions about the true nature of things, and as such, their state of indisputability lends a totally *static*, if temporary, character to the accumulated scientific truth they represent. It is against this static background that 'normal science' is checking its hypotheses and conducts its experiments in 'puzzle solving'. Even if one does not fully endorse Kuhn's account, it is nevertheless imperative to admit that at any given point scientific research must rely on some more or less fixed concept of truth against which it can check further hypotheses, even if those hypotheses will result finally in the adoption of a different concept of truth based on the results of the research undertaken to prove them.

By contrast, efficiency in the technological context is constantly re-defined with respect to each and every one application that any technological process seeks to accomplish. In the eyes of the engineer, there is no absolute, static, or standard concept of efficiency to which s/he must conform. Whatever action, method, or tool is judged as efficient in the course of any technological process is judged so on the sole basis of the immediate context of the application at hand. Even across applications of a similar class, efficiency is never an unquestioned standard, but a goal defined within the present context of action. Therefore, efficiency is not a rigid, timeless faculty to which all technological actions must conform, just like all scientific inquiries eventually must serve the uncovering of the truth about the world, but a dynamic concept which in every technological task acquires specific content according to the nature of the undertaking. Once defined within the immediate context of the application at hand, it is used to assess and aid in the pursuit of the *optimal* solution to the technological problems under investigation. To understand this more clearly, and to be able to assess technological efficiency (and the process of optimization) in a more elegant way, we need to focus on the set of criteria that dynamically define efficiency in the course of a technological process.

First of all, let us sketch the contours of the setting in which a technological process is deployed. Any given technological process is initialized after the emergence of a relevant technological problem, the solution to which this process seeks to produce. Now a technological problem may emerge in a multitude of social and cultural contexts and under the influence of many determining factors, such as economic or political decisions and actions, etc. We cannot discuss here in detail the diverse complex of conditions within which a technological problem arises; however, one thing remains constant every time a technological problem is presented: the technological problem is put forth by an agent which can be a person or a number of persons acting separately, a social group, a community, a corporation, or a complex of corporations, a social institution, or political organization, etc. This agent may also be a technical body or organization seeking to improve on previous solutions to technological problems in order to achieve greater efficiency and/or to incorporate new state-of-the-art techniques into the established body of technical knowledge. In general, the solution to the technological problem will fulfill a need and/or desire of the agent putting it forth, or bring about a desired state of affairs for this agent. The agent assumes the responsibility to cover the cost of solving the technological problem posed and therefore decides the extent of what s/he is willing to pay to bring about the solution ('cost' in this sense should be considered broadly: financially, socially, politically, environmentally, etc.). Given that the technological problem thus put forth is deemed technically and/or socially acceptable[9] (that is, that a possible solution is considered to lie within the range of a predictable and acceptable degree of success under the means and resources available), a technological process is initialized aiming toward its resolution (either in a private or in a public domain, according to the nature of the agent posing the problem). Arageorgis and Baltas (1989) furnish some helpful remarks on the complexity of this setting:

> [T]he state of affairs that the solution of the technological problem will realize is, by definition, desired by a *particular* agent in response to a *particular* need. Other agents may not have this need, they may not consider this state of affairs as desirable or even they may be strongly hostile to it. Accordingly, the solution of a technological problem may encounter important social (and political) resistance and thus involve important social (and political) costs . . . . [A]s no single science is in the position to take into account and control all aspects of the phenomena involved in the solution of a technological problem, adverse side-effects (technological accidents, environmental hazards, etc.) may come about which are unintended and, on the basis of the existing knowledge, unpredictable. The process of

solving a technological problem should take into account such risks as well (Arageorgis & Baltas, 1989, pp. 220–221).

As stated previously, this process has as its first task to define the specific form of efficiency pertaining to the application at hand. The technological process, in other words, must determine what should be considered as 'efficient' within the immediate context of the technological problem posed (i.e. which are the specific characteristics of an efficient solution to the problem).[10] To determine this, a set of distinct criteria of efficiency are used: we can classify them (a) as 'internal' criteria, dealing with what will be referred to as 'technical efficiency', which arise from the inner workings of the technological process, and (b) as 'external' criteria, dealing with economical efficiency in the broad sense, which arise from the agent posing the technological problem. This double set of efficiency criteria guides the whole of the technological process that aims at the realization of the intended application. The 'internal' criteria consist in mainly technical criteria regarding the properties of the available means, the handling of the materials and the accepted technological rules and standard practices which are relevant with the application at hand; the 'external' criteria are broadly understood economic criteria which control the various types of cost involved in the process according to the acceptability of these costs by the agent posing the technological problem and his/her willingness to cover them.[11]

A technological process essentially consists in a large number of decisions and ensuing actions which are contemplated, assessed, and executed in the light of the above double set of criteria. Let us assume that we have the opportunity to inspect a technological process from a hypothetical vantage point which allows us a total perspective on all of its aspects. What we would see from such a vantage point could be described as a finite series of feedback loops. Some of those feedback loops are independent, merely performing a certain number of necessary cycles to produce an output valid in itself within the context of the process and/or used as input to a further loop. The largest part of these loops, however, are interconnected in such a way that the output of one (or many) is immediately compared with and judged according to the output of another (or many).

At the initial stages (cycles) of the work, the outputs produced present a high degree of divergence and even contradiction with one another. Then, as the process continues, the diverging aspects of the issues concerned are used as feedback information to the next cycle of the design process, together with any new data found relevant to the problem at hand. They can be introduced at any such stage and as deemed necessary. The aim of the process is to reduce the level of contradicting elements (and the degree of the contradiction itself) across the network of the loops. At some stage a combination of outputs from each loop is produced which is judged to be satisfactory, and the process comes to an end. The time frame of the successful stage depends entirely on the nature of the application pursued by the technological process as well as on the available resources (human, natural, and artificial) and their proper management. The efficiency criteria are operating in a crucial manner at every step of the above process. Each feedback loop is associated with one or more of the criteria involved, and regulates its operation in order to fulfill it (them). Also, the relative assessment of each cycle's output, as well as the final assessment of the result of the whole process, is performed on the basis of the degree of satisfaction achieved so far for all the criteria which define efficiency in the context of the application pursued.

As can be easily understood, it is absolutely impossible to satisfy all the criteria simultaneously in a maximum way, since many of them represent pairs of mutually annulling demands – the most frequent of these pairs being the seemingly inevitable 'reduce production costs to the minimum/improve quality to the maximum' bifurcation. A maximum solution satisfying every criterion to the fullest is virtually non-existing in this sense. What actually takes place, instead, is a trade-off between the various demands and limitations that the efficiency criteria represent. The process described above is what is often referred to as *optimization*, and it is the operation which essentially achieves the 'ideal' efficiency as defined by the set of criteria used in the process, always within the immediate context of the intended application. The problem of optimization consists in satisfying the technical requirements set by the 'internal' criteria, relative to the simultaneous meeting of the demands set by the 'external' criteria, in such a way that the total value function for the specific application is maximized.

Because the process involves compromising between different and quite frequently opposing alternatives, it follows that the elements considered must fall within a specified, for each element, *range of* (appropriate) *values* in order to allow for the compromise to take place. That is, the technological elements (materials, resources, devices, etc.) considered in the process, must necessarily allow some room for the satisfaction of demands to advance, while at the same time the opposing limitations are restricted, and vice versa. This fact of the optimization process helps to unveil the inner reason why technology and engineering do much of their work by approximations rather than seeking precise solutions. In technology and engineering, one is not so much interested with precise results than with the compliance with specified and agreed-upon ranges of *admissible* values. 'Admissible' in the engineering sense is any range of values of a specific parameter or technical element which (a) adheres to what is at any given time the technical state-of-the-art on the functioning of this parameter or element in a specific class of applications (regulated in the process by 'internal' criteria, e.g. current technical standards), and/or (b) adheres to what is at any time externally regulated as the desirable level of quality (and the 'cost' to achieve this quality) of any given process which includes this parameter or element in a crucial way (regulated by 'external' criteria, e.g. safety regulations). Engineering standards, regulations, and specifications serve exactly this necessity: they are a codification of agreed-upon ranges of admissible values for a variety of technological elements in various classes of similar applications. These codifications incorporate both the compliance to 'internal' criteria specific to various classes of applications (for example, reinforced concrete technology regulations), as well as useful conclusions and proposals for action derived from practical experience acquired through previous work in those classes of applications. They also contain standards and specifications conforming to 'external' criteria, as imposed by the agent assuming the cost of the technological solution (for example, environmental quality standards).[12]

All this boils down to the fact that the technological process results in a dynamically defined *compromising configuration* between the various criteria used, which is achieved through the process of optimization. The configuration that results from the interplay of the different criteria as an optimized solution to the technological problem, manages, in some degree of success that differs from case to case, to both satisfy the demands and succumb to (or perhaps, surpass) the limitations imposed.[13] This configuration is specific to that particular application brought-about and pertains only by a certain amount to other technological problems of the same class; no two technological problems are exactly and in every way the same.

Let us recapitulate what has been said so far, before moving on to the primary objective of this paper: any technological process aims to realize an object and/or a state of affairs that will solve a relevant technological problem put forth by an agent who will assume the cost of its production and for whom the solution to the technological problem will be of importance. To accomplish its task, the technological process begins by defining the concept of efficiency specific to the application it pursues, within the immediate context of this application. To define this specific concept of efficiency, the process utilizes a double set of criteria which regulate both the technical specifications and the properties of the means and resources available ('internal' criteria of technical efficiency) and also regulate what is determined externally (by the agent posing the technological problem) as the desirable quality in resolving the problem, as well as the acceptable cost (in the broad sense) to achieve this quality ('external' criteria of economical efficiency). Having defined the specific concept of efficiency for the application at hand, the technological process engages in the process of designing and producing the object and/or state of affairs it intends to realize using the criteria mentioned to control and check inputs and outputs in the design and production process. A feedback operation is carried out resulting in a compromise between the divergent demands and limitations, which the application at hand has to meet and satisfy. These demands and constraints are represented in every step within the process by the set of efficiency criteria. The resulting solution has a degree of success, which is defined by the relative satisfaction of demands and the overcoming of the limitations imposed, and is always specific[14] to the intended application.

3. A DEMARCATION SCHEME BETWEEN SCIENCE AND TECHNOLOGY

We arrive finally, after this perhaps lengthy but necessary account of the epistemological status of technology and its process (which considered the nature of technological knowledge and the guiding principles of this activity) to the principal issue of this chapter. The statements made earlier are used in an attempt to demarcate between science and technology, with specific emphasis given to the role of engineering in this demarcation scheme. The demarcation scheme considered is shown graphically in Figure 1. Needless to say, of course, this is purely a schematic illustration of the argument presented in what follows and is provided in order to help clarify the positions held. Let us examine first the topology of the elements shown in the figure.

The left-hand side is the locus of the scientific domain, and is showing two partially overlapping rectangles depicting Pure Science and Applied Science respectively. The relationship between those two 'components' of science is understood here in terms of a distinction supported by James K. Feibleman and I. C. Jarvie. Applied science is identified by both Feibleman and Jarvie as a process more closely related to science than technology. In Feibleman's own words: "Applied science consists in a system of concrete interpretations of scientific propositions directed to some end useful for human life." In this respect, the applied scientist differs in function from the technologist: "The applied scientist fits a case under a class; the technologist takes it from there and works it out, so to speak, *in situ*"[15] (Feibleman, 1972, p. 36). Similarly, I. C. Jarvie contends that "Applied science is far more like pure science than it is like invention. [Applied science] is the applying of abstract theories in the world. 'Applying' here means deducing from scientific theories, with the help of some statements of fact, consequences that can be tested and applied"

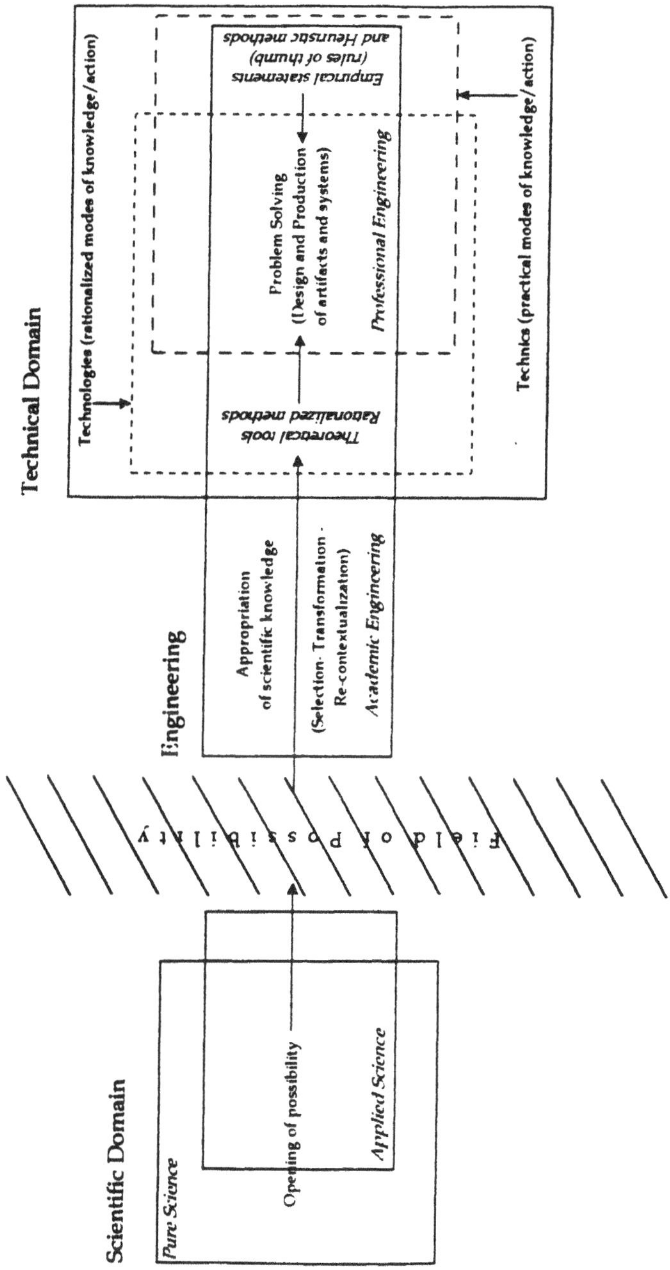

Fig. 1. The cognitive relationship between Technology-Engineering and Science

(Jarvie, 1972, p. 57). In practical terms, the scientific domain of Figure 1 can be seen as a combination of basic research conducted in science institutions and universities (pure science), and applied research conducted in industrial research and development corporate departments and state agencies or laboratories (applied science).

On the opposite end of the diagram stands the technical domain (Technology in the broad sense of the word, or the technical phenomenon in general). The technical domain is shown to incorporate both 'technics' (understood as traditional techniques and modes of practical action) and 'technologies' (understood as modern technological rationalized methods). Within the technical domain, technical activities based solely on traditional techniques and modes of practical action continue to flourish in a variety of human settings. Such activities occur with a remarkable consistency throughout human history and they still endure today, usually in a local or small scale (e.g. small constructions, various manufacturing family businesses, wine-making, etc.). These activities remain constantly outside of any technological process in the modern sense. They rely on practical experience and tradition to carry on their tasks, and they seldom interact with the scientific domain and the possibilities it opens to the technical side. In the diagram these activities occupy the partial space within the dotted line denoted as 'technics', which does not overlap with engineering.

By contrast, technical activities implementing modern technological means are abundant in the contemporary world. However, there are plenty of such activities which also do not overlap with the work of engineering. The operators of such activities have usually received some formal technical education and training that cannot be reduced to simple practical knowledge (e.g. technicians of all kinds; operators of advanced machinery; advanced computer users such as CAD designers, etc.). In the same manner as with technics, then, these activities occupy in the diagram the partial space within the dotted line denoted as 'technologies', which does not overlap with engineering.

Engineering now, is shown to transverse the whole of the technical domain: it extends across both technics and technologies, since both are used in engineering. These elements of the technical domain are described in the diagram in terms of their respective technical knowledge-content, i.e. as practical and rationalized modes of knowledge/action respectively, since we are interested here in the usage of these elements as technical tools and in the corresponding modes of technical action that their knowledge-content prescribes. In this respect, technics are associated with empirical statements (rules of thumb), heuristic methods which have no rigorous scientific standing, and modes of action which stem from previous practical technical experience in various areas (e.g. construction practices, handling of tools and materials). In the same manner, technologies are associated with theoretical tools and rationalized methods, that is, elements of technical knowledge of scientific origin, formulated into useful tools for problem-solving (more on this later). Engineering is also shown extending out of the technical domain (this is actually an extension of it, not a break up) toward the scientific domain. The part of engineering shown to extend outside the main rectangle representing the technical domain in the figure represents the academic engineering community: university departments, research laboratories, engineering academics societies, journals, conferences, etc. The part of engineering lying inside the main rectangle is described as professional engineering: it is the everyday problem-solving engineering activity (either involving design operations or not) carried out by professional engineers across our technological world. The above distinction does not prohibit the consideration of engineering as an essentially uniform type of activity, entirely belonging to the technical domain.

Now let us look at the relationship between the elements depicted in Figure 1. The interplay between the two domains as seen here, evolves as follows: Science, aided by its applied component, opens up a *field of possibility* for practical applications. Accordingly, the same field of possibility is opened up to science by technology, identifying possible items for scientific research. This field is represented by the shaded area standing between the two domains. It is not certain, much less necessary, that all the practical possibilities existing in this field will in fact be pursued in a technical way, and the same goes for the theoretical possibilities opened up by technology to science. The reverse influence of technology on science, through the identification of problems for theoretical research and the manufacturing of precise experimental instruments and apparatuses, is a long-standing argument of the historians of technology against the one-dimensional view of technology as 'applied science'. This argument has been well articulated and presented in numerous works in the past, to the point where it has now been widely accepted. In this paper however, and in the analysis that follows, we want to focus on the science-technology relationship as viewed in the science-toward-technology direction, because it is in this direction that confusions of all sorts still endure.

The decision to pose a technological problem lies entirely with an agent to whom the solution to the particular problem will be of importance. Engineering (its academic part) stands out as a receptor of possibilities opened up by science. Given that a technological problem (or a series of technological problems) has already been posed by an agent or agents willing to assume the cost of the solution(s), a technological area of interest is defined (which could already belong to, or develop into, a class of applications). Engineering then proceeds in the identification and the subsequent *appropriation* of scientific knowledge judged to be relevant and useful in resolving the problem(s) posed. This appropriation consists in selecting among results of scientific research, and in transforming the selected scientific knowledge in technological terms. This selective and transformative process exercised on scientific laws and concepts constitutes a fundamental reason why engineering (and technology in general) is not 'applied science'. (We shall examine more closely the process of appropriation in a moment.) Then, the appropriated scientific knowledge is passed on to the professional engineering problem-solving activity as a useful theoretical tool (a technological rationalized method or an array of methods) to be used in resolving the specific problem posed as well as similar problems in the area of technological interest.

In the course of dealing with a particular problem, professional engineers utilize both rationalized methods and tools and empirical techniques (rules of thumb and heuristics) in problem solving. It should be noted that the available theoretical and empirical tools and their combinations may prove inadequate to resolve a particular problem and hence, additional research work may be required to produce an efficient solution. This additional research may either be of a theoretical nature, thus improving on the existing theoretical tools, or of an experimental nature directed toward gathering sufficient data for design, resulting in some empirical or semi-empirical tool (e.g. nomographs used for specific design purposes). A situation like this will usually engage in some way the whole of the engineering community (academic and professional), thus justifying in practice what we said earlier about the essential unity of engineering.

Since this interaction between the two domains, scientific and technical, is continuously repeated in time, a body of appropriated scientific knowledge turned into theoretical tools consisting in a series of rationalized technological methods is accumulated, which at every given time represents the bulk of established engineering theoretical knowledge available for problem solving within the technical domain. These theoretical tools are conveniently organized into engineering theories by the academic engineering community in order to facilitate their teaching in engineering education and to enable the formulation of more precise specifications concerning their usage.

We need now to discuss the process of the appropriation of scientific knowledge by engineering. We have already noted that many writers on the subject overestimate the significance of the scientific elements embedded in engineering, to the point that they consider them to manifest a scientific attestation of engineering practice. We feel that the reason for this attitude is the failure to appreciate the structural significance of the process of appropriation which is elaborated in what follows.

The process of the engineering appropriation of scientific knowledge begins with a selective operation: If one were to adopt the notion of technology as 'applied science', as propounded by traditional approaches to the science-technology relationship, one would have to claim that technology and engineering 'apply' whatever is found within the field opened up by science (given that a relevant technological problem has already been posed by an agent willing to assume the cost of its solution). In reality what happens is that technology and engineering do not simply pick out of the scientific garden whatever fruits are graciously offered there in order to 'apply' them to solve the technological problem posed. Given the technological problem and therefore given the scope of the application that will solve it, according to what has been previously said in Section 2, the technological process has as a first task to come to terms with the *specific* set of efficiency/economy criteria (i.e. 'internal' and 'external') which will guide the process toward realizing the intended application. Having to conform to these criteria (having as a task to achieve the dynamic compromise between all demands and limitations) the process can be nothing other than extremely selective: the technological process already under way has to find among the tools offered in the scientific garden (and also among other, *non-scientific* tools), those most suitable for the task that lies ahead; in other words, the technological process carefully *selects* (and does not simply pick up) particular items from the field of possibility, guided by the concept of efficiency specific to the application it pursues. And these items are nothing else than scientific cognitive elements (theories, laws or equations, explanations) judged to be possibly useful in formulating theoretical tools for solving the technological problem at hand. Henry Petroski makes a similar point when claiming

> [t]here is hardly an artifact of engineering or technology that can be separated from its physical appearance, and thus it is not surprising that engineers and technologists think and create in terms of pictures. It is for this reason that the naive view of engineering as 'applied science' is simply not valid. It is actually the theories and equations of science that are applied to the object of the engineer's imagination – once there is a picture of what is to be theorized about or analyzed with equations – and so science is really used as theoretical engineering (Petroski, 1990, p. 215).

One might rightfully ask: Is this enough? Does not engineering still 'apply' scientific laws and ideas, even if it does choose among them in relation with its specific goal? The answer is that technology and engineering do not merely choose their theoretical tools, but they also *transform them* in the same movement. We shall articulate this transformative operation using relevant examples. The transformation of a given scientific theory selected as useful theoretical knowledge by engineering, results in three particular operations acted upon the theoretical structure:

(a) The abstract, generalized context of the scientific theory is drastically reduced to those aspects that serve the particular intentions manifested within a given class of technological problems. That is, the general description of the way the world is structured, as furnished by scientific theory, is limited only to those aspects of reality in which we are interested in a technical way. For example, in the civil engineering theory concerned with open channel hydraulics, the abstractions of the underlying scientific theory, namely fluid mechanics, are reduced to what concerns one-dimensional water flow in artificial channels of specific cross-sections only (trapezoidal, rectangular, etc.), since we are not technically interested in fluid flow in general. Faced with the generic scientific description, the requirements of the related class of applications (e.g. water works, irrigation systems, drainage systems, etc.) circumscribe what is relative and what is not with respect to this class within the scientific theory. On the basis of this delimitation to practical aspects, the laws, theoretical principles, and fundamental equations of the underlying scientific theory are manipulated to a (usually simpler) mathematical form which is relevant with those aspects of interest.

In our open channel hydraulics example, the basic equations concerning the flow (continuity equation, momentum equation, etc.) are reduced to what concerns the one-dimensional flow (as opposed to the three-dimensional scientific discussion) of water (and not of any other fluid). The generic considerations of the scientific theory are converted into practical case-studies relevant to problems usually presented within the specific class of applications. Established technical frameworks are formulated and used in the study of such problems. For example, open channel hydraulics engineering theory sets up typical case-studies for a variety of hydraulics problems, e.g. flow over a weir; flow underneath gate structures; flow in channels with variable cross-sections; flow in erodible and non-erodible channels, etc. These typical case-studies constitute assessments of human technical problems that do not form part of the natural phenomenon which the scientific theory aimed to explain and describe in the first place. It is their *interaction* with the natural phenomenon addressed by science that makes them eligible for assessment using the (reformulated for that specific purpose) findings of the scientific theory.

(b) The theoretical material, thus reduced to practical aspects, is then organized in a new way, in order to become more manageable in 'standard' implementation when solving technological problems of the same class. Previous engineering experience reveals those 'standard' courses of action for given classes of technological problems. This does not mean that any particular technological process is not still directly and indispensably linked to the application pursued by it; some preliminary steps in the process though are common for a given class of technological problems, or even across classes of technological problems. In chemical engineering, for instance, the study of unit operations is used to treat a variety of operations within industrial processes which are similar (dilution, filtration, mixing, crushing, and grinding, etc. – see Buchholz, 1979). Also, the hard-to-digest differential equations are replaced by easier-to-use nomographs (e.g. the Moody nomograph used in hydraulics problems) and convenient quantities are introduced which along with semi-empirical charts and equations allow the calculations of certain numerical values needed in given classes of problems. On the basis of this re-organization specific methodologies dealing with certain types of problems are formulated and used in a variety of similar applications (for example, statistical methodologies for handling specific data types used in various engineering applications).

(c) Finally, the transformation is completed by incorporating purely empirical tools and techniques (rules of thumb and heuristics) to the picture. Returning to the example of open channel hydraulics, one of the main tools of channel design is the famous Manning formula, which is a formula of empirical content, derived from experimental data.[16] Many such empirical or semi-empirical formulas are encountered in all the engineering disciplines. These empirical tools and heuristics[17] occupy crucial places within the structure of an engineering theory and they are not simply 'filling in gaps' left over by the transformation process. This means that specific tasks in problem solving are undertaken solely by the use of such empirical rules of thumb on the basis of their simplicity, ease of use, and satisfactory yielding of results. These tools are preferred due to their high degree of efficiency within the context of specific classes of applications. They are sometimes still preferred over more theoretical approaches (engineering theoretical tools) because, simply, they get the job done, provided that their results are deemed satisfactory within the bounds of what is considered to be an admissible answer to the problem at hand (again, it is the 'internal' criteria specific to the application studied with those empirical means that will indicate the degree to which they can be utilized). For example, the semi-empirical unit hydrograph method is still used extensively in hydrology studies, despite the acknowledgment in recent research that its application may sometimes lead to severe over- or under-estimations, and despite the development of more elaborate (and more complex) tools which can be used in its place (linear and non-linear distributed reservoir models, physically-based distributed models, etc. – for a description of these see, for example, Shaw, 1983).

The above transformative process results in the *engineering appropriation of scientific theoretical knowledge*, that is, in the formulation of theoretical tools for use in solving technological problems. The resulting engineering theoretical tool is a specially formulated tool for the accomplishment of given tasks and is altogether different in scope and content from the scientific theory that was used as a primary source of material, along with other experimental sources as well. The scientific cognitive elements included in the engineering theory have undergone a drastic transformation and purely empirical propositions have been incorporated inextricably with these transformed scientific elements. Far from simply being 'grounded' in scientific knowledge then, the theoretical tools of engineering are uniquely *techno-logical*: they are the product of a process aiming at practical ends, which using different sources (scientific, experimental, empirical) produces a tool that is completely *re-contextualized and re-structured* compared to all those original sources. It is a tool combining all those different aspects in a rationalized way, to be implemented in solving particular problems.[18]

This selective and transformative process of appropriation of scientific knowledge described briefly above is taking place within the extended academic part of engineering indicated in Figure 1. Engineering academic research aims at formulating better methodologies, better theoretical tools and techniques, which are then passed on to the professional world of engineering to be used in the everyday engineering activity of solving technological problems. It is useful to note here that although the internal organization of the engineering academia bears remarkable resemblance to its scientific counterpart, its scope is entirely different in content. Engineering academic research enjoys a certain autonomy from external demands and is able to put forth research activities and define areas of interest that may depart from the immediate resolving of given technological problems. However, this partial independence is still necessarily elaborating on issues that are important within the current technological framework. An improved methodology or theoretical tool developed from internal engineering academic research must still pass a crucial test that is virtually unknown to any scientific research result. It has to be accepted and endorsed in common engineering professional practice as a contextually useful tool in resolving certain technological problems, however accurate or correct its results may prove to be. Simplicity, ease of use, and the ability to be used in a wide spectrum of engineering environments (e.g. when problematic data or ill-defined situations are involved), while at the same time yielding satisfactory results, are qualities demanded from any engineering theoretical tool. Therefore, engineering academic research cannot in any case be even partially disconnected from the workings of the professional engineering activity. Primarily, this kind of research focuses on 'second-order' problems, that is, "problems concerning the systematization of the processes of solving technological [i.e. 'first-order'] problems" (Arageorgis & Baltas, 1989, p. 224).

In the concluding section, we will examine again the questions set in the course of this work about technical knowledge and 'technoscience' in light of the relationship laid out between science and technology.

4. CONCLUSION: THE ARGUMENT REVISITED

The discussion so far enables us to revisit our statements on the nature of technical knowledge and the development of a technological process. We are now in a position to clarify the crucial link between any such process and the faculty of *knowing-how*. We had earlier set the question of where and when exactly in the process of a technological endeavor, actions, procedures, or decisions relying on *knowing-how* cognitive abilities are indeed manifested. It becomes evident from the previous discussion that technical knowledge in its entirety is an amalgam of many different contributions. Engineering theoretical tools embed in a technological process elements of scientific knowledge, transformed into useful frameworks for solving various problems belonging to many classes of applications. At the same time, engineers and technologists utilize frequently purely empirical or semi-empirical tools which cannot be derived in any way from scientific statements addressing the same aspect of reality grasped by those tools. They also use heuristic strategies as part of their problem-solving activity, which are not reducible to any kind of strict formalization. So, technical knowledge comprises both technics (empirical techniques) as well as technologies (rationalized technological methods and tools). These diverse cognitive elements are combined in the process of design that brings it all together, under the guiding principle of efficiency, toward achieving the intended goal.

Our initial question is now essentially transformed to the question of whether the process of design can be said to exhibit the cognitive characteristics of *knowing-how*. It is crucial at this point to remember that the human quality of *knowing-how* is manifested in the actual deployment of any process associated with this mode of knowing, and not in the formulation of the rules, canons, or criteria used in the course of this process. As it can be easily understood, such a process exhibits a fundamentally non-analytical character. We cannot offer an exhaustive a priori logical assessment of the train of events within such a process, in much the same way that we cannot give useful directions (other than in a very generic sense) to a person who is trying to learn how to ride a bicycle. Consequently, we cannot offer an a priori assessment of the intrinsic value of any of the rules, canons, and criteria that will be used in this process with respect to the successful realization of its purpose. In other words, since *knowing-how* processes are not aiming at successful formalizations but at achieving concrete results through *action*, prior to the actual outcome of any such process, logical constructions of its deployment can only be more or less generic in nature. Hence, the degree of success in carrying out such processes relies heavily on training and acquired experience.

Engineering conceptualization and design clearly exhibit the characteristics of such a process: following Ryle's exposition of *knowing-how*, design is undoubtedly an intelligent performance and is certainly learned through apprenticeship, through doing actual work, through strict training and experience gained through subsequent design efforts. Design is certainly 'not a single-track disposition like a reflex or a habit'. Furthermore, 'its exercises are observances of rules or canons or the applications of criteria' (i.e. engineering design usage of theoretical and empirical methods and tools and of the criteria defining efficiency), 'but they are not tandem operations of theoretically avowing maxims and then putting them into practice', since no prior detailed assessment of the relative efficiency of any design tool or method is possible outside the immediate context of the application being designed. The exact specification of the desired efficiency of any application goes part-and-parcel with the design process which aims to realize it.

The *knowing-how* character of a technological process is of an undoubtedly higher level than that on which simpler everyday performances based on *knowing-how* capabilities are manifested; it involves diverse cognitive elements and requires advanced skills in manipulating them. However, the process does not owe its higher status to the *knowing-that* cognitive elements that it includes, but rather in the technically-specific and inventive way these are re-contextualized and incorporated into the engineer's toolkit. In spite of the usefulness of analytical techniques employed in engineering, analysis plays a "necessary but secondary role in engineering design" (Ferguson, 1992, p. 163). The following quotation from a report prepared by the MIT Committee on Engineering Design in, 1961, sums up nicely the complex character of engineering design:

> The designing engineer . . . finds himself making only a small fraction of his decisions on the basis of numerical analysis. [Analysis] does not answer all or even a majority of the questions an engineer must answer in a typical design problem, particularly a new one. It seems unlikely that numerical analysis will ever answer more than a small proportion of

> these questions. The remainder of the questions must be decided on the basis of ad hoc experiment, experience (the art of applying knowledge gained by former experiments on the same or similar problems), logical reasoning and personal preference. The subconscious reasoning process, based on experience, which we call intuition, can play a large part (MIT Committee on Engineering Design,, 1961. Quoted in Ferguson, 1992, p. 164).

One may seek theoretical assistance in the courses on design taught within the curriculum of engineering departments or in the growing literature on design and its principles, but formal instruction can never replace the actual work. This is the reason why, in most countries, the young engineer is required to pass through a stage of training-on-the-job before s/he is granted his/her professional license as registered engineer. Formal university education is rightfully not considered enough as a basis for someone to start a professional career in engineering, and so the graduate engineer is required by law to be a trainee first in order to become a professional later. Despite the over-scientific tendency of engineering education, legislative and regulative authorities around the world generally acknowledge the need for the future engineer to acquire 'hands-on' experience before assuming the responsibility of carrying out an engineering project on his/her own. In most countries this period ranges from 3 to 5 years, and the training engineer is required to do work in various areas of the profession so as to acquire skills in all levels (in design and consulting, in construction or production processes, in management and organization, etc.).

Indeed, the engineer learns how (to conceptualize, design, optimize, and manage) by practice, "schooled indeed by criticism and example," as Ryle would have it, but also quite *aided* by 'lessons in the theory'. Nevertheless, efficient design 'precedes the theory of it', and the competent as well as the innovative engineer is a person primarily *experienced* and only on a lesser level well educated in a formal and analytical manner. In other words, an experienced engineer is capable of conceiving solutions that a well educated but inexperienced engineer can not. Engineers among themselves assess a colleague's expertise, valuing above all previous practical experience in his/her area of the profession. Therefore, the competent engineer knows primarily *how* to play *and* at the same time knows to cite well and accurately the rules of the game.

Concluding, the cognitive relationship between technology and science indicated by the proposed demarcation scheme reveals the close interrelations of the two domains while at the same time provides a tool for discerning effectively the cognitive contribution of each one domain in any techno-scientific task. We may outline the main conclusions of this discussion as follows:

(a) The *knowing-how* character of technology and engineering is primarily manifested in the actual deployment of the conceptualization and design process of any technological object or state of affairs. It is the knowledge of combining diverse cognitive elements, parcels of previous practical experience as well as heuristic and resource management strategies, in transforming reality toward a predetermined goal, subject to diverse criteria regulating technical and economical demands and limitations (i.e. subject to the principle of efficiency), and often within ill-defined situations and amidst problematic data. It is the knowledge of bringing all those diverse elements together; that is, the knowledge of *how* to conceptualize, design, optimize, and produce a technological goal. Within the context of a technological process, *knowing-that* becomes a useful tool in the hands of *knowing-how.* Although engineering involves a wealth of abilities and knowledge of a much higher level than those needed for riding on a bicycle, one thing remains constant between these two *knowing-how* performances: one just has to learn by doing it.

(b) Technology-engineering and science are interconnected by a web of relations existing prior to any given technological endeavor; however, this web is *activated* only at the emergence of a technological problem put forth by an agent to whom the solution to the problem will be of importance. Prior to each emergence of a technological problem, science and technology are just accumulating mutual potentialities to each other, in a common bi-directional field of possibility. The interaction between the two domains may, however, be activated many times in the course of a technological process, especially on 'big technology' projects spanning many years of research and development. Or, it can be simultaneously activated in multiple instances when the process is carried out by many technological agents in competition with each other. It is therefore our belief that 'technoscience' does not drive history, nor does it drive around on its own in any way whatsoever. It consists, in every instance we are inspecting, as an unexplored potential waiting to be activated by agents having reasons (and the necessary power) to exploit it. However, we must bear in mind that the mere existence of this potential does not in any way prescribe the result of its possible elaboration in a technical way. A possibility is only a possibility: the actual work to exploit it and the outcome of the process to realize it, may differ considerably from any benevolent prophecies inspired by it.

(c) The blurring of the boundaries between technology-engineering and science is primarily due to their interrelation within a framework of production; this means that the concreteness of technoscience, as observed from outside, is primarily manifested at the stage of bringing-about a desired state of affairs, at the production stage of the systems and artifacts that modern society produces in response to its (either well-founded or ill-defined) needs and desires. Accounts on the status of technoscience and the directions it takes usually lump all relevant considerations to the finishing line, so to speak, thus losing sight of what happens *during* the race. To the average person the workings inside the techno-scientific complex may be impenetrable and/or they may not present any particular interest; s/he may only be interested in the outcome of these workings and they way they affect everyday life. To the philosopher, however, the internal relations within the complex need not be obscure: the discussion here shows that we can discern effectively the limits of the interaction provided that we maintain a strict focus on the particularities exhibited in each domain. It is in this sense also that the development of a philosophy of technology proper is of major importance.

(d) In line with the latter statement, the discussion presented here attempts to subscribe to a philosophy of technology *per se*, in that the argumentation and the line of thinking followed does not depend on any particular metascientific approach or philosophical elaboration in order to assess technology and engineering. In this respect, epistemologies of science ranging from realist to strongly pragmatic may be accommodated in the demarcation scheme proposed without further ado, since the cognitive dimensions of the technical domain are taken to be radically different in scope than those of the scientific domain. The cognitive elements of science delivered into the field of possibility, are appropriated (i.e. re-contextualized) and manipulated in a way specific to technical considerations, regardless of their epistemological account in their original scientific context.

Finally, the demarcation scheme proposed supports the concept of engineering as an essentially hybrid formation of knowledge and activity, although entirely belonging to the technical domain. Engineering, besides technical knowledge, also comprises important artistic elements manifested in design and the visual way of thinking specific to it; it requires important managerial abilities; it involves purely empirical modes of practical action; and heuristic strategies of proceeding with complex, often ill-defined tasks. As such, and given also the important role engineering plays in the science-technology relationship, it is revealed as a primary object for future research in the philosophical elaboration of technology.

ACKNOWLEDGMENTS

This work was carried out within the context of the Graduate Program in Philosophy and History of Science and Technology, jointly run by the

Department of Engineering Science of the National Technical University of Athens and the Department of Philosophy and History of Science of the University of Athens (Greece). I wish to express sincere gratitude to my supervisor, Professor Aristides Baltas, for his guidance and his valuable commentary throughout the course of this work. I am also indebted to Professors Pantelis Nikolakopoulos, Theodossios P. Tassios, Vassilis Karasmanis, and the two anonymous referees of this journal for making many helpful suggestions. Finally, many thanks are owed to Professor Carl Mitcham for his interest, encouragement, and support. An early version of this paper was presented in a series of lectures on 'Philosophical Problems of Technology' organized by the Program in 1997, where I was able to gain useful feedback. I wish to thank all the participants for generating a fruitful discussion.

NOTES

1. Arageorgis & Baltas contend that it is this fuzziness which 'has led some philosophers to hold that a clear-cut distinction between science and technology is theoretically impossible, misleading or insignificant' (1989, pp. 214–215). Examples of philosophers in favor of this view are Mario Bunge, Hans Jonas, and Jacques Ellul.

2. We shall refer to knowledge associated with technology as *technical knowledge* since, as Carl Mitcham points out (1991, p. 81), the term *techn-ology* etymologically implies knowledge. Thus, the purpose of this paper is to demarcate between technical knowledge and scientific knowledge, which is an aspect of the overall demarcation between technology and science implied. But the term *technical* is broader than *technological* in this sense, since it may also connote practical knowledge, which is an important part of knowledge associated with technology and engineering. The structure of technical knowledge will become more clear in what follows.

3. Although the process of engineering design and the various engineering practices followed during the engineering elaboration of a technological problem will be discussed in this paper, they deserve a philosophical treatment of their own, which is outside the scope of this work. We intend to return to this issue with a discussion specifically devoted to design and other engineering practices and the way they relate to both technical knowledge and the actual deployment of a technological process.

4. Mario Bunge's philosophical elaboration of technology as 'applied science' (Bunge, 1972, 1985) is a typical example of thinking on technology from the standpoint of a philosophy of science (in this case, logical positivism). Bunge does not credit practical knowledge as knowledge in the pure (that is, scientific) sense. For Bunge, technology cannot claim any access to knowledge, since practical success does not entail access to the truth. An efficient practice does not necessarily provide a true account of the aspect of reality on which it is exercised. However true this may be, technology does not care to access the truth about the world, which is precisely what Bunge fails to appreciate. Instead of true accounts of the state of things, technology seeks efficient solutions to practical problems. In the course of resolving a problem, more often than not, descriptions of reality in the form of empirical rules are employed

on the sole basis of their yielding satisfactory results, however inaccurate or false they may be in the strict sense ('quick and dirty' solutions, as the engineering jargon goes).

5. There is an obvious and close relation between the question of the epistemological status of technology and the laying out of the science-technology relationship. As Walter Vincenti notes, "viewing technological knowledge as autonomous leaves the relation between science and technology still open to specification" (1990, p. 5). On this subject see also Barnes, 1982; Wise, 1985; Staudenmaier, 1985.

6. The following exposition will necessarily address many issues in a short space. As a result, the reader may feel that some of the topics addressed should be treated more extensively. However, one must keep in mind that here we are trying '*to blaze a trail, rather than cover the terrain*', as the expression goes.

7. In the context of a pragmatic epistemology, Joseph C. Pitt (1983) argues that no valid distinction can be drawn between these two supposedly different types of knowledge. Based on the premises that "knowledge is a tool we use to get around in the world," and "knowledge is for the sake of action," Pitt argues for one basic form of knowledge and a single criterion to define it, namely, action: "In the face of action all explanation and knowledge claims are born equal" (1983, pp. 89–91). Nevertheless, a distinction is still justified if one considers *knowing-that* and *knowing-how* as two different *modes of knowing* rather than two separate kinds of knowledge (hence the use of gerund). Pitt's thesis ultimately reduces all scientific inquiries to purely pragmatic considerations, but, as we are going to see, the demarcation scheme proposed can accommodate even such a strongly pragmatic epistemology of science.

8. The choice of the term *transformation* aims to capture the broad context of the effects of technology on the world, and is not chosen in line with other grandiloquent-style engineering definitions of technology and of the role of engineering itself. Nevertheless, the idea that technology actually transforms nature to the benefit of humanity is central to many traditional and contemporary engineering accounts of technological action. For instance, in the recent report of an international engineering association on environmental issues regarding dam projects, we read: 'The impact of dams and reservoirs on the environment is inevitable and undeniable; land is flooded, people are resettled, the continuity of aquatic life along a river is interrupted, and its runoff modified and often reduced by diversions. Thus, dam engineers find themselves confronted with the basic problems inherent in the transformation of the natural world into a human environment' ('Position Paper on Dams and Environment', *International Commission on Large Dams*, Paris,, 1998, p. 2. Available on-line at http://genepi.louis-jean.com/cigb/chartean.htm). The same idea is echoed in the following definition of engineering by G. F. C. Rogers: "Engineering refers to the practice of organizing the design and construction of any artifice which transforms the physical world around us to meet some recognized need" (1983, p. 51).

9. It is evident that the social and technical acceptability of a technological problem as a legitimate claim on the part of any social agent, is subject to many determining factors. Technical acceptability is greatly influenced by the prevailing technical ideal at the time the technological problem is presented. Social acceptability of a technological problem is often an issue of dispute between social groups, or between society and the agent posing the problem. Social dominant values and ideologies shape decisively the outcome of such disputes.

10. The specification of efficiency within the context of the intended application, is usually conducted via the preparation of a *feasibility study*, which usually precedes the

actual design and realization of the project. It is possible that this study may prove the project to be non-feasible (that is, the costs involved are not deemed acceptable by the agent posing the technological problem), thus leading to the abandonment of the project (cf. Radnitzky 1978, p. 1012; Arageorgis & Baltas, 1989, pp. 226–227).

11. A somewhat similar account of the elements controlled by the double set of efficiency criteria is given by Herbert Simon (1981). Simon, identifying the said elements, distinguishes between "command variables" (means or inputs), "fixed parameters" (laws or rules), and "constraints" (ends or outputs).

12. Few of the 'external' criteria defining efficiency within the immediate context of the intended application can be incorporated into engineering standards and specifications. Many of those criteria relate to qualities extremely variable in time and subject to many socially and/or historically determining factors. For instance, the financial efficiency of any given technological project is a notion defined primarily under the influence of current conceptions of proper investment policy, market research orientations, etc.

13. While their importance in shaping technological outcomes does not escape our field of view, this discussion does not deliberately address the influence of the engineer's personal values (social, cultural, political, or otherwise) on his/her work. Engineering has often been described as a value-driven activity. However, ethical implications arising from the person(s) engaged in this activity are not at issue here; we need to maintain a strict focus on the technical and cognitive inner workings of the technological process.

14. The term 'specific', in this sense, may be understood in reverse as follows: it is always the intended application as the *projected embodiment* of all the technical and economical prerequisites, demands, and limitations to which it must conform and which it must satisfy, that sets decisively the scope of the process aiming at its realization.

15. Feibleman's only mistake is that he proceeds, elsewhere in his paper, to reduce the technological operations to a mere *modus operandi* level, which does great injustice to the actual work being performed in a technological context.

16. It is interesting to see how attempts from the academic engineering community to give these empirical tools a more firm theoretical basis are often self-regulated in view of their current usage in practical applications. Such attempts are also criticized on the grounds of their success in maintaining or improving on the simplicity and ease of use of these tools (in other words, in maintaining or improving their level of efficiency). We can witness this using a recent discussion that appeared in vols. 118–119 of the *Journal of Hydraulic Engineering* (ASCE), involving the classic Manning formula mentioned here as an example. Professor Ben Chie Yen (vol. 118, no. 9,, 1992) proposed a new dimensionally homogeneous form of the Manning formula, which in his view could overcome the long-standing problem of the dimensions of Manning's roughness coefficient *n* in relation with the system of measurement units used (reference can be made in relation to the Manning formula and the dimension problem to many standard hydraulic engineering texts, e.g. Chow, 1959). Professor B. A. Christensen (vol. 119, no. 12, 1993), discussing Yen's paper, remarked that the "price paid" is "a complication of matters" and that "the new coefficient [suggested by Yen] is not easier to estimate than the old." Professor Yen replied in the same issue defending his version of the formula (the technical particulars are not important here) and stating that "The main objective of the [original] paper is to make the Manning formula *more acceptable from the fluid-mechanics viewpoint without changing the general appearance of the*

formula, which is extensively used in the profession" (emphasis added). Therefore, efforts to make empirical tools more theoretically sound pay respect to the established use of these tools in engineering practice. Also, the outcome of these efforts is judged not on the basis of their successful scientific 'grounding' of the empirical rule, but on the basis of their improving the efficiency of this rule in common practice.

17. For an account of the importance of heuristics in engineering see B. V. Koen (1991). Koen gives the following definition of a heuristic: "A heuristic is anything that provides a plausible aid or direction in the solution of a problem but is in the final analysis unjustified, incapable of justification, and fallible. It is used to guide, to discover, and to reveal." A heuristic has the following characteristics: "(a) A heuristic does not guarantee a solution; (b) It may contradict other heuristics; (c) It reduces the search time in solving a problem; and (d) Its acceptance depends on the immediate context instead of an absolute standard" (1991, pp. 35–36). Koen is right when he stresses the role of heuristics in engineering practice; but then he proceeds to identify the engineering method itself with the use of heuristics, thus failing to appreciate the importance of other cognitive and methodological tools available to engineering.

18. Walter Vincenti's important remarks on the theoretical tools of engineering are quite similar to our argument here: Theoretical tools are distinguished by Vincenti into two classes: mathematical methods and theories, and intellectual concepts. Mathematical methods and theories involve "purely mathematical tools having no physical content in themselves," tools derived from reformulation of "essentially physical" knowledge of scientific origin, engineering theories limited to specific phenomena or devices (and otherwise useless), and "phenomenological theories" lacking rigor and scientific standing and even "demonstrably faulty in some degree," which are used solely for engineering calculation and "because they work, however imperfectly, and because no better analytical tools are available." In a similar manner, intellectual concepts range from "highly scientific to intensely practical, from (though not correspondingly) specifically mathematical to explicitly physical" (Vincenti, 1990, 213–216).

REFERENCES

Arageorgis, A., & Baltas, A. (1989). Demarcating Technology from Science: Problems and Problem Solving in Technology. *Zeitschrift fur allgemaine Wissenschaftstheorie*, *20*(2), 212–229.

Barnes, B. (1982). The Science-Technology Relationship: A Model and a Query. *Social Studies of Science*, *12*(1), 166–171.

Borgmann, A. (1984). *Technology and the Character of Contemporary Life*. Chicago: University of Chicago Press.

Buchholz, K. (1979). Verfahrenstechnik (Chemical Engineering) – Its Development, Present State and Structure. *Social Studies of Science*, *9*(1), 33–62.

Bunge, M. (1972). Towards a Philosophy of Technology. In: Mitcham & Mackey (Eds), 62–76.

Bunge, M. (1985). Technology: From Engineering to Decision Theory. In: *Treatise on Basic Philosophy*, vol. 7, *Epistemology and Methodology III: Philosophy of Science and Technology*, part 2, Life Science, Social Science and Technology. (pp. 219–311). Boston: D. Reidel.

Carpenter, S. (1974). Modes of Knowing and Technological Action. *Philosophy Today, 18*(2), 162–168.

Carpenter, S. (1983). Alternative Technology and the Norm of Efficiency. *Research in Philosophy and Technology, 6*, 65–76.

Castoriadis, C. (1990). Technoscience. In: *Omilies stin Ellada* [Lectures in Greece]. (pp. 133–167). Athens, Greece: Ipsilon.

Castoriadis, C. (1991). Technique. In: *Les Carrefours du Labyrinthe*, Zisis Sarikas, trans. (pp. 269–302). Athens, Greece: Ipsilon.

Chow, V. T. (1959). *Open-Channel Hydraulics*. International Student Edition. Tokyo, Japan: McGraw-Hill Kogakusha.

Christensen, B. A. (1993). Discussion of Dimensionally Homogeneous Manning's Formula by Ben Chie Yen. *Journal of Hydraulic Engineering, 119*(12), 1442–1443.

Feibleman, J. K. (1972). Pure Science, Applied Science, and Technology: An Attempt at Definitions. In: Mitcham & Mackey (Eds), pp. 33–41.

Ferre, F. (1988). *Philosophy of Technology*. Athens, GA: University of Georgia Press. Reprinted, 1995.

Ferguson, E. (1992). *Engineering and the Mind's Eye*. Cambridge, MA: MIT Press.

Gaines, B. R. (1976). Foundations of Fuzzy Reasoning. *International Journal of Man-Machine Studies, 8*(6), pp. 623–668.

Gilles, B., Fel, A., Parent, J., & Russo, F. (1986). *History of Techniques*, vol. 1, *Techniques and Civilizations*, vol. 2, *Techniques and Sciences*. New York: Gordon and Breach.

Hess, D. J. (1995). *Science and Technology in a Multicultural World: The Cultural Politics of Facts and Artifacts*. New York: Columbia University Press.

Jarvie, I. C. (1972). Technology and the Structure of Knowledge. In: Mitcham & Mackey (Eds), pp. 54–61.

Johnston, S., Lee, A., & McGregor, H. (1996). Engineering as Captive Discourse. *Society for Philosophy and Technology Quarterly Electronic Journal, 1*(3–4). URL: http://scholar.lib.vt.edu/ejournals/SPT/v1n3n4.

Jonas, H. (1972). The Practical Uses of Theory. In: Mitcham & Mackey (Eds), pp. 335–347.

Koen, B. V (1991). The Engineering Method. In: P. T. Durbin (Ed.), *Critical Perspectives on Nonacademic Science and Engineering*. (pp. 33–59). Bethlehem, PA: Lehigh University Press.

Kuhn, T. S. (1983). *The Structure of Scientific Revolutions*. 2nd ed. Trans. V. Georgakopoulos & V. Kalfas. Thessaloniki, Greece: Synchrona Themata.

Layton, E. T. Jr. (1974). Technology as Knowledge. *Technology and Culture, 15*(1), 31–41.

Layton, E. T. Jr. (1976). American Ideologies of Science and Engineering. *Technology and Culture, 17*(4), 688–701.

Layton, E. T. Jr. (1984). Science and Engineering Design. *Annals of the New York Academy of Sciences, 424, Bridge to the Future: A Centennial Celebration of the Brooklyn Bridge*, 173–181.

Layton, E. T. Jr. (1991). A Historical Definition of Engineering. In: P. T. Durbin (Ed.), *Critical Perspectives on Nonacademic Science and Engineering*. (pp. 60–79). Bethlehem, PA: Lehigh University Press.

McGinn, R. E. (1990). What is Technology? In: L. A. Hickman (Ed.), *Technology As a Human Affair*. New York: McGraw-Hill, pp. 10–24.

Mitcham, C. (1991). Engineering as Productive Activity. In: P. T. Durbin (Ed.), *Critical Perspectives on Nonacademic Science and Engineering*. Bethlehem, PA: Lehigh University Press, pp. 80–117.

Mitcham, C. (1994). *Thinking Through Technology: The Path Between Engineering and Philosophy*. Chicago: University of Chicago Press.

Mitcham, C., & Mackey, R. (Eds) (1972). *Philosophy and Technology: Readings in the Philosophical Problems of Technology*. New York: Free Press.

M.I.T. Committee On Engineering Design (1961). Report On Engineering Design. *Journal of Engineering Education, 51*(8), pp. 645–660.

Petroski, H. (1985). *To Engineer is Human: The Role of Failure in Successful Design*. New York: St. Martin's Press.

Petroski, H. (1986). *Beyond Engineering: Essays and Other Attempts to Figure without Equations*. New York: St. Martin's Press.

Petroski, H. (1990). *The Pencil: A History of Design and Circumstance*. New York: Alfred J. Knopf.

Pitt, J. C. (1983). The Epistemological Engine. *Philosophica, 2*(2), pp. 77–95.

Radnitzky, G. (1978). The Boundaries of Science and Technology. In: *The Search for Absolute Values in a Changing World* (Proceedings of the Sixth International Conference on the Unity of the Sciences, San Francisco, Nov. 25–27,, 1977). New York: International Cultural Foundation, pp. 1007–1036.

Rogers, G. F. C. (1983). *The Nature of Engineering: A Philosophy of Technology*. London, England: Macmillan.

Ryle, G. (1990). *The Concept of Mind*. London, England: Penguin Books.

Shaw, E. (1983). *Hydrology in Practice*. London, England: Van Nostrand-Rheinhold.

Simon, H. A. (1981). *The Sciences of the Artificial*. (2nd ed.). Cambridge, MA: MIT Press.

Skolimowski, H. (1972). The Structure of Thinking in Technology. In: Mitcham & Mackey (Eds), pp. 42–49.

Staudenmaier, J. M. (1985). *Technology's Storytellers: Reweaving the Human Fabric*. Cambridge, MA: MIT Press.

Vincenti, W. (1990). *What Engineers Know and How They Know It: Analytical Studies from Aeronautical History*. Baltimore: Johns Hopkins University Press.

Walentynowicz, B. (1968). On Methodology of Engineering Design. In: *Aktendes XIV Internationalen Kongresses fur Philosophie*, Band 2 (Wien 2–9 September, 1968). Wien, Germany: Verlag Herder, pp. 586–590.

Wise, G. (1985). Science and Technology. *Osiris*, 2nd ser., vol. 1, pp. 229–246.

Yen, B. C. (1992). Dimensionally Homogeneous Manning's Formula. *Journal of Hydraulic Engineering, 118*(9), pp 1326–1332.

Yen, B. C. (1993). Closure of discussion of Dimensionally Homogeneous Manning's Formula' by Ben Chie Yen. *Journal of Hydraulic Engineering, 119*(12), pp. 1443–1445.

THE CONCEPTION OF TIME AND ITS RELATIONSHIP TO TECHNOLOGY

Theodore John Rivers

It is no longer feasible to attempt an analysis of technology without making reference to time, an observation applicable both to the philosophy and history of technology. It is not so much that the phenomenon of time has changed but our understanding of its historical relevancy that makes it necessary to include time whenever discussing technology, particularly, our understanding of the historical-cultural conception of time. By conception we mean the manner in which different ages, cultures, and societies utilize time, how they give an accounting or reckoning of it, and how they assign a value to it. As a fundamental condition of human reality, time needs to be considered seriously as a major factor whenever humanity makes an effort to modify nature as an attempt to improve human existence.

Before we briefly outline what we will discuss in this chapter, we need to discuss the role that 'artifice' plays when dealing with technology. In fact, just as it is no longer possible to discuss technology without including time, it is also no longer possible to discuss technology without the inclusion of artifice to which it is appended.

Various thinkers, particularly José Ortega y Gasset and Martin Heidegger have noted the importance of artifice, or of ideas similar thereto, in modern technology.[1] As used here, artifice refers to a framework or matrix, however skillful or crafty it may be; it is a device for action, a plan or method for application. An artifice is a type of scheme upon which all manner of things are appended, from information to methodology, and from technical expertise to cognitive perception. It imposes its own order and maintains its own

Metaphysics, Epistemology, and Technology, Volume, 19, pages 215–231.

ISBN: 0–7623–0681–5

organization. In fact, the organization of information, which is a large part of its methodology, is the fulcrum of artifice. It guides the addition of new information and helps in the retrieval of old information.[2]

As a result, an artifice is ordinarily composed of several parts, which support a phenomenon. One component of technology's artifice is secularization; others are the subject-object dichotomy and its conception of time. Being long in the making, technology's artifice is historical or cultural rather than metaphysical in nature.[3] Although we have mentioned only a few components of technology's artifice, the most fundamental component is time, which entails a discussion of its ontological variants (past, present, future), its linearity (Christianization), its reckoning by mechanical means (the clock), and the effect of mechanization upon time (time-consciousness). We will discuss all of these features in what follows.

Technology may be described in various ways, but most commonly it is equated with knowledge or skill, to which we can add activity, the latter being the ability to do something. Technology may also be defined as the effect of knowledge or know-how because objects are made through its application. Because technology is both a means and an end, it is the cause of the ability to do something as well as the result of this ability, and it is found in many types, from the technologies of production (associated with Marx) to the technologies of the self (associated with Foucault). Technology deals with the means by which nature is modified, thereby imparting a knowledge of technical skill. Ultimately, all technologies allow us to take control over something as a means of attaining certain ends.

Technology and its artifice act in conjunction to form a particularly distinct society, for which there may be no better description than 'the technological age': a social milieu unlike any that preceded it, and most likely, unlike any that will follow. Although all people and all cultures have technology, only modern technology has an artifice; that is, only the technological age has technology accompanied by an elaborately and technically-devised framework as well as a framework supporting technology. But to describe artifice as technique when the latter means the totality of all techniques is an understatement, because technique (its totality) is only one part of artifice.[4]

Modern technology is different from, although it shares characteristics with, pre-modern technology. Certainly, the manner of improving well-being is one common characteristic, and openness to assimilation is another. But modern technology's artifice places it apart from all other types of technology because pre-modern technology was considered the end of practical applications, whereas modern technology is merely the means of their application. A notable example of the difference between pre-modern and modern technology is the construction of the Egyptian pyramids, which most likely were conceived and erected for one major purpose – a place to contain the earthly remains of the pharaohs or members of the royal family. Modern technology could not conceive of the pyramids without either devising ways of improvement (which would lead to a change in structure) or utilizing the pyramids as a means for other ends (which would lead to a change in purpose).

Technology in pre-modern times played a subordinate role in society because ends were the only intended purpose. An example of the difference between pre-modern and modern technology can be seen in the development of the medieval public market. The medieval market, contrary to its modern counterpart, was no mere abstraction. It existed only because it was devised for specific days at definite times and places. The amount of time allotted for a market was specifically assigned for buying and selling goods within limited hours (usually in the morning) with protection from violence (usually from thievery and assault), but only during these specified times.

All of the above is merely preparatory to a discussion of time, which, as we said above, is fundamental to everything that we understand about technology. This relationship between time and technology is applicable not only to technology as we now understand it, but also to its historical evolution.

1. TIME IN GENERAL: PAST, PRESENT AND FUTURE

We need to discuss time first before we can discuss the role it plays with technology, since there are different interpretations of the meaning of time and of the concept of eternity. In the Old Testament (Genesis 1:1), time is believed to have a beginning, which is attributed to the occasion when God created heaven and earth. This interpretation also agrees with Plato who believed that time could not be eternal because it represented being's mutability as defined by space, which supposedly existed before time. Plato regarded time as having a beginning.[5] But Greek culture and most Greek philosophers perceived time as an attribute of the natural world that was thought to be uncreated, that is, eternal, as in Aristotle's explanation of time.[6] Eternity is allegedly time's timelessness. Although we should be aware of these differences of interpretation, we do not have to be overly concerned with them. Empirically, there is no eternity of time because none of us can ever experience it. We cannot know the beginning of time – just a beginning – and we cannot know the end of time – just an end.

We can begin by saying that time is a condition of being. Time is evident in the universe, in its creation and expansion, on the earth, and in all living things. We cannot know being without time; we can only know when time accompanies being, that is, time must be present in order to know being just as being must be present in order to know time. Time is bound to being, and being is entwined with time.[7] More than being indispensable to each other, they are inconceivable without each other. Nevertheless, time is perpetually changing because constancy is not a characteristic of the universe, an idea made famous by Heraclitus. Time is not a void or an emptiness waiting to be filled up by activity. It is not in things, but, in a sense, flows through them. Time is a phenomenon encompassing three basic variants: that which has been, that which is, and that which will be. Time has a reversible destiny: it is both a destroyer of being and a creator of it. Aristotle described time as the measure of motion.[8] The flow of time itself, or at least the human experience of it, may well be an illusion, that is, a misinterpretation: an imposition we make upon the world because we are conscious of time and because we need time in order to understand reality. But there is no universal time for the individual, just one's own time, although the latter also includes memory and possibility, that is, the object of the past and the essence of the future, respectively.[9]

Time is a phenomenon whose referents are prevailing conditions. More than a period of duration, which is an interval between events or occurrences, time is conditioned by and is related to all other doings and beings, both preceding and subsequent. Human reckoning makes time divisible into three periods, although these three divisions are more presumed than perceived: the past, which is real (in so far as it has a relationship to the present), but not actual; the present, which is real and actual; and the future, which is neither real nor actual. The principal terms here are reality and actuality. When both reality and actuality are manifested, we have a presence. When neither is manifested, we have a future. When neither is manifested, and the memory of being spent has a relevance to the present, we have a past.

It should be self-evident that the past is a memory because the object of memory is the past. Even when we attempt to memorize something, that is, to commit it to memory, we attempt to assign a place for it so that it may be recollected. Although the past is a record of remembrance, it may be an inaccurate record; the past does not lie ready to be dredged up and made living again simply because we have need of it. Because the past has slipped away beyond the reach of the present, it is time that was once filled with activity that is now filled with silence; it is time that has lost its sense of activity. The past is time that has been consumed, denuded of sound, motion and meaning, although still visible when light travels through space-time; yet once perceived, it, too, is consumed by itself.[10] Although non-existent, the past to us retains in large part the essence we have worked so hard to create, and, in this sense, the past means our own individual past – not the past of heavenly bodies, but of our own bodies.

Interpretations of the past vary, but the past itself is fixed and determined. Thus, the constitution of the past is necessity. Constrained by its own inevitability, the past does not exist. If it were to retain a presence, it would not yet have a pastness, and if it has a pastness, it would not yet have a presence because the past and the present are mutually exclusive. The past represents time severed from being, drifting farther and farther away into nothingness; and memory makes the past real in the sense of reflection, but it cannot actualize it. Nevertheless, there are things (such as various literary works and ideas) that seem to transcend time. Although created in the past, they seem to defy time's variants because they are capable of existing in the future as well as the past. These defiers of time have attained what Mikhail Bakhtin has called 'great time', that time in which some things can exist for centuries because they resist the traditional limitations of time (Bakhtin, 1986, p. 4).

Far too often we think that the past is purely individual, that is, that there is no common past, just our own past. Although it is true that we all have a past, even if some of us live only for a day, there is also a past of our families, our culture, as well as the universe, because at one time there was a presence as a condition of their being. The past is not only subjective, it is also objective. It is not only individual, it is also universal. The past is more than a memory; it is also historical evaluation. But the past must not be confused with history, for history is not the past per se, but its interpretation. History is not a recollection, but an analysis.

Nevertheless, not everything from the past is historical; only a small part of it is. The commonly used expression that the past is history is a misnomer. More accurately, we should say that the past is expended time. The past is a larger circle into which the smaller circle of history fits. History is really the present conception of the past, and that conception changes. Although facts, which are derived from the past, never change because they are frozen in time, this condition is not binding upon history. So-called 'good history' gives an accurate and fair interpretation of the past (in the light of all time), and so-called 'bad history' is written to mimic the conditions of the day (in the light of this time).[11] And as we already said, the past is real, but not actual. If it were not real, there would be no need for historians, and no reason in trying to learn from them. Herein is the paradox: the past is real because it has a relationship to the present, but it has ceased to exist.[12]

What we can say about the present can now be compared with what we have said about the past. First of all, it seems readily apparent that the present is an actuality as well as a reality. The present is an actuality because it is not a possibility. The present is actual because it is neither virtual nor theoretical nor ideal. Although we live in the present, our minds have the additional characteristic of both anticipating the future and remembering the past. The mind is able to traverse what some would like to call time barriers, which allegedly give a basis for visions and premonitions proclaimed by prophets and mystics, but in fact we have no such ability because what the mind does is to project into what it perceives to be occasions other than the present by either imagining the future or recollecting the past. The mind cannot really traverse time because the person to whom the mind is connected must always be in the present in order to be. Dreaming or hallucinating is one thing, but actually being in another time is something quite different. Since a fundamental characteristic of being is defined by time (and time-consciousness), this characteristic means that each being has its own time in order to be.

Despite its self-evident manifestation in our lives for always being readily at hand, we cannot say that there is basically only one segment of time, the ever-present present. But we can say that the relationship between the past (the time behind us) and the future (the time before us) is paramount to existence because the use of the future, especially its uncertainty, is the means through which we actualize and make true the present, which is wedged, if you will, between the past and the future, between time as it slips into nothingness and time as it comes into being. Possibilities may reside in the future, but they are manifested in the present.

The present defines the future by setting limits to its possibilities, although the future itself remains an illusion, that is, the future cannot come into being unless the present wills it, making it pointless to assume that the future is real because it is just around the corner. The future does not reside in a nebulous place somewhere waiting to be actualized. It is not drawn from a pool of undefined time, nor is it, if we were to make a comparison to Greek mythology, formed out of chaos. The future has no suffering and no pain, but it also has no hope and no desire. It is filled completely with nothing, except the dimension of possibility.[13] Denuded of reason and emotion, the future is conditioned more by uncertainty than likelihood; that is, something can be possible, but not likely, an example being worldwide peace – a possibility, but not a likelihood.

The future is contingent upon the present, not because it is subject to change – since the present is also – but because it is dependent. Lacking any control over its own destiny, it can neither come into being nor remove the shroud of uncertainty (that is, possibility) unless the present acts. The future is really a conception we have in the present of what we think the future might or should be; but it cannot exist unless it has acquired a presence, and if it does acquire a presence, it loses its futureness. Just as the present and the past are mutually exclusive, so too are the present and the future. All other time variants are excluded from the present. The future is contingent not because it is free from necessity but because it is uncertain. It is, to use a word appropriated from Sartre, merely 'possibilized' (Sartre, 1956, p. 129).

We should avoid the assumption that what is true today will also be true tomorrow, since it is dangerous to assume that the truth of today will continue in a straight line into the future, although it may be equally dangerous to deny constancy when it is apparent. The future is unreal, indeterminate, and possible, but it is not predictable. Assumptions that we make about the past are different from the assumptions we make about the future, for no matter how much (or how little) we know about the past, we know it far better than we can ever know the future. Predictions about the future are projected without consideration of natural or human-made catastrophes, and therefore look quite different from the hardships of reality, to which we can add that an age such as our own that makes assumptions of constancy between the present and the future can only be disappointed.

2. TIME RECKONING: CYCLIC VS. LINEAR

Although the present is the only variant of time we can experience, the future is undeniably the principal variant of technology, which is peculiar because the future is neither real nor actual. The future is paramount for technology not because it is uncertain when characterized by possibility, but because it has been incorporated into linear time. This incorporation lies at the foundation of technology's artifice. It would make no sense if technology were preoccupied with the past, or even, the present, which we might find hard to believe. Not only is the future the place to which technological innovation points, but it is also the only mode of time that can concern technology. Although pre-modern technology may have dealt with the present insofar as ends of specific goals were achieved, modern technology deals with the future insofar as means as an avenue for other means are perpetuated. The future as a mode of time and technology's preoccupation with means as its main contention link technology's use of the future with the linearity of time.[14]

The three variants of time acquire additional meanings when we consider their place in time reckoning, and by the latter we mean not the existential-ontological awareness of time as it passes from the future through the present into the past, but the idea of the cultural accounting of time. Time reckoning is society's assessment of time, a perception made manifest as time is used by different peoples and different cultures, a process or description in which time-awareness is molded and shaped by distinct peoples at particular times with specific goals in mind. Because all people are born into a distinct time demonstrating a history unlike any other, time reckoning is a way of comparing and contrasting one people or culture with another.

One such way of time reckoning important for Western culture and for the westernization of the world was that devised in Greco-Roman antiquity, which perceived that time moved in cycles, that each individual life, or age, or the world itself would be repeated endlessly without beginning or end.[15] Since cycles vary in duration, they could last a lifetime, or several generations, or aeons. Cyclic time to the Greeks and Romans meant growth as well as decay, and fortune as well as misfortune. Although the Greeks and Romans were not the only people who had a cyclic view of time, their conception began to change when Christianity introduced the idea of the linearity of time. Because cyclic time is grounded in nature, be it the passing of day into night, or the phases of the moon, or the seasons of the year, it expresses the rhythm of nature. By its very definition, cyclic time means that no event is unique, that everyone and everything will be repeated again and again without end. Cyclicity signifies that there will be a return to the same state of being without end, a view governed by necessity, not choice, but there was nothing to emulate in this return except a commitment to life itself. Of course, time associated with cyclicity would contain all of the follies of humanity. It would contain the strife and struggle of life in general, the horrors of war and pestilence, the petty intrigues and senseless absurdities of pride and greed, lust and anger, that is, things far more painful to remember than to forget. But cyclicity also signifies the regeneration of the world by utilizing past time for the benefit of present time.

The cyclic conception of time underwent a transformation when Christianity was established; indeed, Christianity's conception of time is its greatest legacy, its most enduring aspect. If all its doctrines, dogmas, and perceptions were erased, its conception of time would continue to exert tremendous influence because it lies at the basis of Western culture. Of the three fundamental influences on the development of Western culture as it evolved through the Middle Ages – namely, the Greco-Roman, the Germanic ethnicity, and the Christian – Christianity is by far the most significant, with its greatest characteristic being its linear, progressive conception of time.[16] It is a legacy evident in everything that we do because we perceive time to be linear, a succession of continuous and unending increments progressing onward within a continuum. This conception of time is defined from the first coming of Christ (his life and crucifixion) to the eventual second coming at the end of the world. Christianity's conception of time contrasts with the Greco-Roman conception of time as repetitive and fateful. This does not mean that individuals and cultures before Christianity had no conception of progression or could not even recognize a non-cyclic pattern in nature; rather, it means that in pre-Christian times there was no need to make linear time a major feature of culture.

We must keep in mind, however, that the cyclic view of time is really the view of philosophers (and theologians), and not the view of non-philosophers. As we noted above, Plato thought time had a beginning, whereas Aristotle thought time was eternal. What ordinary Greeks thought we do not know, but we can assume with some confidence that they followed the direction of their culture and their setting within time, whatever that was.[17] As far as the cyclic view is concerned, both Anaxagoras[18] and Epicurus,[19] for example, denied it. Furthermore, the view of philosophers must be distinguished from those of historians, since the latter look at the passing of time quite differently because of their profession. Greek historians, notably Herodotus and Polybius, did not support the cyclic view of time.[20] As far as the cultural accounting of time is concerned, it is evident that Greco-Roman antiquity was well acquainted with the idea of cyclicity, which was criticized and challenged by Christianity as being opposed to the unique occurrence of the Crucifixion and the coming of the Messiah. Cyclicity signified paganism, linearity Christianity.

Christianity's conception of time has given Western culture its progressive dimension, its excessive optimism about a better world, its reason for struggling hard in the pursuit of perfection. It needs to be emphasized that the technological age could not exist without this linear view of time. Even the Renaissance, which relied so heavily upon the past, looked forward because it, too, accepted Christianity's linear view of time. Christianity is unlike other religions because of its rapid and consistent expansion throughout the centuries. Although there were other cultures, societies, or religions that also viewed time linearly, none had an influence in any way comparable to that of Christianity.[21] Even the way Christianity divided time into historical epochs, introduced by the abbot Dionysius Exiguus in the early sixth century, is indicative of this linear, forward progression: BC representing time before the Christian era and AD (*Anno Domini*) representing time after Christ's birth. Everything in our culture indicates the idea of linear succession, from a simple *line* of thought to the *progress* of career goals, from the study of a *continuity* of events to an *update* of the news, from birth through baptism through Holy Communion through marriage through extreme unction to death.

But the linearity of time does not ignore nature's cycles; rather, it subordinates them. It denigrates nature's ebb and flow to an insignificant aspect of the world, to a presence with little bearing on human destiny. And when tragedy strikes by earthquake or hurricane or through other natural calamity, we are always at a loss for an explanation because we have forgotten that nature has its own rhythm.[22] Linearity is insensitive to the complexity of natural time because it renders the latter unworthy of consideration.

3. THE MECHANICAL CLOCK

Without a doubt, Christianity should be credited with having the greatest influence on Western culture, which is especially evident in the Middle Ages, because the latter surrendered itself so easily to a conception that permanently altered history and the shape of things to come. The Middle Ages firmly established Christianity as the religion of the Western world, and it gave everything, including time, a religious connotation. It ultimately made time a divine characteristic, which served the Middle Ages well until the thirteenth and fourteenth centuries, when a secular or non-religious meaning began to appear (Leclercq, 1976, p. 145). This laicization or secularization occurred when economic growth and urban development began to accelerate as European society was rapidly expanding economically, politically, and intellectually, when governments built up their bureaucracies and when their loyalties rivaled the powers of the church. As a result, the thirteenth and fourteenth centuries gave time a more worldly meaning, a meaning with stronger ties to the earth, not levitating in the clouds.

Contemporaneously with Christianity's influence on time and time's gradual secularization, the invention of the mechanical clock must be considered yet another step in the long process of time reckoning. The development and use of clocks had a direct bearing on the reckoning of time because clocks have made us far more conscious of the presence and passage of time than any other invention. Time, of course, has long been measured by non-mechanical inventions, such as the water clock or sundial, but none of these, all of which were evident in antiquity, had the impact of the mechanical clock. First believed to have been invented in monasteries in the late thirteenth or early fourteenth centuries, the mechanical clock was the most important time reckoning device since the introduction of the idea of linear time, and was responsible for creating the perception of time's uniformity.[23]

The daily life of medieval monasteries in which the discipline of work, study, and prayer became commonplace presupposes the keeping of time by some means of reckoning. The rising and setting of the sun, which determines the amount of daylight, as well as the changing of the seasons, were the first regulators of time. Because the latter are natural sources of time reckoning, it is likely that not every monastery needed the artificial time-keeping provided by clocks, so we must not overemphasize their importance. But regardless of the place that time reckoning played in medieval monasteries, the reckoning of time itself became an increasingly important phenomenon. In the process the natural conception of time was altered to meet unnatural and pragmatic uses by means of the mechanical clock. Simultaneously with the use of mechanical clocks in monasteries, clocks were beginning to be introduced into European cities in the fourteenth century, just as the Middle Ages showed evidence of evolving into an industrial (that is, non-agrarian) society.

Ultimately, the mechanical clock led to the mechanization of production, the means by which working conditions were industrialized. In conjunction with the mechanization of time went the regulation of raw materials, the use of mechanical means (if there were any) in production, and the form the finished product took. Regardless of the promulgation of numerous laws that prescribed the conditions of production, the result was an increase in the mechanization of everything. Time became a commodity. It could be studied, regulated, and now manipulated. Once this occurred, the mechanical clock lead to the industrialization of time, creating a methodical conception in which each resounding second rang out with an ubiquitous ticking, like a metronome – audible, perceptible, and discernible. Although the full industrialization of time occurred later in the nineteenth century, it had its origins in the Middle Ages.

4. TIME-CONSCIOUSNESS

Regardless of the source of its power, the clock achieves more than a division of time into a continuum of a forever flowing presence; it also makes possible the consciousness of time, that is, it makes time-consciousness conscious.[24] The clock has made us so conscious of time that it is nearly impossible for us to imagine time without it. We are hounded by time's presence and cajoled by its influence. With the clock, we 'see' time, we feel its linearity, acknowledge its progression, but we are also blinded to its alternatives. Not only does the clock make time visible, it also makes it accessible, malleable, and manageable so that time becomes yet another phenomenon subject to domination. Although time is a natural and innate condition of the world, the units of time (seconds, minutes, or hours) are artificial creations. These units exist only because the clock exists, which combined time measurement (which the sandglass measured) and time reckoning (which the sundial reckoned).[25]

When we wish to ask what time it is, it is expressed differently from one language to another. For example, in German, it becomes "How late is it?" (*Wie spät ist es?*), and in French "What hour is it?" (*Quelle heure est-il?*). Similarly, in Spanish, *¿Qué hora es?* means not only "What time is it?" but also "What hour (*hora*) is it?" When we ask "What time is it?," "How late is it?," or "What hour is it?," we presuppose a time-measuring device of some sort. Even in

Mandarin Chinese, when we ask "What time is it?" (*Gay dim jung?*), the literal translation is 'What time clock?' All of these examples indicate that there can be no accounting of time without the clock. Even in the expression 'It is one o'clock', the 'o'clock' is simply a contraction for 'of the clock', when specifying the hour of the day.

For purposes of discussion, we need to clarify the difference between time-consciousness and temporality. Simply put, time-consciousness is an awareness of time because we are temporally-conscious beings and because we have made time extremely important; but although time-consciousness has considerable influence on us, it does not qualify as a condition of being. The manifestation of time that is a condition of being is known as temporality: naming not time-conscious but the time-bound. Temporality as a phenomenon of being is determined by the presence of time, by time's influence on all that is. Temporality is not conditioned by awareness, but by the parameters of being. To qualify as a condition, the latter must be a prerequisite; it must be essential to being and to being's presence in the world. As a condition, it must be so fundamental as to lie at the base of its structure. Time-consciousness is not innate in being, but is a reflection of it, an observation of time because we give an accounting to it, an accounting is a basis for time reckoning. Of course, the clock does not contain time; it merely measures it. In fact, time is nowhere to be found. It exists as a condition of being and only as a condition.

We must look deeper into the ideas of temporality (time-boundness) if for no other reason than because Heidegger attributes temporality as "the reason for the clock."[26] But contrary to Heidegger, the clock is not the result of temporality just because time is a condition of being; rather, the clock is the result of human inventiveness and the influence of social change in history. The sundial, for example, preceded the invention of the mechanical clock by many centuries. The sundial was invented when the cyclic conception of time was the acceptable manner of time reckoning, whereas the mechanical clock was invented in Europe when the linear conception of time under the influence of Christianity had replaced cyclicity.[27] It does not seem likely that there would be a need for the mechanical clock, which reckoned time linearly, in a non-linear culture, but there was a great need for the sundial in a cyclic culture, which day after day and year after year reiterated the consistent and ubiquitous presence of the sun in daily and annual cycles. Had the mechanical clock been invented in antiquity it would have been far ahead of its time; in fact, it would have been irrelevant, because there would have been no need for a device that reckoned time linearly in a culture characterized by a cyclic conception of time.

The invention of the clock has little effect on the importance of the existing present. Likewise, the past, that is, our memory of it, would continue despite the invention of the clock. But it is a different matter for the future. Despite the lack of reality and actuality in future time, their absence does not discourage technology's use of the future as its primary variant of time, nor does their absence undermine technology's optimism or even its idea of progress. Blindly pushing ahead, technology uses the future as its agent to define what we all know is indefinable because the future is neither real nor actual. It is for this reason that we can say that the use of the future by technology is an illusion. We are not concerned with pre-modern technology's preoccupation with ends that are executed in the present; rather, we are concerned with modern technology's preoccupation with means (its methodology) that are projected and extended into the future. Modern technology's use of the future deals with a variant of time forever shrouded with uncertainty, and uncertainty means unpredictability. Much of the blame for this false sense of security is attributable to the mechanical clock and to the clock's mechanization of time. The clock gives the impression that it can control the future simply because it makes time visible through mechanical means. The movement of the hands of a clock seem to do more than just symbolize the passing of time; they seem to be time itself. We are easily fooled by a machine that seemingly depicts reality.

When understood in relationship to linearity and mechanization, time means one thing when it comes to non-technical endeavors, such as law, religion, or literature, but another when it comes to technology. There is a distinct difference between the influence of time on these different endeavors. This difference is apparent, insofar as the influence of time on law, religion, and literature is conditioned by the fluidity of its presence, regardless of the variants of time (past, present, and future), but the influence of time on technology is conditioned by only one variant (the future), and is therefore unpredictable. Although time has a bearing on all aspects of being, it is especially revealing for technology because the latter is concerned with only one time variant: the most uncertain one.

5. CONCLUSION

What does all this mean for technology and for the technological age? First of all, we must differentiate the existential awareness of time from the cultural accounting of it. We are concerned with the assessment of time made available by society and conditioned by history, not with internal time-consciousness, such as that analyzed by Husserl phenomenologically or Heidegger onto-logically. And this cultural accounting of time conceived by Western culture has permanently altered technology. It has affected not only Western culture, but also wherever westernization has had an influence. It lies at the foundation of modern technology and technology's artifice. It is one reason why the West has prospered technologically, which perhaps is why the rest of the world emulates the West. It is also a cause for the blind optimism associated with the idea of progress.

In daily existence, the present is all we have. It is the variant of time we live and work in, but the present is losing its importance as we look far into the future. As we said above, this conception is originally attributable to Christianity. Now it is attributable to the technological age, which extends its Christian underpinning, however secular or irreligious it may seem. To put it simply, Christianity's linearity of time, reinforced by the methodical ticking of the clock and its resultant time-consciousness, has made the future the measure of all reckonings. It has made the unpredictability of the future a certainty, and the uncertainty of the future a predictability.[28]

NOTES

1. José Ortega y Gasset, in his *Toward a Philosophy of History*, speaks of a "supernature" (1941, pp. 152–154, "Man the Technician"). Ortega y Gasset was the first professional philosopher to write about the philosophy of technology. Martin Heidegger, in his *The Question Concerning Technology and Other Essays*, speaks of a framework or "enframing" (*Gestell*) (1977a, pp. 19–35). Also see Zerubavel, 1981, cited in note 25 below.

2. A well-known example of artifice as a plan of action is the exploitation of nature by Europeans when they settled in the New World and of the aboriginal American cultures they encountered, which is so vividly described by Frederick Turner in *Beyond Geography: The Western Spirit Against the Wilderness* (1980).

3. Although this is not the place to discuss Heidegger's description of technology as a metaphysical phenomenon, suffice it to say that the use of metaphysics as a means of categorizing technology does not explain what happened to technology before the beginning of metaphysics, which is attributable to the Greeks. As a result, we are at a loss for an explanation for any technological undertaking that predated the Greeks, from early forms of architecture and navigation, for example, to major engineering feats, such as the pyramids at Giza, to say nothing of simple inventions as the wheel, inclined plane, and pulley. For Heidegger's view, see Heidegger, 1977a, and his 'Letter on Humanism', in Heidegger, 1977b, pp. 189–242.

4. Technique in the sense of totality is a key element in the thought of Jacques Ellul. See his *The Technological Society* (1964).

5. Plato, *Timaeus*, 38b 5–10.

6. Heraclitus, fragment 30, cited in Kirk, 1970, p. 307, with commentary, pp. 308–324. See also Aristotle, *Physics*, VIII, 1 (251b 11–19).

7. Also see the analysis in Paul Ricoeur, *Time and Narrative* (1984), vol. 1, pp. 7–12, regarding Augustine's reflection on being and time (*Confessions*, book XI, chaps. 14–28).

8. Aristotle, *Physics*, IV, 12 (220b 33).

9. Similarly, see Ricoeur, 1988, vol. 3, pp. 86–91, which presents an analysis of Heidegger's distinction between ordinary time and phenomenological time.

10. See the analysis in Whitrow, 1980, pp. 273–274 and 347–348.

11. Paul Ricoeur describes a type of 'temporal distance' in which extraneous ideas from the past preserved in the present have become more significant than a survival from the past in the present (1988, vol. 3, p. 149).

12. Although Henri Bergson tells us that the past is an idea (1911, p. 74), he also adds that it has not ceased to exist (p. 193).

13. See Bakhtin, *The Dialogic Imagination: Four Essays*, where 'adventure time' in Greek romances is described as lacking everything because time is empty – the world remains the same because nothing changes, and heroes are as they have always been and will always be (1981, p. 91).

14. The close relationship between technology and the future is also noted by Lorenzo C. Simpson in *Technology, Time and the Conversations of Modernity* (1995), especially on pages 43–60, where he says that technology causes time to be 'domesticated', that is, tamed, so that the future (the variant of technology) can be rendered manipulable. Unfortunately, Simpson supports the illusion that the future is controllable by saying that technology attempts to make the future predictable, but it is uncertain how much control we can exert over the future. Although technology attempts to achieve stability, this attempt fails ultimately because the future will always be unpredictable, uncontrollable, and unknowable. How can we as mere mortals control that which is not yet?

15. For example, see Ovid, *Metamorphoses*, XV, 153–479. Cyclic time is a recurring theme in early Christian writers, but as a theme for ridicule, not praise. See Tatian, *Oratio ad Graecos*, VI; Origen, *Contra Celsum*, IV, 68 and V, 20–21; and Augustine, *De civitatis Dei*, XII, 11–13. For cyclic time in general, see Eliade 1965, pp. 118–123.

16. Although there is a cyclic conception in the Old Testament (for example, Isaiah 65:17–20) where death will be conquered and the world renewed, this view is superseded with the coming of Christ as the Messiah and the acceptance of Christianity as a religion. The Bible is a more or less continuous narration of history, but see the criticism of the Judaic conception of linear time in Whitrow 1988, pp. 51–54.

17. Bakhtin raises an interesting point that 'everyday time', that is, time in Greek romances, is non-cyclical. It is unlike the cyclicity of time evident in nature and mimicked in myth (1981, pp. 127–128).

18. Anaxagoras, fragment 4 in Simplicius, *In Physicorum*, 157,9, cited in Kirk & Raven, eds., 1976, 389–390. Also see Aristotle, *Physics*, 250b 18–26.

19. Epicurus, *Epistula ad Herodotum*, 38–39, cited in Bailey, 1926, pp. 21 and 179–180. Also see Lucretius, *De rerum natura*, III, 945–952, where a linear view of life is presented.

20. Herodotus, *The Persian Wars*, I, 207; Polybius, *The Histories*, VI, 51. Also see Momigliano, 1966, pp. 10–13.

21. One possible exception is Islam, but this religion has a Judeo-Christian background.

22. Nature's rhythm is described quite pointedly by Erazim V. Kohák (1984, pp. 5–6, 15–17), who also adds that 'artifacts and constructs' obscure time (p. 78).

23. In general, see Dohrn-Van Rossum, 1996, pp. 45–123 passim. Dohrn-Van Rossum's criticism of Lewis Mumford's excessively mechanistic perception of

medieval monastic life (p. 33) seems well-founded. For Mumford, see his *Technics and Civilization* (1934), pp. 12–18.

24. A similar idea is evident in Heidegger's *Jetzt-Zeit* ('now-time'), his description of world-time evident in clocks (1996, p. 386).

25. Zerubavel, 1981, p. 38. We should note in passing that Zerubavel's book is an attempt to discuss what he calls the "sociology of time," which concerns how time regulates the social world. Although deliberately avoiding an analysis of time from a historical-philosophical perspective, this book has much to offer.

26. Heidegger, 1996, p. 380. Also see Heidegger's discussion (p. 383) regarding the history of clocks, which seems to conflict with what he says about temporality.

27. The earliest known clock that worked mechanically was built in China between 713 and 741 (usually attributed to 715) by I-Hsing and Liang Ling-Tsan. See Needham et al., 1998, vol. 4, pp. 470–471. Although this clock was powered by water and not by a flywheel (verge and foliots) as in Europe, it was the first known escapement clock. I-Hsing's invention was put to different uses than that of the European mechanical clock, since the former was devised as a mechanism to be used solely by the Emperor for astrological purposes, whereas the European mechanical clock was devised as a mechanism to be used by monks for far more practical purposes. On the nature of Chinese cyclicity, see *The Cambridge History of China* (Twitchett et al., eds., 1986–1991), vol. 1, pp. 688–692. Also see Dohrn-Van Rossum's *History of the Hour*, where the author notes that the principle of escapement employed in China was different from that principle employed in Europe, since the latter used "the centrifugal force of an oscillating inert mass" (1996, p. 87).

28. I wish to thank Dr. R. Dean Ware of the University of Massachusetts (Amherst) for helpful criticisms of an earlier draft of this essay.

REFERENCES

Bailey, C. (1926). *Epicurus: The Extant Remains, With Short Critical Apparatus, Translation and Notes*. Oxford, England: Clarendon Press.

Bakhtin, M. M. (1981). *The Dialogic Imagination: Four Essays*. M. Holquist (Ed.) and Trans. by C. Emerson & M. Holquist. University of Texas Press Slavic Series, vol. 1. Austin, TX: University of Texas Press.

Bakhtin, M. M. (1986). *Speech Genres and Other Late Essays*. C. Emerson & M. Holquist. (Eds) and Trans. by Vern W. McGee. University of Texas Press Slavic Series, vol. 8. Austin, TX: University of Texas Press.

Bergson, H. (1912). *Matter and Memory*. Trans. by N. M. Paul & W. S. Palmer. London, England: G. Allen. Originally published in French as *Matière et mémoire* (1911).

Dohrn-Van Rossum, G. (1996). *History of the Hour: Clocks and Modern Temporal Orders*. Trans. by T. Dunlap. Chicago: University of Chicago Press.

Eliade, M. (1965). *The Myth of the Eternal Return, or, Cosmos and History*. Trans. by W. R. Trask. 2nd printing with corrections. Princeton, NJ: Princeton University Press.

Ellul, J. (1964). *The Technological Society*. Trans. by J. Wilkinson. New York: Vintage Books. Originally published in French as *La Technique ou l'enjeu du siècle* (1954).

Heidegger, M. (1977a). *The Question Concerning Technology and Other Essays*. Trans. by W. Lovitt. New York: Harper and Row.

Heidegger, M. (1977b). *Martin Heidegger: Basic Writings from Being and Time* to *The Task of Thinking*. D. F. Krell (Ed.). New York: Harper and Row.

Heidegger, M. (1996). *Being and Time*. Trans. by J. Stambaugh. Albany, NY: State University of New York Press. Originally published in German as *Sein und Zeit* (1927).

Kirk, G. S. (Ed.) (1970). *Heraclitus: The Cosmic Fragments*. Cambridge, England: Cambridge University Press.

Kirk, G. S., & Raven, J. E. (1976). *The Presocratic Philosophers: A Critical History with a Selection of Texts*. Cambridge, England: Cambridge University Press.

Kohák, E. V. (1984). *The Embers and the Stars: A Philosophical Inquiry into the Moral Sense of Nature*. Chicago: University of Chicago Press.

Leclercq, J. (1976). The Experience of Time and Its Interpretation in the Late Middle Ages. *Studies in Medieval Culture*, *8* and *9*, 137–150.

Momigliano, A. (1966). Time in Ancient Historiography. *History and Theory: Studies in the Philosophy of History*, *5*/*Beiheft 6*, 1–23.

Mumford, L. (1934). *Technics and Civilization*. New York: Harcourt Brace Jovanovich.

Needham, J. et al. (1998). *Science and Civilisation in China*. 7 vols. Cambridge, England: Cambridge University Press.

Ortega y Gasset, J. (1941). *Toward a Philosophy of History*. New York: W. W. Norton.

Ricoeur, P. (1984–1988). *Time and Narrative*. Trans. Kathleen McLaughlin and David Pellauer. 3 vols. Chicago: University of Chicago Press.

Sartre, J.-P. (1956). *Being and Nothingness: An Essay on Phenomenological Ontology*. Trans. by H. E. Barnes. New York: Citadel. Originally published in French as *L'Être et le Néant. Essai d'ontologie phénoménologique* (1949).

Simpson, L. C. (1995). *Technology, Time and the Conversations of Modernity*. New York: Routledge.

Turner, F. (1980). *Beyond Geography: The Western Spirit Against the Wilderness*. New York: Viking.

Twitchett, D. et al. (Eds) (1991). *The Cambridge History of China*. 15 vols. Cambridge, England: Cambridge University Press.

Whitrow, G. J. (1980). *The Natural Philosophy of Time*. (2nd ed.). Oxford, England: Oxford University Press.

Whitrow, G. J. (1988). *Time in History: The Evolution of Our General Awareness of Time and Temporal Perspective*. Oxford, England: Oxford University Press.

Zerubavel, E. (1981). *Hidden Rhythms: Schedules and Calendars in Social Life*. Chicago: University of Chicago Press.

BAKHTIN'S CONCEPT OF INDIVIDUALISTIC IDEOLECTS AS A PROPITIOUS MODEL OF COMMUNICATION AMONG HUMANS AND BETWEEN HUMANS AND COMPUTERS[1]

Karina Stokes

Linguistic communication is all too frequently misunderstood among individual human language users; how much more difficult, then, would it be to create artificially intelligent systems with the capability of effectively communicating via natural human languages? This is a question that is frequently dismissed as one which will automatically resolve itself once the technology of computer systems advances enough to make high-speed, accurate phoneme recognition and lexical parsing more feasible. It is, however, a theoretical issue, rather than a technical void, that prevents communication between human and machine from becoming a reality. It is plain to many observers that communication difficulties can cause discord in relationships among persons who are well acquainted with one another, and often prompt breakdowns in the exchange of information among co-workers who are less than intimately connected. Why, then, do researchers in artificial intelligence who seek to create language using/understanding systems continue to ignore research by philosophers, cognitive psychologists, linguists, rhetoricians, and literary critics who seek more

Metaphysics, Epistemology, and Technology, Volume, 19, pages 233–252.
2000 by Elsevier Science Inc.
ISBN: 0-7623-0681-5

accurate understandings of the process of linguistic communication among humans? For that matter, why do performance technologists and other training and development professionals who strive to enhance on-the-job achievements fail to benefit from relevant insights into the nature of communication as well?

Both issues can be addressed by examining how and why communication breaks down and by determining how interpersonal communication can be enhanced between individuals. The relationship between two communicating human beings can then be extrapolated to encompass the interaction between any two intelligent language users, natural or artificial. An examination of problems encountered in communication among culturally diverse people readily illustrates the need for a dialogic perspective in the workplace; since the cultural rift between human and machine is more extensive than that between any two people, this same examination may clarify why the traditional linear and analytical model of linguistic communication must be abandoned if intelligent computer systems are to achieve the ability to use and understand natural language.

The contemporary workplace is no longer comprised only of White, Euro-American males; now men and women of many colors, cultures, and creeds work together as professional teammates. It is obvious that communication and team-centered problem solving become more difficult when members of a team speak different languages; what is not often recognized is that the more culturally and personally different two people are, the more divergent their ways of using the same language and their ways of understanding the same events will be. Among performance technologists (those who seek to optimize productivity in professional settings), feedback has been identified as an essential part of workplace communication, because performance indubitably misses the mark when directions, standards, and expectations are not sufficiently comprehended by performers. According to William Deterline, feedback serves two functions: "correction or improvement of unacceptable performance and acknowledgment and maintenance of satisfactory performance." He also admits that "[n]either function is *intuitively* performed as effectively as it could be by managers dealing with subordinates" (Deterline, 1992, p. 299). The question arises: why do managers fail to instruct performers on how to improve unacceptable performance or to affirm workers' progress when improvement is evinced or a task is performed well? Current strategies for effective communication in the workplace do not address this dilemma because their practices, which are based on faulty theories, produce anomalous rather than beneficial results.

WHY THE OLD THEORY CANNOT COPE IN THE CURRENT CLIMATE

The most commonly held views of the communication process espouse a linear model of human communication and rely on a correspondence-based representational epistemology; such a view is clearly detailed in the writings of rhetorician James Moffett. The analytical model of communication includes a speaker, a hearer, and some message about an objective reality which is encoded by the speaker and decoded by the listener. This analytical model treats discourse as a product of a single speaker's singular, unified, rational mind; it implies that the symbolic processes involved in communication can be explained by various techniques of analysis which rely on logics and grammars. Learning, according to the traditional model of communication, can be accomplished by the transfer of knowledge, habits, and values from the mind of a teacher to that of a student; miscommunication can be described as a breakdown in one of the steps in this linear process.

Traditional View of Communication

(1) The initial state is the inception of ideas based on sensory input in the mind of a speaker/writer.
(2) Ideas are formed into linguistic symbols according to rational rules.
(3) A physical act produces an utterance or writing.
(4) Transmission occurs via sound waves or written letters.
(5) The transmission is physically heard or read.
(6) Symbols are logically processed, mentally sorted, and comprehended.
(7) An idea in the mind of a listener/reader is the result.

The analytical communication model, being entrenched in the correspondence-based representational paradigm, treats a message as some 'thing' in an objective reality and suggests that a message may, under proper circumstances, be encoded and decoded without significant transmutation. Symbols are assumed to hold a strict correspondence with ideas which are treated as 'things' in an objective reality. The cognitive being, in this view, acquires information about an objective reality in such a manner as to preserve the distinct identities of 'real objects' in the act of cognitively grasping them. Thinking, then, is the manipulation of the collected symbolic representations (as David Hume claims), and language is based on the idea of correspondence; so statements are true or false depending upon their correspondence with external reality. This correspondence epistemology makes communication a matter of correctly matching linguistic signs with their references in the 'objective world', and of comparing uttered propositions with 'reality' in order to discern their meanings. Communicating, then, is a matter of correctly encoding one's understanding of reality into linguistic signs so that the information can be transferred from one mind to another.

More recent philosophers have noted that an 'objective reality' cannot be discerned, let alone compared with ideas in one's mind. Further, one's perceptions are apt to change as one gains new experiences. Thus, the notions of a static, external 'reality' and of an exact duplicate 'representation' of this reality in the minds of perceiving subjects are myths that make it easier to discuss human thinking and communication in a concrete manner while obfuscating the real nature of such complex events. From a dialogic view, sending and receiving messages without deviation is obviously not the normal course for human communication nor for thinking; rather, discourse, like thought, depends upon *interpretation* and *context*.[2] Language specialists have recently pointed out that conversation always involves more than one participant, even when only one speaker is present, as in talking to one's self, because of the interactive nature of language (Schegloff, 1982, pp. 71–72).

In talking to one's self, a language user formulates his or her ideas into linguistic utterances and presents them as if talking to an other. This activity can be seen most poignantly in those who are trying to weigh two sides of an issue in order to make a decision. Such persons may even apply persuasive rhetoric in an effort to convince themselves to adopt some course of action or another and may, in this process, conceive of new alternatives heretofore unconsidered. The analytical view of communication does not account for this sort of interactive thinking nor for many other common human uses of language. Linguistic communication is a dynamic activity, not a linear procedure with fixed rules and rigidly defined responses; nonetheless, this human activity is the only tool available for communicating ideas to other persons and to oneself.

It is a well established maxim among human performance technologists that to enhance performance in the workplace, it is necessary to communicate to task performers what performance is desired of them and then provide feedback which lets performers know how close to (or far from) the desired performance outcome they are. Deterline depicts this process with a simple feedback loop diagram consisting of a person, an action performed by that person, a result of that action, and feedback to the acting person; this feedback "consists of information about the nature of the action and its result, in relation to some criterion of acceptability" (Deterline, 1992, p. 294).

Unfortunately, the process of communicating is never as simple as the traditional "sender giving message to receiver" models would imply. For one thing, the cognitions of one mind cannot be 'encoded' into language and then be 'decoded' by another mind into identical cognitions. Interpretations are the necessary result of humans possessing separate minds which must communicate intentions and ideas through a fluid medium of language. Furthermore, personal understandings of verbal feedback can provide or deprive a performer of the needed motivation to continue to improve; feedback can be a positive or negative reinforcement of behavior depending on how the message and its delivery are interpreted. So the matter of providing accurate feedback in an appropriately encouraging manner becomes a daunting responsibility for supervisors whose aim is to enhance the productivity of performers exhibiting demographic characteristics (and concomitant modes of perceiving and interpreting) which manifest extensive diversity.

The traditional model of communication, which is presumed by most human performance technologists to correspond to reality, unfairly places a greater onus than necessary on those who are responsible for improving the yield of performers. One expert states that:

> [t]he sender has primary responsibility, since the message and its intent originate with him. Clarity, completeness, simplicity, and other characteristics that facilitate understanding can be designed into a message only at the sender's end .... The receiver has the responsibilities of attending to, attempting to understand, interacting with, and remembering the details of the message (Deterline, 1992, pp. 296–297).

This human performance authority further states that "[b]oth responsibilities are best met through the use of feedback" (Deterline, 1992, p. 297). He then proceeds to outline his idea of the most effective method of ensuring that messages are sent and received properly by all parties involved in a communicative endeavor; his formula requires an awkward array of "tell me what you think I just said" drills. Such an approach ignores the fact that communication does not, in reality, proceed in the mechanistic fashion delineated by theorists who crave neat and tidy explanations of messy and multifaceted processes.

Analytical (linear, objective, hegemonic) theories of communication dismiss obvious anomalies to the linear view of linguistic knowledge transfer, or they blame individual communicators for failing to express or interpret ideas properly. The more 'anal retentive' of the adherents to these approaches strive to capture linguistic meanings by incorporating the notions of stereotypical (the-way-they-do-it) uses, associations, and modes of employment for words; they often propose more and more elaborate models of the communication process in order to control the flow of information and document 'information transfer' procedures. Reliance on rules to ensure proper transmission of messages results in a linear analysis of the path of a message from a sender to a receiver in order to determine who or what is to blame when miscommunication occurs. Typically, the person with the lower status or less powerful position (or the more marginalized culture or gender) is the one who is faulted for failing to communicate 'properly' with a supervisor or more powerful person. When communication styles fail to interact productively, it is assumed that one individual needs to learn to communicate 'better', which usually is to say that the lower status individual needs to learn to communicate in the manner of the higher status person. Typically, the higher status persons in business today are phallocentric, eurocentric, analytical communicators. This covertly enforced hegemony of communication style in the workplace limits creativity and thwarts problem solving efforts.

A useful paradigm for promoting creative thinking is called 'analoguing', which involves using metaphors in communication. When Timothy Thompson studied organizational creativity in an Ohio newspaper news room, he discovered that several analogies were productive, that analoguing "helps avoid excessive categorization and separation by looking for unities and patterns in phenomena rather than for divisions" (Thompson, 1986). Thompson's research concludes that creativity can be seen as "a muddling dialectic between stability and flexibility, an image of messing things up before recombining" (Thompson, 1986). He then recommends that groups might better solve problems if they were to pursue diverse paths to a solution rather than insisting on a single sanctioned procedure; they should strive to create analogies for their problems in order to gain fresh insights.

In order for such an effective strategy to be utilized, those who form an organization's culture must establish "an atmosphere of experimentation, acceptance of mistakes, high-variety solutions through chaotic action, loose goal setting and limited formal planning, time for wandering and incubation of ideas, rewards for random and unjustified variations in problem making and problem solving, as well as other creativity motivators like a sense of challenge and achievement, freedom, playfulness, and a system of mutual support" (Thompson, 1986). Clearly, this sort of novel, non-hierarchical, contemporary workplace needs a dialogic, non-linear, non-analytic communication strategy for its creative problem solving as well as for informing and motivating workers within a professional culture which encourages mutual respect and individual innovation.

Typically, approaches to training via initial instruction or post-performance feedback and plans for enhancing motivation in the contemporary workplace focus on modifying/manipulating behavior with rewards and punishments, or they strive to create appropriate stimulation levels while making topics relevant and bolstering confidence and satisfaction in performers concerning their ability and desire to accomplish required tasks. Unfortunately, the behavioral and ARCS[3] models of learning assume that communication is a matter of correctly applying logical formulae to communicative exchanges. These methods fail to profitably exploit the dynamic features of human communication which are involved in learning and teaching; they also neglect the whole of the persons who comprise one's partners in a verbal interchange. Just as linear logics and mathematics only depict a portion of naturally occurring processes which they attempt to model, so too, traditional, analytical approaches to the use of language in communicative acts fail to capture the entirety of the dynamic and dialogic act.

Many mathematicians have turned to fuzzy logics and other alternative methods in order to develop better models of physical processes; modern linguists have, likewise, begun to explore the insights of dialectical philosophies and dialogic methods of criticism in order to develop a better understanding of the dynamic social process of communication. Dialectical methods of depicting language use and of understanding discourse are difficult to master and implement for those who are accustomed only to linear thinking; however, these methods are worth the additional effort because they provide a richer sensibility and account for intuitively obvious phenomena which can be observed in any human interaction. Further, a dialogic view of communication accommodates diversity among communicators; it allows for a sharing of meaning amongst non-homogeneous intelligent beings.

BAKHTIN'S THEORETICAL BASIS FOR A NEW SOLUTION

What is shared between persons allows for greater subtlety in communicative interchanges, while what is different and unknown can cause misunderstandings. Polysemy is a phenomenon of language that occurs frequently in human discourse and is often exploited for the sake of humor. Shared phonological and morphological relations allow for puns. The reason that puns are humorous is that the same phonemic combination signals two very different lexemes with their associated concepts. Humorous ('punny') remarks can be perceived in multiple valid interpretations. It is the ambiguity resolution process that tickles the funny bone and provides intellectual pleasure (Morreall, ed., 1987). Sometimes, however, a speaker creates a pun unintentionally, yet a listener recognizes it and derives pleasure from its humor. Clearly, in such cases, the 'decoding' process is not a reflection of some linear 'encoding-sending-decoding' operation. Rather, a transmutation, an active interpretation, of the linguistic message occurs. This alteration in the message is not a result of transmission problems, or signal noise, but is derived from the context of the speech act, the ideolect of a listener, and the social interaction of a speaker and a listener. Traditional analyses often treat such phenomena as flukes of language, and persist in claiming that linguistic signs are stable (at least within a static time frame along the historical development of a language), and so misunderstandings are seen as problems to analyze and solve. Unfortunately, linear procedures for conducting an objective analysis cannot properly identify the dynamic causes of miscommunication, let alone forge a feasible solution. What is needed is a cognitive revolution, a paradigm shift.

A viable theory of human communication can be found in the writings of a literary critic, Mikhail Bakhtin, who views the construction of meaning as a dialogic process that involves the entirety of a context. Context includes (perhaps among other things): the linguistic environment, the behavior of the speaker and of the listener, the situation common to both, and the horizon of reality surrounding the speech situation (Johnson, 1988, p. 170). Meaning is determined by context (Richards, 1955). A person perceives incoming information, compares it with previously acquired information, experiences a dialectical interaction between new and old data, and thereby attaches meaning to an utterance. Instances of words or symbols are merely the stimuli, whereas previous experiences with other, possibly similar, contexts and with the language (as a whole system) constitute the basis for interpretations of these stimuli. Since every human has unique past experiences, everyone attaches slightly different meanings to symbols. Meaning is thus individualized.

The individualization of meaning need not cause insurmountable communication problems, though, because, according to dialectical philosophy, all individuals are essentially one at a higher level of understanding. This notion of underlying unity is also prevalent in Jungian psychology and in contemporary rhetorical theory. As Kenneth Burke puts it, differences are culturally formulated and will disappear when driven into the background. To stress opposition or to define differences is to mistake cultural convention for true reality. The whole purpose of rhetoric, for Burke, is to overcome the misunderstandings inherent in the individualized nature of language and to seek a common ground, a reconciliation, an identification so that true communication becomes possible. Once the dialectical identification of the speaker and the hearer is achieved, then cooperation can be induced by the use of linguistic symbols (Burke, 1966). This dialogic act of communication, which relies on context and cooperation, is alien to analytical, linear approaches to communication.

A speaker's current context influences the form which his or her communication takes. Mood, perceptions, and beliefs at a particular moment and place determine a speaker's word choice, intonation, and other expressive gestures. "The linguistic significance of a given utterance is understood against the background of language, while its actual meaning is understood against the background of other concrete utterances on the same theme, a background made up of contradictory opinions, points of view, and value judgments — that is, precisely that background that, as we see, complicates the path of any word toward its object" (Bakhtin, 1981, p. 281). The context of a communicative act is based on the speaker's current perceptions of the physical and emotional situation as well as the speaker's perceptions of the listener and of the subject matter.

While the speaker's context influences the manner in which he or she formulates an utterance, the listener's context delineates the interpretation which is given to the perceived communication. This context encompasses the listener's current perceptions of the situation, of the speaker, and of the subject matter. For instance, a father may take his child to a hot air balloon show and enthusiastically describe all the features of each flying apparatus to a bored and inattentive audience. If a vendor selling helium-filled balloons passes by, the child might utter, "Daddy, I want a balloon." The father, who is focused on the grand spectacle of the show, would likely reply that he could not afford to buy such a thing, but he might take the remark as a sign that his child shares his enthusiasm for these flying machines. In the midst of a birthday party or a county fair, the same statement would elicit a very different response, and the child might well receive a little balloon on a string. The context in which the utterance occurs constrains the possible meanings which can be associated with it.

One of the fundamental principles for Bakhtin is "that meaning is context bound but that context is boundless" (Clark & Holquist, 1984, p. 218). The number of situations in which one might communicate are innumerable; however, the possible interpretations or meanings which can be ascribed to an utterance are constrained by the context in which the act of speech occurs. The mutable rules of language are founded on a social logic and, therefore, can change over time according to the needs of the social community. The primary function of language is communication, as a form of social interaction with other language users, and not merely as a transmission of a message. "The only way that words can mean is to be understood" (Clark & Holquist, 1984, p. 213). The capacity of words to mean is distinct from definitions. Discourse is action; it is not the sequential transmission of encoded signals.

While the current context affects the interpretation given to an utterance, so too does one's horizon. Each individual has previous experience from sensory input and beliefs (some signified, which is associated with a linguistic signifier) about that experience. The sum total of this experience and stance toward it constitutes what Martin Heidegger calls one's horizon; and this horizon shapes one's interpretations of further experience and communication. For instance, an older man may tell his grandchildren about his experiences in World War II and mention that he thought one of the computers he met during the war was very attractive and personable. The kids, unfamiliar with the, 1940s usage of the term 'computer' to refer to a (typically female) person who performed arithmetic computations would be confused by Grandpa's attraction to a machine like the Pentium PC which sits on Mom's desk at home. These people do not share the same background knowledge and awareness of technological changes. Once the grandfather explains wartime decoding practices, the horizon of the grandchildren will be changed so that they are capable of interpreting the word 'computer' as "a person who performs arithmetic computations" in connection with pre-desk-top-computer experiences of Grandpa. The specific contexts in which a word is uttered in the future will combine with the horizon developed in the individual to create a new particular meaning for the uttered word at any given time.

In addition to the linguistic context on several levels, listeners bring to the task of interpretation their own personal biases. Their belief systems (that is, their sets of interrelated symbols or signifiers which are associated with sensory experiences or with other symbols) will open up multiple possible interpretations and close off others, especially in the case of more deeply held beliefs. If previous experience shapes the interpretation given to an utterance, then every new utterance of a word will derive a slightly different meaning, since experience constantly flows and changes from moment to moment. Taking this fact of communication into account allows one to be more tolerant of individual differences and thereby to communicate more effectively. Incorporating this concept into an accurate account of interpersonal communication requires a dialectical understanding of the act of communication and the active participation of all communicators.

An utterance of the same word is never given the same interpretation twice, because of the dialectical nature of communication that requires an interplay between the speaker, the listener, and the subject matter which are all situated in contexts. Anyone who has repeated a joke which left everyone rolling on the floor yesterday only to find that it is met with raised eyebrows and a few half hearted smiles today has experienced the "you had to be there" contextual nature of communication. Accounting for this dialectical process goes beyond current studies in pragmatics; not only do contexts, beliefs, and personal relationships have causal roles in determining meaning in an utterance, but utterances also act upon one's interpretation of the situation, one's understanding of one's horizon, and one's feelings toward the speaker. One's world view can constrain one's interpretation of new information.[4] Also, new information can change one's beliefs, subtly or drastically.[5] So the process works in both directions as a dialectical interplay. A linguistic sign can depict a multitude of meanings, or significances, depending upon the context in which it is uttered, the horizon of the listener, and the intention of the speaker. Bakhtin views communication as an act of interpretation, rather than as a procedure of encoding and decoding representations; his dialogic view does not separate the act of communication from the communicating actors.

Comprehension and production are, furthermore, intimately connected aspects of the dialogic act of linguistic communication. Speakers are likewise listeners because they monitor their articulations and may begin an utterance over or may continue an utterance further in order to clarify perceived ambiguity. Meaning is not transmitted by speakers to listeners; they co-create meaning in a time and a place, all of which can never be re-created exactly the same again. Bakhtin insists on the mutability of linguistic symbols. He opposes the formalist's view of language as possessing definite, predictable structures; he focuses, instead, on context as the main factor involved in the ascription of meaning. He concludes that language is dialogic in nature and that one need only examine the process of discourse to see the dynamic flow that occurs in this human activity. At the same time, from any speaker's perspective, words appear to be "always already there" because of the obvious fact that the vocabulary and grammar pre-exist. Yet one's words are also "never ever before" because they are spoken in a novel and unique context for the individual speaker (Clark & Holquist, 1984, p. 217). The rules of the language predate the individual in time; yet it is the individual who shapes the meaning of words, or the significance associated with the signs, in a particular present moment. At one and the same time, speakers bend language to their needs and are shaped by the language which has been previously molded by their ancestors. The social contexts of communication constantly fluctuate over time; and thus, the words of the language have meanings ascribed to them in a fluid manner.

For Bakhtin, communication is a unifying process that resolves oppositions and brings participants into a dialogic exchange with one another. In any real act of communication, both a speaker and a listener experience a change in their understandings of the world, and even in their very selves. An utterance which has no effect upon a listener, or a speaker, does not constitute an act of communication. Furthermore, the nature of communication necessitates that speakers formulate an utterance in conformity with their ideolects (ways of understanding and using the language system) and that listeners interpret an utterance according to their own context bound ideolects. The individual ideolects of any particular speaker and listener constantly change according to the experiences which each individual encounters.

Dialogic theory points out that linguistic communication, as well as human cognition, takes place in a social context as a social construct that is based on the interpretations of social beings. Such a theory of communication thus shares much of the dynamics included in general systems theory. Communication becomes productive only when language users experience enough agreement about the significance of linguistic signs, sufficient overlap in their horizons. That is, people who share more background information, beliefs, biases, etc., have a greater chance of understanding one another via language. Of course, no two people share the exact same horizon. Fortunately, as Bakhtin notes, language is elastic enough to accommodate human language users to some extent, so that exactitude is not necessary.

Language is a dynamic tool for communication. Its regularity is based on some degree of unconscious or unrecognized agreement among language users, yet its supple nature allows for dynamic, individualistic employment. Each person acquires and uses language in accordance with his or her unique personal experiences; and to the extent that individual humans share common experiences, their horizons, and thus their ideolects, intersect and thereby make communication possible. Genuine dialogue requires that each speaker look at him or herself through the eyes of the listener, and that every listener put him or herself in the place of the speaker. It is through this context-bound, dialogic activity between self and other that communication takes place. Real communication, therefore, requires an intimate sharing of horizons, a consubstantiation between the participants.[6] A mutual understanding of experience (or, at least, some shared background assumptions about the way in which things operate in the world) is essential for any productive communication.

WORKPLACE APPLICATIONS FOR DIALOGIC COMMUNICATION THEORY

Tolerance of the fact that every speaker uses language in a slightly different manner is essential to effective communication. Also, cognition (or communicating with one's self) can be better grasped with Bakhtin's perception of language use, which views thinking as communication among the many voices within one's self, the many personality aspects and moods which every individual exhibits. Recognition of the vast resources available within every human mind allows individuals to approach every problem from multiple perspectives and thereby enhance the probability of finding a viable solution. Unfortunately, those who remain refractory in their analytical thinking negate their own creativity and sociability; they reject the dialogic paradigm just as Ptolemaic astronomers refused to accept Copernican views.

A supervisor who understands the ideolects of workers is able to enhance the effectiveness of his or her communication with them. A dialectically savvy manager may approach new employees by first assessing their past experiences and social/cultural backgrounds and then determining what is needed to create cohesion and understanding among all parties. Establishing a common ground is an essential first step to sharing information and providing meaningful feedback. Team-building activities among work groups are moving in the right direction regarding the dialogic concept of establishing mutual understanding. The first priority must be to create a shared horizon within the workplace. This framework for cooperation can only be produced if workplace leaders recognize the existence of diverse ideolects and distinct horizons within any social group.

It is not suggested here that managers should make excuses for why employees of different cultures fail to perform; rather Bakhtin's theory indicates that managers must understand the process of communication, learn to understand the ideolects of all of their co-workers, and provide communication aimed at reaching each individual. In order to accomplish this goal, it may be necessary to provide workplace experiences designed to create some sharing of ideolects and merging of horizons among co-workers. Creating shared experiences, or common ground, among a group of people can help them to perform more effectively, because shared experiences make their communications with one another more productive.

A decision which seems completely innocent and devoid of bias can, in fact, "carry an enormous amount of ideological and philosophical baggage, which is in turn communicated to [individuals] either directly or indirectly" (Nieto, 1997, p. 389). For instance, culturally dominant groups are portrayed as the norm in educational materials, business practices, and office etiquette; anyone who is different from this norm is presumed to be culturally or intellectually disadvantaged. Such messages can be internalized by individuals from subordinated cultures when such seemingly minor things as office celebrations of holidays focus on White, Euro-American traditions alone. The subtle but powerful message is that "their cultures, families, languages, and experiences, have low status, and they may consequently learn to feel inferior" (Cummins, 1989; Darder, 1991) (Nieto, 1997, p. 392). As a result of this sort of passive racism, all members of an organization suffer. Likewise, unconsciously maintained gender bias can hinder half of an organization's work force from contributing their optimal performances.

Communication reform aimed at empowering all individuals to perform at their peek must beware of the subtle prejudices lurking in the dark corners of one's horizon like unnoticed insects that wait for opportunities to spread the diseases which they passively carry. Workers should be encouraged to share their experiences and perspectives in an open and accepting environment so that all horizons may receive equal validation and appreciation, and so that all individuals may contribute to broadening the collective horizon of their organizational systems. According to a 1989 study by education specialist, J. Cummins, students who are encouraged to develop positive cultural identities via interactions with teachers derive a sense of control over their own lives and the confidence and motivation to succeed academically (Nieto, 1997, p. 398).[7] Such confidence can be incorporated into the workplace by promoting an attitude among supervisors and workers alike that biases and prejudices need not be tolerated; they can be changed. A positive perspective is one which acknowledges that all cultures and individuals have something valuable to contribute. A negative view presumes that some groups or individuals have nothing to offer and should not even try. This latter vision is a pessimistic fallacy that prompts individuals to give up, stop trying, and thereby fail in accordance with such a self-fulfilling prophecy. The dialogic and tolerant perspective assumes that all individuals can make a difference, that all actions produce a result, and that all persons have value and worth; this principle is empowering.

Performance improvement efforts must focus, first, on adopting a new vision of communication so that supervisors may provide accurate and usable information, so that harmony is maintained among diverse workers, and so that innovative solutions can be encouraged rather than discouraged. Instructions that spew forth from a supervisor do not constitute communication with individuals. Every genuine act of communication entails a sharing of horizons, a merging of individual ideolects into a unified understanding. No matter how clear instructions seem to their author/speaker, the purpose of the communicative endeavor is to co-create meaning with an *other*. Knowing the people with whom one works is essential to establishing a consubstantiation of being, a co-creation and sharing of meaning. An organizational culture conducive to such vital communication supports efforts to maintain harmony and treats discord as a communal problem to be solved rather than as a fault of some individual(s).

To the good old boys who claim that communication in the workplace was easier when the workforce was dominated by men with the same cultural perspectives, the obvious response is: "That's right!" Indeed, hegemonic discourse works well enough with a homogeneous community. But the contemporary workplace is filled with diverse horizons and innumerable ideolects; the multitude of perspectives and voices in the workplace today presents both a challenge and an opportunity. While increasing diversity makes it impossible to continue doing business-as-usual, it also makes it possible for an organization to innovate and adapt by exploiting the vast array of perspectives inherent in a heterogeneous population. A reasonable view of the situation, an examination not clouded by fear of the unknown, could recognize that the twenty-first century marketplace requires the inclusion of multiple problem-solving strategies and multifarious communication styles in order to "do it better and faster with less."

ARTIFICIAL INTELLIGENCE RESEARCH APPLICATIONS

A dialogic theory of communication may also bolster artificial intelligence research concerning the issue of natural language comprehension in two ways. First, the members of a research team may become more productive and creative in solving research problems if they free their own minds from the tyranny of linear thinking and hegemonic discourse styles. The practice of giving individuals room to think and expand their horizons not only helps business organizations compete more effectively for financial gain, it also helps scientists to overcome technical and theoretical problems. Secondly, applying a dialogic understanding of communication to the research problem itself may open new avenues by which natural language understanding by artificial intelligences might successfully be approached. Once it is understood that a machine intelligence must be as dynamic and individualistic as a human intelligence, then 'errors' in comprehension and production of speech may be viewed as a normal part of the process of acquiring facility with language rather than as indications of failure. A research team organized around dialogic principles of communication and thinking has the best chance for success in producing an artificial intelligence with its own horizon and its own individualistic and dynamic ideolect, with which to communicate with human intelligences.

While horizons and ideolects are dynamic, correspondence-based rationalistic theories of truth are bound by predefined realities. Rule-based computer models of language comprehension which adhere to these theories are subject to the limitations inherent in them. Once a rule-based program for addressing some domain is instantiated, then encoded rules will be followed with tireless precision. Even programs that allow for periodic updates to the rule set must rely on further rules to execute the alteration procedures. But human interaction with the world is not a strictly rule-based procedure; rather it is a dynamic, dialectical activity. One's horizon (one's current understanding of the world) is changed by new experiences, just as new experiences are shaped and interpreted according to one's horizon. Rule-based models are not capable of dialectical interaction, but are limited to the linear logic which directs them. Human communication involves a dialogic sharing between two beings via individualistic ideolects (ways of using language); these ideolects change according to one's exposure to communication just as horizons shift with experience. In order for an artificially intelligent machine to communicate with humans in the manner of humans, it must acquire language and experience simultaneously as humans do. A dynamic 'feel' for the language and an ever-changing 'conception' of reality must be developed before any intelligent individual, natural or artificial, can communicate with another intelligent individual.

In considering a rationalistic view of the nature of truth, one finds that most of these theories of truth are, in fact, theories about the judgment and verification of truth propositions, and not about the human experience of truth itself. Thus, rule-based computer models, which rely on static, black and white statements about what *is* true and what *is* false, cannot communicate with individual humans in a human-like manner because they lack the ability to form and maintain dynamic, unique horizons with which to interpret the many shades of gray found in human experience and human linguistic utterances. The rationalistic paradigm limits artificial intelligence research as well as discourse analysis precisely because it ignores the dialectical involvement of an individual's horizon in matters of interpreting utterances and evaluating their veracity.

An expert system is the epitome of artificial intelligence based on rationalistic prescriptions. Using an expert system to model natural language use may be likened to a pantomime creating an illusion while human language use may be depicted as an action within an environment. In the case of a pantomime, a person may appear to be leaning against a wall; however, no wall actually exists and no interaction between a person and an object actually occurs. A real action involves an interplay between a person and a physical or mental environment. The pantomime model of reality resembles aspects of the richer experience of an action. The mime interacts with an imaginary world which exists only in the mind of the viewer; program users may assume the existence of some intelligence in an expert system, just as an audience believes in an invisible wall portrayed by a mime. Real intelligence exhibited in goal-directed actions must interact with experienced phenomena in the midst of a flow of action. The mime artificially constrains actions in order to achieve a verisimilitude with real action. A mime could, after all, extend his or her hand beyond the invisible wall, but refrains in order to mimic a real interaction with a physical object. Natural human actions typically are constrained only by physical laws and not by such artificial choices. Likewise, language use involves a dialogic interaction with social contexts, linguistic strings, and speakers and listeners. Logical models of language use do not achieve this interaction; therefore, rationalistic models fail to capture the essence of human communication via language.

Some current researchers claim that logic or grammar alone (especially fuzzy logic or modified transformational grammar) can adequately model human cognitive functions. But even expert systems based on fuzzy logic rely upon predefined rules which freeze the world view of a machine and prevent it from adapting to unique experiences in life. Fuzzification only relaxes the artificial constraints of logic, it does not eliminate them. Likewise, additional levels of transformational rules extend the usefulness of grammatical analyses; but they do not fully encompass the act of linguistic communication. Attempts to encode context (as in the restaurant scripts of Schank) have also not been able to mimic a dialectical interplay between strictly defined contexts and incoming novel data (Anderson, 1985, pp. 130–131). A reliance upon symbolic representations insures that expert systems (traditional rationalistic artificial intelligence techniques) can never capture the dialectical nature of cognition and communication. The rationalistic research programme is based on a misinterpretation of the nature of communication and cognition. It fails to account for the social and dialectical characteristics of language.

Rationalistic/analytic approaches to simulating natural language comprehension strive to capture linguistic meanings via models that incorporate stereotypical uses, associations, and modes of employment for words. Such notable authors as Terry Winograd & Fernando Flores (1987) and Hubert Dreyfus & Stuart Dreyfus (1986) have promoted the claim that viable research in artificial intelligence cannot continue with the rationalistic model of human communication and cognition which they deem to be inadequate. The foundation of their assertion is the Heideggerian notion that communication is a dialectical process requiring the active participation of all parties involved.

Their Heideggerian views lead them to conclude that while computers may be employed as tools in more productive ways than are available at present, no further optimism for the endeavors of artificial intelligence research is warranted.

While these criticisms of the traditional representational view are germane, their conclusions are unduly pessimistic. Further analysis of Heidegger's concept of horizon and Bakhtin's related ideas about individual ideolects in human cognition and communication provide insights that may lead to a more optimistic research programme for artificial intelligence in general and for understanding human communication (and miscommunication) via language, in particular. So long as researchers insist on working from within a rationalistic paradigm which views language as a system of static signs, then the predictions of Winograd and Flores and the Dreyfus brothers will be born out, since a static model cannot depict the entirety of a dynamic act. Alternative research approaches, which exhibit an adaptive learning ability, show promise for modeling the dialectical acts of human cognition and linguistic communication.

Back-propagating neural networks, for instance, have been employed to achieve speaker-independent phoneme recognition (involving minimal pair distinction) without pre-defined rules; the training for such a system took only minutes and required no prior knowledge of what kinds of speakers' accents would be used as input (Stokes, 1991). Extending this result to allow machines to learn to connect semantic associations with phonemic input would be far less taxing than developing a system of rules which will be able to accommodate all potential future experiences with linguistic utterances. To focus only on the logical considerations of denotations and correct grammatical structures is to exclude what is essential to the human *use* of a language system.

Philosophers, psychologists, linguists, rhetoricians, and literary critics examine the nature of communication from a variety of perspectives. The importance of context and horizon inspires a good deal of research in these fields. Unfortunately, it is often the case that relevant research in one field is not made available, or remains unconsidered, by others. This has been the case with Heidegger's dialectical concept of horizon and Bakhtin's notion of a context-bound dynamic ideolect. Linguistic communication becomes much more interesting and complex when viewed as a dialectical activity which conjoins a speaker and a listener in a relationship. Any model of communication which hopes to be viable must account for this dialogic interaction among participants, for this interaction is an essential part of the nature of the human activity of communication.

NOTES

1. Mikhail Bakhtin wrote a great deal about literary criticism during the early part of the twentieth century in Russia; however most of his writings were suppressed by, or were hidden from, the communist government until their recent revival in the Western world.

2. A fairly comprehensive examination of the role of context can be found in the writings of H. P. Grice.

3. Focuses on generating: Appropriate stimulation, Relevance, Confidence, and Satisfaction.

4. Further details on the role of context can be found in the writings of H. P. Grice.

5. See Quine & Ullian, 1978 for details.

6. See I. A. Richards (1955) and Kenneth Burke (1966) for further discussions of mutual sharing and consubstantiation.

7. Nieto (1997) cites the original study as Cummins, J. (1989) *Empowering Minority Students*. Sacramento, CA: California Association for Bilingual Education.

REFERENCES

Anderson, J. R. (1985). *Cognitive Psychology and Its Implications*. (2nd ed.). New York: W. H. Freeman.

Bakhtin, M. M. (1981). *The Dialogic Imagination: Four Essays*. M. Holquist (Ed.) and Trans. by C. Emerson & M. Holquist. University of Texas Press Slavic Series, vol. 1. Austin, TX: University of Texas Press.

Burke, K. (1966). *Language as Symbolic Action: Essays on Life, Literature and Method*. Berkeley, CA: University of California Press.

Clark, K., & Holquist, M. (1984). *Mikhail Bakhtin*. Cambridge, MA: Harvard University Press.

Deterline, W. A. (1992). Feedback Systems. In: H. D. Stolovitch & E. J. Keeps (Eds), *Handbook of Human Performance Technology: A Comprehensive Guide for Analyzing and Solving Performance Problems in Organizations*. (pp. 294–311). San Francisco: Jossey-Bass.

Dreyfus, H. L., & Dreyfus., S. E. (1986). *Mind over Machine: The Power of Human Intuition and Expertise in the Era of the Computer*. New York: The Free Press.

Garcia, S. S. (1992). *The Pervasiveness of the 'Psychology of Marginalization': Barriers to Institutional and Personal Development in a Multicultural Education Environment*. ERIC database document no. ED 369 371.

Grice, H. P. (1975). Logic and conversation. In: P. Cole & J. L. Morgan (Eds), *Syntax and Semantics*, vol. 3: *Speech Acts*. (pp. 41–58). New York: Academic Press.

Johnson, M. L. (1988). *Mind, Language, Machine: Artificial Intelligence in the Poststructuralist Age*. New York: St. Martin's Press.

Moffett, J. (1983). *Teaching the Universe of Discourse*. Boston: Houghton Mifflin.

Morreall, J. (Ed.) (1987). *The Philosophy of Laughter and Humor*. Albany, NY: State University of New York Press.

Nieto, S. (1997). School Reform And Student Achievement: A Multicultural Perspective. In: J. A. Banks & C. A. McGee-Banks (Eds), *Multicultural Education: Issues and Perspectives*. (pp. 387–407). Boston: Allyn and Bacon.

Quine, W. V. O. & Ullian, J. S. (1978). *The Web of Belief*. (2nd ed.). New York: Random House.

Richards, I. A. (1955). *Principles of Literary Criticism*. New York: Harcourt Brace Jovanovich.

Schegloff, E. A. (1982). Discourse as an Interactional Achievement: Some Uses of 'uh huh' and Other Things That Come Between Sentences. In: D. Tannen (Ed.), *Analyzing Discourse: Text and Talk*. (pp. 71–93). Washington, DC: Georgetown University Press.

Thompson, T. N. (1986). *Analoguing Creativity and Culture: A Method for Metaphors*. ERIC database document no. ED 282 263.

Winograd, T., & Flores, F. (1987). *Understanding Computers and Cognition: A New Foundation for Design*. Reading, MA: Addison Wesley.

White, A. (1984). Bakhtin, Sociolinguistics and Deconstruction. In: F. Gloversmith (Ed.) *The Theory of Reading*. (pp. 123–145). Totowa, NJ: Barnes and Noble.

THE NEGATIVE TELEOLOGICAL METAPHYSICS OF HANS JONAS

Ton van der Valk

Twenty years after 'The Imperative of Responsibility',[1] five years after his death, we may ask what the lasting contribution of Hans Jonas and his philosophy will be to the ethics of technology in the twenty-first century and after. However, to answer the question of its merits, a comprehensive interpretation of his work is required. This task is complicated by the fact that Jonas spent a lifetime writing only essays. *The Imperative of Responsibility*, especially, needs to be viewed in connection with the previous volumes *The Phenomenon of Life* and *Philosophical Essays*, to understand and appreciate its bridging function between metaphysics and ethics – that is, not only between 'is' and 'ought', but also between the foundations and the applications of responsibility.

In this chapter, Jonas's work will be presented as a philosophical system characterized by a *negative teleology*, both with regard to its metaphysical assumptions and its ethical implications. The keyword, 'teleology', must be understood in the Aristotelian sense, which is not synonymous with 'consequentialism'. But in contrast with Greek teleology, and in accordance with a Jewish-Christian perspective, Jonas would describe the human telos not as 'happiness' but as 'responsibility'. And where Aristotle felt quite able to show the road to happiness, Jonas strongly rejects every attempt to fix the human essence, or, as he prefers to say, "the image of man." We can easily recognize a threat to the human image, but we cannot and should not try to define the greatest good for man. This is why his teleology needs to be conceived in a negative sense.

Metaphysics, Epistemology, and Technology, Volume, 19, pages 253–279.

ISBN: 0–7623–0681–5

With regard to the second keyword, ‘negative’, some comparisons may be made between Jonas’s negative teleology and the tradition of negative theology. But in the first place, Jonas’s system is not a theology, since he deliberately wants to develop metaphysics in the language of natural science. Secondly, in his few explicitly theological essays, he insists on the at least partial intelligibility of God and the order of creation. Jonas’s works may in fact be read as an attempt to offer a naturalistic translation of the idea of an ordo creationis, meant to save its moral purport from our declining faith in revealed religion. In this translation the notion of God is exchanged for that of freedom which, on the one hand, is objectively visible in nature but, on the other, can be defined only negatively, as lack or want of freedom – or through the concept of mortality. But we must not focus too much on the apologetic meaning of such a ‘translation’. In his search for a foundation of ethics Jonas purposively votes for a metaphysics able to stand on its own rational feet. Therefore its ethical implications must be taken as independent of any religious traditions.

The first part of this chapter (sections 1–3) will be devoted to the interpretation of his work, the second (sections 4–6) to the evaluation of its merits as an ethics of technology.

1. COLLECTED ESSAYS

Great philosophers have always watched over the unity of form and content in their works. For Jonas too, the essay may have served a special, presumably critical purpose, but this conceals the larger coherence of his thought. After his pre-world war work on ancient Gnosticism,[2] which, although important as a background, should not be regarded as part of his philosophy, but as a contribution to the history of religions and religious thought, he published four volumes of essays.[3] *The Imperative of Responsibility* appears to form an exception to this practice, but on closer inspection this book too appears composed of material from several separately published essays.[4] Although according to the introduction to the original German edition, Jonas intended to present an extensive, compelling argument (Jonas, 1979a, pp. 9–10), the attempted integration has not been fully elaborated.

Instead of summarizing *The Imperative of Responsibility*, I will review the different essays out of which the book has been composed, identify their separate messages, and indicate in parentheses the places where they have been inserted in the book. Clearly the material has sometimes been altered and supplemented, but the stories have hardly been changed. Jonas’s thinking shows a major coherence over time.[5]

(1) In 'Technology and Responsibility: Reflections on the New Tasks of Ethics'[6] (chapter 1), Jonas argues that new possibilities and long-term consequences of modern technology call for a revision of ethical theory, preferably incorporating some source of wisdom and respect.

(2) In 'Responsibility Today: The Ethics of an Endangered Future'[7] (divided over chapters five, two, and six),[8] he introduces his 'heuristics of fear', issuing in the practical guideline to give prevalence to the bad over the good prognosis.

(3) In 'The Concept of Responsibility: An Inquiry into the Foundations of an Ethics for our Age'[9] (chapter 4),[10] he pleads for a forward-looking, caring kind of responsibility, in which the obligating authority coincides with the object of responsibility.

(4) In 'Reflections on Technology, Progress and Utopia' (chapters 5 and 6),[11] he argues against the Marxist utopia that there is neither a perfection in the image of the human, nor any kind of objective progress in the history of humankind (on which an ethics for the future could draw).

(5) Finally, the essay 'Socio-Economic Knowledge and Ignorance of Goals'[12] may be mentioned as a preliminary study for chapter 3, including the last part of the preceding and the first part of the succeeding chapters (Jonas, 1984, pp. 44–86), spanning the full metaphysical foundation of *The Imperative of Responsibility*. As Jonas concedes himself, it may even be regarded the native ground of the entire book.[13] In 'Socio-Economic Knowledge and Ignorance of Goals' Jonas argues that the study of economics is not indifferent to its object, as is the case in the natural sciences, but carries an immanent normativity comparable to the study of medicine or law. In chapter 3 of *The Imperative of Responsibility* Jonas stretches this argument further in an attempt to convince the behavioral scientist and the Darwinian biologist of the same point, which adds the major philosophical problem that in these two cases, normativity and teleology, even if the latter could be established beyond doubt, no longer necessarily coincide.[14]

As we can now see, *The Imperative of Responsibility* is more like a combination than an integration of separate essays. The following questions remain unanswered:

(a) Why does humankind have to exist?[15] Even granting Jonas's contention that the presence of purpose in the universe is better than its absence (Jonas, 1984, p. 80), still the conclusion does not follow "that there be a mankind" (Jonas, 1984, p. 43).

(b) How does humankind have to exist?[16] Even if we accept the imperative character of humankind's existence, we still do not know anything about its better or worse existence. The heuristics of fear may provide a clue here, but fear is very subjective, and Jonas's ethics was meant to be objective.

(c) How can Jonas deduce from the highly general imperative, "that there be a mankind," his very specific applied ethical judgements on brain death and cloning human beings?[17]

(d) If there is teleology in nature, then why not in humankind and history? Or the other way around: if humans, as rational beings, are truly free and independent, why should they obey any naturalistic ethics?[18] This point is important with regard to Jonas's criticism of progress and Utopia.

(e) Why not simply update subjectivist ethics?[19] In the first chapter of *The Imperative of Responsibility* Jonas should have, but has not sufficiently explained why only his 'restoring [of] the category of the sacred' (1984, p. 23) could solve the ethical aporias of modern technology.

As it turns out, most questions do not concern the foundations, but are related to the applications of Jonas's ethics. The kind of responsibility Jonas has founded seems too abstract and general.

This abstract character of Jonas's "responsibility for the future" may be due to the fact that his efforts were directed in the first place at providing new foundations for (technological) ethics. In the epilogue of 'The Phenomenon of Life' Jonas declares that his 'philosophical biology' already contains an implicit ethical theory, based on a unified conception of nature (1982, p. 282). It only needs to be explicated. In the prologue to *Philosophical Essays* this intention is invoked again (Jonas, 1974, p. xvii). This volume already bears testimony of Jonas's turn toward ethics, but still in the form of separate essays. The task of putting together the pieces is postponed to *The Imperative of Responsibility*.

Jonas is well aware that founding an ethical theory would call also for applications. However, the metaphysical foundation, bridging 'is' and 'ought', consumes all his attention. In the original German version of *The Imperative of Responsibility* Jonas promises the follow-up of an 'applied part'.[20] The subsequent volume of essays, *Technik, Medizin und Ethik*, published only in German, carries the subtitle 'Zur Praxis des Prinzips Verantwortung' [Applying the Imperative of Responsibility], but in fact does not contain even a single reference back to the metaphysical foundations established in the preceding book. This omission was implicitly acknowledged in the English edition of *The Imperative of Responsibility* by skipping all (but one[21]) references promising an applied sequel. It seems best to regard *Technik, Medizin und Ethik* as the German translation of the ethical part of *Philosophical Essays*. In this case the 'applied part' was in fact already written, which confirms the task of *The Imperative of Responsibility*: to knit together two domains of Jonas's thought which were already present. Following this lead we may find much more coherence and comprehensiveness in Jonas's thinking than when reading *The Imperative of Responsibility* separately.

2. FREEDOM IN NATURE

Although no interpretation of Jonas's works may disregard his lasting predilection for the essay, it is tempting to present him as one of the latest great philosophical system-builders. His philosophical counterparts are Descartes, Spinoza, Leibniz, Kant, and Hegel. Jonas's mission is to reconcile the irreconcilable: he endeavors to found a deontological ethics on top of a teleological metaphysics. With Kant he shares the conviction that human freedom is the alpha and omega of morality. Every definition of the human telos would be a denial of the intrinsic value of the individual; the categorical imperative would change into a hypothetical, and humanity's autonomy would slide back into heteronomy.

But Jonas shares with Aristotle the conviction that value cannot be conceived of in a dualistic way, independently and over against nature.

> The objectivity must really stem from the object. Thus all proofs of validity for moral prescriptions are ultimately reduced to obtaining evidence of an 'ontological' ought (Jonas, 1984, p. 130).

In short, Jonas needs some kind of natural purpose to found objective value in nature, but at the same time such a purpose of nature would be his greatest enemy because it would contradict human freedom and responsibility and turn all intrinsic value into instrumental value. As we have seen, Jonas has solved this apparent contradiction by attributing objective value only to general purposiveness, that is, not to any particular purpose, but to living nature in general. Each particular organism owes its intrinsic value to its quality of being a purposive, i.e. living, being (Jonas, 1984, pp. 77, 80).

Jonas is thus on the verge of committing a naturalistic fallacy. But this is what makes his ethics interesting. He deliberately challenges modern, subjectivist ethics by crossing the boundary between 'is' and 'ought' (Jonas, 1984, pp. 44–45). His arguments against subjectivism are that the logical account of the naturalistic fallacy hinges on a 'value-free' concept of nature that has been neutralized beforehand, and that every thinkable moral theory will inevitably have its metaphysical presuppositions. The problem is how to select the least unattractive ones. However, anyone who expected a clear case of naturalism would come away disappointed. If recognizing objective value in general purposiveness is regarded as a naturalistic fallacy, then, at least, one has to concede, that Jonas is not (yet) pushing some kind of totalitarian interpretation of "the natural purpose of history." Or, put differently, at *le moment suprême*, Jonas avoids the naturalistic fallacy by not concluding anything specific. In fact, when the critique of naturalism is motivated by fear of totalitarianism, then Jonas can be counted in the first ranks of critics. What is the verdict on the naturalistic fallacy then, but a different formulation of: "We must simply not try to fixate humanity in any image of our own definition"? (Jonas, 1974, p. 181).

The concept of a negative teleology,[22] in which the telos is deliberately not specified and only general natural purposiveness is pronounced to be good, will enable us to discover the unity of Jonas's philosophical system.

At first sight *The Imperative of Responsibility* gives the impression of a far too general kind of "emergency ethics of the endangered future" (Jonas, 1984, p. 140), which could at best support a moral obligation to keep some parts of this planet inhabitable.[23] In one of his last essays, 'Matter, Mind and Creation', Jonas merely reinforces this abstract impression by seriously entertaining the question of whether the presence of some kind of Martian morality elsewhere in the universe might make any difference to our responsibility for rescuing planet earth (Jonas, 1996, pp. 196–197; it would not). True, it is a thorny question for anyone who wants to build human morality on the goodness of natural purposiveness; Jonas will have to face the fact that natural purposiveness will dwell on and on, and that this particular humankind is just a stage in the infinite flux of becoming.[24] Nevertheless such questions remain rather abstract.

However, revisiting Jonas's metaphysics after an inquiry into his ethics, we sense a more substantial reason behind the restriction of his attribution of objective goodness of being to general purposiveness only: this generality follows from the negative approach required by his concept of intrinsic value. The intrinsic value of all forms of life stems from the freedom of all striving nature which unfolds itself in an encompassing evolution, and in each and every organism; this freedom is overwhelming and unfathomable. Any description of this ultimate freedom would be inadequate, and therefore diminish, limit, and injure it. Thus the teleology of nature admits of a negative revelation only; that is, one may recognize the inadequacy of any description of this freedom without being able to offer a better description.

When viewed as parts of a deliberately negative teleology, Jonas's ethics and metaphysics show remarkable consistency. The critical questions mentioned in section 1 may now be addressed.

(a) Why humanity has to exist if general purposiveness is good becomes a tautological question when the concept of purposiveness turns out to be derived from the model of human freedom. In the essays 'Is God a Mathematician?'

and 'To Move and to Feel: On the Animal Soul', both contained in *The Phenomenon of Life*, Jonas characterizes the direction of evolution as an increase in freedom (Jonas, 1982, pp. 80–86, and, 192ff). In his last essay, 'The Burden and Blessing of Mortality', he claims that every gain of freedom affirms itself, in spite of the fact that the other side of the coin is an ever increasing "fear of death." For instance, the faculty of sight is not merely a means in the service of the survival of the organism, but becomes part of the value that the organism attaches to its own life, so its fear is increased by a fear of going blind. "Even the sickest of us, if he wants to live on at all, wants to do so thinking and sensing, not merely digesting" (Jonas, 1996, p. 93). So the intrinsic value of life, or at least the phenomenon of subjective purposiveness, intensifies during the course of evolution, culminating in humanity's self-conscious freedom and morality.[25] Nevertheless humanity's existence may have resulted from 'blind chance' (Jonas, 1984, p. 100). The hierarchy of evolution is due to mere self-affirmation of being in its ever more sophisticated resistance to mortality. This does not yet found the undeniable value of "this mortal coil,"[26] but it does establish, as requested, a moral hierarchy among the species. We have a right to eat the pig, because we are more afraid to die.

(b) The general object of responsibility consists of everything intrinsically valuable, which by definition is equal to all transient or vulnerable things (Jonas, 1984, pp. 125–126). At the same time, Jonas admits that inevitably, "each purpose can be realized only at the cost of other purposes" (1984, p. 81). Regarding the resulting question of *how* we should be responsible, the general formulation of Jonas's answer is: "Only the idea of Man, by telling us *why* there should be men, tells us also *how* they should be" (1984, p. 43). Or put negatively: "No condition of future descendants of humankind should be permitted to arise which contradicts the reason why the existence of mankind is mandatory at all" (1984, p. 43). When humanity is conceived of as the only potential bearer of responsibility (1984, pp. 98–99), and responsibility is regarded as the most explicit form of natural purposiveness, then humanity must be in order for responsibility to continue to be:

> The *presence of man in the world* ... has itself become an *object* of obligation: the obligation namely to ensure the very premise of all obligation: that is, the *foothold* for a moral universe in the physical world – the existence of mere *candidates* for a moral order (Jonas, 1984, p. 10).

Let this be 'why' humanity has to be; still it does not tell us 'how' we should exercise our responsibility.

The briefest summary of Jonas's ethics is that we are responsible for responsibility, but that the object of responsibility (freedom in nature) is such that it cannot be described. All Jonas can say is that in his conception, responsibility is total, continuous, and future-oriented (1984, pp. 101ff). He opposes 'formal' to 'substantive' (1984, pp. 90–92), respectively, 'contractual' to 'natural' (1984, p. 94) or 'reciprocal' to 'non-reciprocal' (1984, p. 94) responsibility, in order to make clear that true responsibility will always remain open and *unlimited* (1984, pp. 101ff). Every definition of responsibility will have to be regarded as a breach of responsibility: every specification will fence out something, every gradation would entail a degradation. Just as any description of ultimate freedom would diminish freedom, Jonas's ultimate conception of responsibility remains somehow unmentionable. The prohibition of any definition of responsibility itself offers the only clear prescription of Jonas's ethics. Therefore this prohibition will have to serve as the foothold for the elaboration of Jonas's entire applied ethics.

Accordingly, Jonas tries in many different ways to gain "a knowledge wrested from ignorance" (1984, p. 118). So, instead of specifying any particular object, he can only say that the object of responsibility is the continuation of responsibility itself, which demands our protection of the prerequisites of the existence of responsibility in the world, of which one is the presence of the human species, sufficiently fed and educated to give responsibility a chance. "Put epigrammatically: the possibility of there being responsibility in the world, which is bound to the existence of men, is of all objects of responsibility the first" (1984, p. 99). In truth, however, the entire argument remains strictly formal.

Parents and statesmen must pass their burden of responsibility on to the next generation (1984, pp. 107 and 118), but again Jonas cannot express any pedagogical priorities. Nevertheless he explicitly defends himself against the imputation of normative emptiness when he states the general principle that the statesman is also always responsible for preserving the possibility of responsible statesmanship in the future (1984, p. 118). In contrast with the Kantian demand of logical self-consistency, Jonas appeals to an empirical self-consistency in morality, which according to him is by no means empty. Clearly, one may assume the prevention of famines and armed conflicts and the advancement of education to be in line with a care for the preconditions of responsibility. However, such an empirical self-consistency of morality will always presuppose some knowledge of what 'responsible statesmanship' or responsibility in general is about, which itself cannot be merely empirical, since it must function as a critique of empirical reality.

(c) Nevertheless, reasonably convincing and specific prescriptive answers to the question of how to be responsible are produced by the negative procedure of the 'heuristics of fear' (1984, pp. 26–27 and 202–203). We do not need to be suspicious of a deduction of highly specific prescriptions, concerning for instance the cloning of human beings, from highly general premises, such as the goodness of general purposiveness, because no such deduction takes place. Instead, Jonas simply presupposes a shared and common knowledge of evil: "We are not unsure about evil when it comes our way, but of the good we become sure only via the experience of its opposite" (1984, p. 27). This substantial presupposition is not in contradiction with the former formalistic approach. We cannot positively describe the object of responsibility just as we cannot describe human freedom. But knowing that human freedom cannot be described implies at least some knowledge of human freedom – at least enough to be able to recognize the inadequacy of past and future coups d'etat to fixate freedom.

Therefore Jonas can say that threat will enlighten the true image of humanity (1984, p. 26). We must "keep open the horizon of *possibilities*" (1984, pp. 139–140). The critique of Utopia (1984, pp. 186–201) and especially the case-study on eugenics (1974, pp. 141–167) may be regarded as applications of this imperative. In both cases Jonas makes a distinction between fighting illness and misery and creating better people in a better world (1984, p. 172; 1974, p. 152). The first, negative, intention leaves the model of humanity implicit and therefore open; the second intention is based on a notion of humanity's perfection, which evidently fixates the human being in an image of our own definition. Problems will arise, however, in drawing the dividing line between fighting poverty, misery, and illness, and pursuing 'melioristic' progress. In the last section we will return to this question of application.

(d) Instead of necessarily having to choose between either freedom or teleology Jonas wants to have both at the same time. He feels confident of criticizing both Marxism, for introducing (positive) teleology into history,[27] and Existentialism, for its dualistic premises (Jonas, 1984, pp. 87–88; Jonas, 1982, pp. 211ff). His negative teleology is an explicit attempt to give a monistic account of freedom. On the one hand, Jonas rejects dualism by referring to the common experience that there is matter without mind, but no mind without matter (1984, pp. 67 and 209). On the other, he evades final determinism by describing the purpose of nature only as the self-affirmation of being through negated non-being (1984, p. 82). (Jonas is not afraid of causal determinism; the existence of mind, freedom, or subjectivity forms his initial datum. See Jonas, 1984, pp. 205ff.) It is interesting to notice Jonas accusing Ernst Bloch of fostering "a hidden teleology" (1984, p. 199), and at the same time naming the newborn child a 'teleological promise' (1984, p. 134). The difference is subtle: the former kind of teleology, in spite of being hidden, seems to be positively determined; the latter only negatively, e.g. in terms of being 'sound and healthy'.

(e) This question concerns judging the merits of Jonas's conception of responsibility, which we shall do in the last two sections.

3. A SECULAR EQUIVALENT OF CREATION

Jonas never called his philosophical system a negative teleology; he never even called his collected essays a philosophical system at all, but we will do so now. Several times Jonas suggests that he is looking for a secular equivalent to "the order of creation." In his introduction to *Morality and Mortality*, Lawrence Vogel, too, defends the thesis that Jonas's metaphysics should be read as a natural theology (Jonas, 1996, pp. 20–21). The doctrine of the biblical God who, in Jonas's own résumé, created heaven and earth, who saw that his creation was good, who created the human being in his image, and makes known to humans what is good because his word is inscribed in our hearts (1974, p. 177), offers a model for interpreting both Jonas's metaphysics and his ethics. In 'Contemporary Problems in Ethics from a Jewish Perspective', Jonas says of these biblical propositions, that they

> are of course in no way 'refuted' by anything science has found out about the world and ourselves .... [S]ome equivalent of their meaning, however remote of the literalness of their statement, must be preserved if we are still to be Jews and, beyond that special concern of ours, if there is still to be an answer to the moral quest of man (1974, p. 177).

In the rest of the essay Jonas explains, as we have seen in section 3, what the ethics of technology could learn from the doctrine of creation, with regard both to nature and to humanity.

How the desired metaphysical alternative was to be conceived becomes revealed in the epilogue to *The Phenomenon of Life*. Here again we find an explicit reference to creationism:

> Ontology as the ground of ethics was the original tenet of philosophy. Their divorce, which is the divorce of the 'objective' and 'subjective' realms, is the modern destiny. Their reunion can be effected, if at all, only from the 'objective' end, that is to say, through a revision of the idea of nature. And it is becoming rather than abiding nature which would hold out any such promise. From the immanent direction of its total evolution there may be elicited a destination of man by whose terms the person, in the act of fulfilling himself, would at the same time realize a concern of universal substance. Hence would result a principle of ethics which is ultimately grounded neither in the autonomy of the self nor in the needs of the community, but in an objective assignment by the nature of things (what theology used to call the *ordo creationis*) – such as could still be kept faith with by the last of a dying mankind in his final solitude (Jonas, 1982, p. 283).

Apart from the essay 'Contemporary Problems', which is exceptional in its exclusive address to a Jewish audience, Jonas always carefully refrains from appealing to religious authority. Even though, as we will see below, in his view faith and reason are still compatible, he does not want to create the wrong impression. For his primary concern is not the authority of religion, but the reverence for striving nature. The apologetic significance of his work is a by-product of his revision of the idea of nature. Only in much later work does he lose his hesitation to name 'creation' the final object of our responsibility: "At stake is the preservation of Earth's entire miracle of creation" (1996, p. 112). The last word of the last sentence of his last public address summarizes what it was all about:

> The latest revelation, from no Mount Sinai, no Mount of the Sermon and no Bo-tree of Buddha, is the outcry of mute things themselves that we must heed in curbing our powers over creation, lest we perish together on the wasteland of what was creation (1996, pp. 201–202).

The promised principle of ethics grounded "in an objective assignment by the nature of things" is provided for in *The Imperative of Responsibility.* Although Jonas repeats "that religious belief has answers here which philosophy must still seek" (1984, p. 45), at the same time he makes clear that he does not want to base his fight against the ecological crisis and the technological threat to the image of the human on mere religious conviction:

> . . . e.g. from the "order of creation," faith can argue that according to God's will men should be there in his image and that the whole order should exist inviolate . . . . Faith in revealed truth thus can very well supply the foundation for ethics, but it is not there on command, and not even the strongest argument of need permits resorting to a faith that is absent or discredited. Metaphysics on the other hand has always been a business of reason, and reason can be set to work on demand (1984, p. 45).

To reach a general public, Jonas wants 'secular reason' (1984, p. x) to take over the teachings of revealed religion. If we can regard this as the final summary of Jonas's intentions, this raises two central evaluative questions. First, is his ethics really secular or, in other words, does he really reach a general public? And second, does his translation of creationism into a negative teleology result in an effective ethics of technology? The first question will be dealt with here; the second must wait until the next section.

In a nutshell, Jonas's secular equivalent is rational, but not compelling. What Jonas means when he talks about 'secular reason' becomes clear from his scholastic dispute with Duns Scotus (Jonas, 1984, pp. 46ff). According to Scotus's theological voluntarism the world is good because God wanted it. However, according to theological rationalism as represented by Jonas, God willed the world because it was good (Jonas, 1984, p. 48). Both theological positions assume God to be good. But Duns Scotus primarily stresses God's free will (Jonas, 1974, p. 42). The entire creation is dependent on God, so God could never be bound by any laws of nature. Therefore God is unpredictable and incomprehensible. Revelation becomes the sole source of ethical knowledge. This implies an ethics in which faith and good intentions come first and autonomous judgment is no longer possible. "Theological voluntarism runs the risk of lapsing into Gnostic dualism," says Vogel in the name of Jonas (Jonas, 1996, p. 20). In voluntarism Jonas perceives the same subjectivism which he fears and despises so much in modern ethical theories; in fact the former prepared the ground for the latter, according to Jonas, so Duns Scotus exactly marks the spot where the history of philosophical ethics went wrong (Jonas, 1974, p. 43).

In his theological essay 'The Concept of God after Auschwitz: A Jewish Voice', included in *Morality and Mortality*, Jonas works out his alternative. To solve the theodicy trilemma, in which one has to choose, given the presence of evil, between God's power, intelligibility, and goodness, he is ready to sacrifice God's power, but not, as Duns Scotus does, his intelligibility (Jonas, 1996, pp. 139–140). In another theological essay, 'Matter, Mind and Creation', Jonas adds that God's limitation in power makes us not only free, but at the same time burdens us with the responsibility for the success or failure of the divine enterprise (1996, p. 191; quoted below). So at the same time Jonas renders the meaning of life intelligible for believers and non-believers alike, and takes away the careless excuse of good intentions.

God's powerlessness, however, does not yet suffice for a general amnesty for all evil, because after all, God still decided (powerfully) to abdicate his power in the act of creation. So Jonas still has to explain intelligibly the goodness of life to justify its creation. As we have seen, Jonas cannot realize this without recourse to an "evidential intuition" of the goodness of general purposiveness (1984, p. 80). This evidence, however, cannot be expected to be generally convincing. Most of the great religions have esteemed post-mortal heavenly existence far above this earthly coil. Both life in the animal kingdom and among humans may be designated, in Hobbes's terms, as "brutish, nasty and short." In fact the counter-arguments are much more generally intelligible than the proposed evidential intuition. Nevertheless, in 'The Burden and Blessing of Mortality' Jonas presents a reasonably convincing answer to the problem of natural evil (earthquakes, lions, and diseases). Fear is the other side of the coin of freedom, and the death of the individual has to be regarded as a necessary condition for the eternal resurrection of the species (Jonas, 1996, pp. 94ff). Regrettably, however, the problem of human evil (greed, murder, war) remains unanswered, as Vittorio Hösle also notes.[28]

The term 'secular' is used by Jonas as a tiny pun, for 'secularization' to us moderns mostly means accepting the premises of reductionistic materialism, whereas he, quite on the contrary, intends to "restore the category of the sacred" (Jonas, 1984, p. 23), be it with modern means. His appeal to theological rationalism, however, need come as no surprise, because modernity simply has nothing intermediate to offer between faith and materialism. Theological rationalism solves this either/or problem because it serves two purposes: the same arguments are open to believers and nonbelievers alike. This leads Vogel to ask what the special benefit of believing any longer then will be (Jonas, 1996, p. 24). According to Jonas there may be consolation for the desperate and encouragement for the righteous in the myth of a God who will remember all forgotten victims in his great 'Book of Life' and whose personal 'Image' is at stake (1982, pp. 271ff).

I will not further deliberate on the theological merits of Jonas's metaphysics, but restrict myself to the question of what he has to offer to the general public. As the following quotation may show, Jonas literally translates a religious message into naturalistic language. In 'Matter, Mind and Creation' he writes:

> [T]he destiny of the divine adventure is placed in our unsteady hands in this earthly corner of the universe and . . . the responsibility for it rests on our shoulders. So the deity, I imagine, must become anxious about His own cause. There is no doubt that we have the power in our hands to thwart the purpose of creation (1996, p. 191).

Compare this with the following lines from *The Imperative of Responsibility*:

> In man, nature has disturbed herself and has only in his moral endowment (which we may still ascribe to her) left herself an unsure substitute for the shattered sureness of her self-regulation. It is a terrifying thought that on this fickle ground her cause should now rest — or let us say more modestly: so much of what man can see of her cause (1984, p. 138).

Is such a translation acceptable? Will it reach a more general public? Or will the audience remain the same or even shrink only further?

Several times, Jonas has compared his own metaphysical foundation of ethics with the ontological argument for the existence of God (1984, pp. 43 and 130; 1996, p. 107; 1991, p. 32). But instead of regarding God's existence as a logically necessary component of his essence, Jonas wants to derive the obligating essence of responsibility from the contingent fact of its existence. This reversed order conforms to his anti-Platonic preference to start from "becoming, rather than abiding nature" (1982, p. 283). According to Jonas "we must seek the essential in transience itself .... Only for that which is changeable and perishable can one be responsible" (1984, p. 125). Therefore, the archetype of Jonas's view of objective responsibility, where the object most clearly coincides with the commanding instance of responsibility, is the newborn infant (1984, p. 134). The responsible care of the parents for their baby is the most explicit form of the struggle for life which, looking backward in evolution, is already present in every kind of metabolism. Both in his understanding and in his valuation of organic nature, Jonas uses the higher forms of life as a model that can teach us about lower forms of life (1984, p. 69). Of course, this choice implies an anthropomorphistic shortcut to evidence which yet would have to be demonstrated. According to Jonas, one has to choose between two different ontologies: either the self-denying reductionism of the modern sciences, or some kind of – however circular – anthropocentrism. Finally, he holds on to the conviction that the second option is the most reasonable (1996, p. 108).

Totaling up accounts, we must say that Jonas's monistic account of freedom in nature is certainly attractive. However, as an ontological foundation for ethics it remains ambivalent. On the one hand, there is the phenomenological evidence of duty in the squealing appeal of the newborn infant; on the other, there is the evidence of violence and terror which is no less real for the victims. If human beings are to be taken seriously in their responsibilities, then they have to be taken seriously in their hatred and negligence too. Repeatedly, Jonas refuses to draw up "a balance sheet of mankind's performance on earth" (1984, pp. 49 and 99; 1996, pp. 93–94). To avoid such a "hedonistic calculus" he has recourse to the phenomenon of duty, which is simply incomparable with happiness (1984, pp. 99). However, when Jonas declares this duty to be intelligible from the order of nature, the unpleasant balancing of human good and evil inevitably returns. The reality of brute violence may still spoil our readiness to accept general purposiveness as a good in itself.

Jonas is determined to reestablish objective discourse on moral values and their metaphysical foundations. As he himself admits many times, however, his own argumentation has no compelling force (1984, pp. x, 45, 80, and 132; 1996, p. 108). It may persuade the reader, and that is precisely what it is meant to do. But one may reply: What difference does it make when "objective metaphysics" remains a matter of choice anyway? After all, arguments of mere taste or prudence could persuade as well. Indeed Jonas's objective moral approach offers no guarantee for reaching a significantly wider general public. It ought to be judged as an attempt to point out a third way between dualism and nihilism. In meta-ethical respects, Jonas has to teach a lesson in modesty to those adhering to exaggerated faith or authenticity, on the one hand, and those expecting incontestable scientific certainty, on the other.

At least as interesting as the acceptability of Jonas's axiomatic moral foundations, however, is the persuasive force of his deduction of practical moral guidelines from his first premises. Presumably, the more general the foundations, the less compelling the applications will be. We have discovered

Jonas's metaphysics to be very general indeed. We will now have a look at the applicability of his ethics of technology.

4. TECHNOLOGICAL DETERMINISM AND THE ETHICAL VACUUM

Before deciding on the merits of Jonas's negative teleology, we have to determine how Jonas conceives of the application of his ethics. In the ethics of technology two more or less opposite approaches can be found, which I will refer to as applied ethics and cultural criticism, respectively. The difference rests with the issue of technological determinism. Technological determinism is by definition incompatible with applied ethics. According to cultural criticism, morality itself is so much influenced by the conditions of modernity that it can no longer take an independent stance toward modern society. Therefore, diagnosing the cultural determinants of technological development becomes the first task of technological ethics. A radical change of mind, accompanied by a social or scientific revolution, is presumed to be needed to restore a more authentic morality.[29] So, generally speaking, cultural criticism may accuse applied ethics of legitimating technological development instead of questioning it. Representatives of applied ethics reply that the critical stance toward modern culture must be very attractive to conservative minorities who can no longer impose their values on others.

Jonas's position is ambivalent. His medical ethics case studies on experiments with human subjects and brain death present themselves clearly as contributions in applied ethics. At the same time, his contention that modern value subjectivism has created an 'ethical vacuum' (1984, p. 22) betrays a severely critical view of modern society. It is not immediately clear whether, in practical respects, *The Imperative of Responsibility* is intended as a critique, meant to heal society from its subjectivist blindness, or just as a preliminary foundation for a set of values that could be used as arguments in a democratic discussion. The crucial question is this: Which is primary in Jonas's ethics, the *meta-ethical objectivity* or the *normative ethical content* of his kind of responsibility?

In interviews, Jonas had to defend himself over and over again concerning his flirtations with the idea of a communist authoritarian government (Jonas, 1984, pp. 142–151).[30] He judges representative democracy to be much too preoccupied with short term concerns to take care of the distant future (1984, pp. 146 and 151). To ensure our physical survival even "a temporary absence of freedom" (1996, p. 112), concerning both the freedom of scientific research and some 'democratic freedoms' (1996, p. 111), may have to be considered.

For a while a "well-intentioned, well-informed tyranny" (1984, p. 147) may have to take over, because "only a maximum of politically imposed social discipline can ensure the subordination of present advantages to the long-term exigencies of the future" (1984, p. 142). It is fully clear that Jonas meant this provocation as a warning, but this still cannot take away the impression that his conception of the realization of his ethics is rather Bolshevist.

Jonas's lack of faith in democracy, however, does not yet make his theory a clear instance of cultural criticism, because not all applied ethics is necessarily liberal. More decisive will be his explanation of the relative autonomy of modern technological development. Jonas is certainly not a technological voluntarist; his entire theory is based on the idea that technological development is out of control in its large-scale and long-term effects, and that we now need a large-scale and long-term ethics to match it (1984, pp. ix–x, 6–8, 23, and 118). If we were able to control the consequences of technology from day to day, no future ethics would be required at all. But clearly, Jonas thinks this not to be the case. However, in *The Imperative of Responsibility* he does not offer a satisfying theory of technological development. In his warnings against the cumulative power of technologies, he merely suggests, or even just proclaims, that modern technology has become autonomous (1984, pp. 21, 32, 140–142, and 203). Moreover, the examples he gives point in different directions.

Consider, for instance, his rejection of the criterion of brain death[31] and his constant warnings against overpopulation (1984, p. 141; 1985, pp. 50 and 155ff). The first example fits into a plain cultural critical, deterministic theory of technological development. Heidegger could have used it to demonstrate our instrumentalist mode of encountering reality. The roots of determinism in this case are to be found in utilitarian value subjectivism: people have no respect anymore even for the sick and dying. Woe to us (1985, pp. 238–239)! The second example belongs to a more modestly deterministic theory, according to which doing good now can lead to bad outcomes later, because of our growing dependency on technology.

In his description of the "apocalypse of the 'too much'" (1984, p. 202), Jonas reveals that he is more afraid of the slowly creeping danger of overpopulation than of a sudden extinction by the atomic bomb. The latter will still be a matter of human action and decision, whereas the former is a matter of macro-economic behavior where no one can decide to stop things anymore (1984, p. 202; 1985, pp. 10–11 and 48ff). The deterministic causes of the apocalypse of the too much are objective, natural factors, like resources and demographic curves. This implies that we could do something about it, provided that we would act on a global scale with special regard to future secondary effects. So the issue of overpopulation is still compatible with some kind of planetary applied ethics, whereas an abolition of the brain death criterion will require a radical change of mind first. In the first case, Jonas only needs to plead for respect for long-term issues; in the second, he needs first to establish a foundation for the very principle of respect.

The diagnosis in the first part of 'Contemporary Problems' is unambiguous: Darwinism, historicism, and psychoanalysis have extended the value-free conception of nature from modern natural science to the human, and now morality is eroded by relativism (Jonas, 1974, p. 170). So, presumably, the philosophy of biology developed in *The Phenomenon of Life*, which made room again for conceiving of value in nature, could be used to repair relativism and grant anew some metaphysical significance to evolution, history, and humanity. In *The Imperative of Responsibility* the diagnosis of relativism as the root of the moral and ecological crisis is left out, possibly in order to reach a more general public, but the message still remains the same: "For the very same movement which put us in possession of the powers that have now to be regulated by norms – the movement of modern knowledge called science – has by a necessary complementarity eroded the foundations from which norms could be derived; it has destroyed the very idea of norm as such" (Jonas, 1984, p. 22). According to this necessary complementarity we might expect that Jonas's re-introduction of teleology into natural science would solve the problem at its roots: a new kind of natural science would result automatically in new kinds of technology. Such a theory would fit nicely into a cultural-critical perspective.

Jonas, however, does not expect a more nature-compatible technology to spring from his re-enchantment of natural science. Instead, he merely uses his philosophy of biology to resolve the apparent contradiction between modern science and the idea of creation (1985, p. 85). He wants to redress the 'crisis' of modern technology, not by changing natural science, but through restoring firm moral boundaries. In the essay 'Toward a Philosophy of Technology"[32] he puts forward his view that technological development is led by the control of materials: from stone to bronze to iron, or from physics to chemistry to biology, or from plants to animals to humans. Technological progress will continue its course unless it is checked by external means. If such a check remains absent, finally "man himself will become an object of his own technology" (Jonas, 1984, p. 17). At least this means that Jonas is less fanatic in his cultural criticism than, notably, his former teacher Heidegger, to whom any claim of an independent morality is idle, as our entire notion of reality is taken to be caught up in our modern-day fate of technological objectivity.[33] Hösle, too, notes that here lies the most important ethical distinction between teacher and pupil.[34]

Jonas would never believe in the recommendation of '*Gelassenheit*'. Instead he calls fatalism "a mortal sin."[35]

Even though Jonas turns out to belong in the moralist camp, still his diagnosis of relativism as the root of technological determinism is not compatible with an open, democratic discussion. A charge of relativism may be correct, but it can never be a moral argument; it will always be *ad hominem*. Further, his theory of science-based technological *inertia*, moving on and on until it is stopped, seems rather outdated after two decades of 'constructivism'.[36] Thus Jonas's theory of technological determinism is not only questionable from a democratic point of view, its concept of technological development is also not very convincing. However, Jonas has presented not only a critique, but a comprehensive normative moral alternative as well. Instead of regarding his ethics as a crusade against nihilism, we may treat it as a defense of a different set of values. In the last sections, therefore, I will focus on the normative content of his negative teleology.

5. THE MERITS OF RESPONSIBILITY

The new questions with which technological development has enriched ethical theory since the 1970s are generally well known: medical technology, future generations, and environmental protection have all posited new challenges. If Jonas's efforts in providing foundations for an ethics of technology really make a difference, these are the cases where it would have to show. In the first chapter of *The Imperative of Responsibility* he has weakened the accusation of relativism, which was prominent in 'Contemporary Problems', to the statement that "all former ethics are not able to meet the requests of modern technology." Now even if this be true, it leaves open the possibility of revising existing theories, apart from the option of discarding all former ethics. So we would have to distinguish the merits of Jonas's theory of responsibility not in comparison with Kant or Mill, but in comparison with their recently upgraded versions put forward by, for instance, Karl-Otto Apel or Joel Feinberg. However, apart from a few brief comments, such a task cannot be undertaken here. I will restrict myself to drawing up an inventory of Jonas's answers to the main ethical aporias of modern technological development.

When we regard the field of medical ethics as being divided between the extremes of patient autonomy and justice, on the one side, and the integrity of the body and the sanctity of life, on the other, then it is clear that Jonas will belong on the last, more conservative side. 'Responsibility' demands first of all a reverence for the image of the human, which is present in every patient. And as said before, every 'definition' will have to be regarded as an insult. This precludes not only positive eugenics, but also cloning and brain death. At any cost Jonas would want to avoid to have to determine explicitly the borderline between worthy and unworthy life.

The field of environmental ethics has been even more obviously divided into anthropocentrism and ecocentrism. By "widening the moral circle," so as to include animals and even plants, some have tried to fill in the gap,[37] but this strategy only conceals the moral dilemma instead of solving it. Jonas seems to have established a more promising synthesis. According to Vogel he would have exploded the very distinction (Jonas, 1996, p. 17). Even though Jonas has invested all living nature with intrinsic value, he retains a moral hierarchy conforming to the stages of evolution, which finally show a rapid increase and intensification of freedom. Humankind's responsibility must be regarded as the self-reflective capstone of this pyramid of freedom. So, finally, Jonas's position would have to be counted as a form of 'stewardship' (Jonas, 1984, pp. 8 and 129–130).[38]

Regarding our obligations toward the future, a variety of theories exist, mostly concerning meta-ethical questions, but the normative question remains twofold: What sacrifices do we have to make for our descendants – how few or many? Jonas, again, occupies a rather extreme position, according to which we are responsible not to future individuals, but to the negative teleological idea of the human, which demands the continuing presence of people on earth to carry on the burden of responsibility (Jonas, 1984, pp. 41–43). What counts first of all, is "to keep open the horizon of possibilities" (1984, p. 140). However, if we are not allowed to limit the future in any way, its demands may grow unlimitedly. Jonas's overexacting ethics may turn into a kind of terror of the future: in order to save humankind from a truly dark ending (see Jonas, 1984, p. 203), he would agree, as we have seen, with a suspension of the freedom of scientific research and of some, not further specified 'democratic liberties'. After all, a tyranny, even when not so benevolent, may be temporary, whereas a global radioactive poisoning could destroy humanity forever (Jonas 1996, pp. 111–112).

Pointing out new dilemmas may be silver, but solving them is gold. One of the foremost reasons not to pay too much attention to any diagnosis of an ethical vacuum standing behind "the crisis of modern technology" is that this tends to ignore the truly problematic character of the aforementioned new moral questions. As far as they all have to do with 'progress', the issue of our obligation to future generations includes much of the other issues too. The question, then, is not whether we ought to preserve the jungle or not, but what to say to the Brazilian poor who want to start a farm. Further, no one denies the self-evidence of the demand that we respect the sick and dying, but what to say to the patient with a severe kidney deficiency who could be helped by an organ donation? The real moral question cannot simply be whether or not humans ought to exist indefinitely. Instead it will always spring from a dilemma: how much are we prepared to offer up to posterity? What measure will mediate between the short and the long term?

The real force and promise of Jonas's ethics of technology lies in the fact that he bridges the recurring split between, to put it simply, the 'subjectivists' and the 'objectivists'. In medical ethics we find autonomy and justice standing over against the integrity of the body and the sanctity of life; in environmental ethics it is anthropocentrism versus ecocentrism; regarding the future, it is justice versus sustainability. As long as these debates remain overshadowed by meta-ethical considerations their outcomes will be predictable. Some mediation is needed to limit the demands of either side and to provide for new answers. Although Jonas most certainly belongs to the objectivist camp, his theory also honors the individualist, traditionally subjectivist principles of autonomy and justice. His ethics is based on two pillars, the 'teleological' one of respect for life and the 'negative' one which lifts the individual out of the species. Jonas wants to combine the Kantian imperative of respect for the human as an end in itself with the Aristotelian sensibility for living nature. Such a two-eyed theory will be indispensable meeting the dilemmas that the twenty-first century will pose ever more vehemently to the ethics of technology.

6. PROGRESS, JUSTICE, AND SUSTAINABILITY: HANS JONAS MEETS THE FUTURE

Jonas's measure to meet the future consists of three useful guidelines, which all follow immediately from his negative teleology. First, his negative concept of progress; second, a general apology for natural evil; and third, a basic egalitarianism inherent in his notion of responsibility.

According to Jonas, as we have seen in section 3, *progress* of society may be realized negatively in the removal of bad circumstances, but not positively in the creation of more optimal circumstances. This view may also be formulated in terms of the distinction of *subsistence* and *existence*. Social subsistence, characterized by poverty and deprivation, admits of an application of the concept of progress, whereas in the case of social existence, which is characterized by sufficiency, this concept becomes meaningless. Beyond the threshold of sufficiency, real progress is no longer possible. Luxury, or 'meliorism' (Jonas, 1984, p. 36; Jonas, 1974, p. 151), cannot count as a form of progress. Two separate societies which are both in a state of 'existence' have to be considered as simply different. There is no measure to decide which one is best. From this distinction follows the rule that progress is morally demanded, whenever it is possible. Or formulated differently, to avoid misunderstandings: fighting subsistence is morally demanded, improving existence is not (Jonas, 1984, pp. 36 and 172).

The big open question which remains, however, is where to draw the line between subsistence and existence, between necessity and luxury? What level of scarcity is tolerable enough to be termed sufficient for genuine humanity? Even with Jonas, feelings of compassion tend to overrule rational morality:

> no skepticism concerning the advent of the good relieves us of the duty to replace bad conditions with better ones to the best of our ability. That which is a moral scandal must be done away with, even if we do not know for sure what in the end we shall get in its place (Jonas, 1984, p. 172).

The point, therefore, is that one could go on settling moral scandals forever. The determination to remove all moral scandals from the world turns out to be no less idealistic than the quest for Utopia. The first, most prudential argument that Jonas brings up against Utopia, its physical unattainability (1984, pp. 186–192), could be put forward against his own views. In other words, in the name of Jonas's negative conception of progress, just as much damage could be done to the environment. So the first guideline alone is yet incapable of accommodating the demands of sustainability.

To add a further point of critique, Jonas's distinction between his own negative ideal of removing moral scandals and the positive ideal of Utopia is in fact at least questionable too. As he admits himself, Marxism is hardly interested in drawing a detailed, positive sketch of communist Utopia; it is already satisfied with a mere proclamation of the advent of the 'new Man' (1984, pp. 16, 156–157, and 179). Even capitalism may legitimize itself negatively by seeing itself as fighting scarcity for the benefit of all instead of seeking abundance for the happy few. In his introduction to the work of Jonas, Vogel writes:

> If nihilism is at the root of our cultural crisis, then the only sufficient response would be a philosophical critique of nihilism. This is precisely the task that Hans Jonas sets for himself (Jonas, 1996, p. 5).

But it is necessary to reply that if nihilism is *not* at the root of our cultural crisis, then a philosophical critique of nihilism is not sufficient. And luckily, Jonas's response is not restricted to a rejection of nihilism and an avowal of modesty. As we have seen in the last section, the examples of brain death and overpopulation pointed in different directions. I repeat that a charge of nihilism does not contribute to clarification of the dilemmas of the future. It is not just the hubris or nihilism of Utopia and capitalism that poses a threat to posterity.

Therefore a plea for humility (Jonas, 1984, p. 22) and modesty (Jonas, 1984, p. 201) will not suffice.

Jonas is also aware of the fact that even doing the good (that is, fighting moral scandals) may have undesirable consequences in the long run. He calls the problem of overpopulation, the problem of having to choose between the needs of the hungry masses and saving scarce resources for the next generations, a "vexing dilemma" (1985, p. 49), and "the axis of an ethics of technology" (1985, p. 50). Thus, in addition to his negative conception of progress, Jonas understands the need to put restraints on what may count as a moral scandal too.

This is where the second guideline comes in. The following passage from the essay 'Medical Experiments with Human Subjects' shows how Jonas complements his rejection of positive progress with a general apology for natural evil:

> The destination of [medical] research is essentially melioristic . . . . Unless the present state is intolerable, the melioristic goal is in a sense gratuitous, and this not only from the vantage-point of the present. Our descendants have a right to be left an unplundered planet; they do not have a right to new miracle cures. We have sinned against them, if by our doing we have destroyed their inheritance – which we are doing at full blast; we have not sinned against them, if by the time they come around arthritis has not yet been conquered (1974, p. 117).

Illness and mortality cannot count as moral scandals which need to be eliminated at all costs. This second guideline is elaborated most explicitly in the essay 'The Burden and Blessing of Mortality'. The looming possibility of death is a continuous threat to all organic existence, but it is a blessing as well, because as a logical counterpart of natality it prevents freedom from ever growing old and dim (Jonas, 1996, pp. 95ff). The finiteness of life does not diminish or taint its intrinsic worth. Mortality is a burden for sure, but not one that we are unable to carry. So if mortality is a constitutive part of life and life, according to Jonas's metaphysics, is good, then mortality cannot be bad. Therefore a general apology for natural evil follows immediately from his negative teleology. Jonas wants to get rid of the kind of dualism that makes our bodily existence appear as an enemy of our spiritual self. (As a matter of fact, no one would maintain to be entitled to a good health.) Natural evil is not at odds with genuine humanity. Nevertheless the overcoming of natural evil through technological inventions represents a form of progress. Not in the positive sense, as simply a matter of luxury, but neither one of strict necessity either. According to Jonas, the successes of modern-day medicine in finding cures for many serious diseases should be regarded foremost as "something akin to grace" (1974, p. 117).

As a second guideline for dealing with dilemmas, a general apology for natural evil is very meaningful to technological ethics, especially concerning medicine. This was Jonas's main argument in the case of 'Medical Experiments with Human Subjects': not taking the demand of informed consent fully seriously means insulting the goal of medical research by the means of achieving it. But the harshest dilemmas of the future still persist. The dilemmas of justice versus sustainability, which will likely be the core problem of the twenty-first century, have not yet been brought closer to a solution. Poverty is not a matter of natural evil. Jonas gets into trouble when he has to choose between saving the rainforest or the children of the poor Brazilian farmer:

> Infanticide is a crime like every murder, but a child's dying of hunger, that is, permitting it starving to death, is a sin against the first and most fundamental of all responsibilities which man can incur (1984, p. 134).

There are, as well, numerous less urgent cases, in which the economic interests (jobs) of one party stand over against a variety of environmental values. As the quotation shows, when the needs of the present are really serious – that is, if they cannot be regarded as meliorism or reluctance to natural evil — Jonas would grant priority to the present over the future after all. Does this mean that the future still is not in safe hands? The fact that Jonas would ultimately put the present first proves at least that it is mistaken to blame him of 'moral terrorism' in the name of the future.[39]

But there is one more guideline which has relevant implications for the future, and that's the issue of egalitarianism. As a third guideline, a notion of equality, applied as a principle of distributive justice, will not solve the theoretical dilemma of justice versus sustainability, but it does draw attention to our present opportunities for countering human misery without infringing on the rights of future generations:

> a socially better, juster, *distribution* of [the] fruits [of our productive capacity] . . . could already with the yield of present technological economy greatly mitigate today's deprivations on our earth, and there is no doubt that the problem is in part not technologic-physical but economic-political' (Jonas, 1984, p. 187).

Although such notions of distributive justice are commonly accepted in our days – and Jonas does not seek an explicit justification for his rule of justice – still it is interesting to contemplate this rule in the context of his negative teleology. We may notice then, that egalitarianism is indigenous to the Jewish-Christian concept of responsibility, in comparison with the Greek concept of virtue. Virtue as a moral concept introduces a measure of excellence, and thereby a measure for comparing people. In opposition to the Greek hierarchy, the Jewish-Christian concept of responsibility is strictly egalitarian. For Jonas "not-being irresponsible" is the only way to explain what it means to be responsible (1984, p. 93). There will be a hierarchy in irresponsibility, varying from slight to grave, but not so on the positive side: directly in accordance with Jonas's negative teleology, there is no excellence in responsibility. When the human 'Good' remains open, there is no measure to distinguish the 'better' either. Someone is either responsible or not. The parents caring for only one child are just as responsible as the statesman who watches over a whole nation, and in that sense they are equals.

As a concluding remark I would like to mention that Jonas succeeds very well in accommodating our common intuitions with regard to both moral objectivity and the normative content of our Jewish–Christian inheritance. Inevitably, philosophy will have to question again and again whether his grounding intuition of the idea of Humanity is justifiable or not. Nevertheless, Jonas's negative teleology offers a most careful and cautious elaboration of this intuition, which for many generations to come will continue to exert its convincing force.

NOTES

1. Jonas, 1979a; English translation: Jonas, 1984.
2. Jonas, 1954; English (shortened edition): Jonas, 1963.
3. Jonas, 1982; German: Jonas, 1973a, 1974, 1985. Jonas, 1992, and perhaps, 1996.
4. See references in Jonas, 1984, p. xi.
5. The only trace of development in his thinking seems to be a slight withdrawal from the once suggested resemblance between modernity and Gnosticism. Concerning this remarkable continuity it must be remembered that by the time he came to New York Jonas was already over fifty years of age and his thoughts had ripened.
6. From Jonas, 1973b; also in Jonas, 1974, ch.1.
7. From Jonas, 1976.
8. Chapter 5. I–II; chapter 2. I–IV.5; chapter 6. III.2. (Jonas, 1984, pp. 136–142, 25–44, and 202–203, respectively.)
9. From Engelhardt & Callahan, eds., 1977, pp. 169–198.
10. Chapter 4, from I.9 onward. (Jonas, 1984, pp. 86ff.)
11. Jonas, 1981, Chapters 5. VI.2 to 6. III.1, except for 6. II.C.1–3. (Jonas, 1984, pp. 162–198 and 200–202.)
12. From Jonas, 1974, pp. 81–104.
13. Cf. Jonas, 1974, p. 95, note 7. The first statements of Jonas's revised categorical imperative can be found here, as well as the examples of 'the hammer', 'the court of law', 'walking', and 'digesting' (Jonas, 1974, p. 86). However, the theory of the existence of objective purpose in nature is not yet fully worked out.
14. With the examples of 'walking' and 'digesting' respectively.
15. This point was raised by Johannes Wendnagel (1990, pp. 24–28) and Richard J. Bernstein (1995).

16. This question has been raised most clearly by Karl-Otto Apel, who demanded more attention to the emancipation of the poor and the deprived (1986).

17. These judgements are established in the essays 'Against the Stream' and 'Biological Engineering: A Preview', both contained in *Philosophical Essays* (Jonas, 1974), and reprinted in *Technik, Medizin, und Ethik: Zur Praxis des Prinzips Verantwortung* (Jonas, 1985). The obscure deduction is criticized by Hans Achterhuis in *De Maat van de Techniek* (Achterhuis, ed., 1992, p. 160).

18. This problem is elaborated upon most extensively by Gilbert Hottois, defending the thesis that technology may be regarded as the next step in evolution (Hottois, 1995).

19. Many authors have tried to answer moral questions regarding modern technology within a subjectivistic meta-ethical framework, with regard to the subject of justice to future generations for instance. Karl-Otto Apel (even in explicit disregard of Jonas' metaphysical approach), Derek Parfit and Joel Feinberg are a few.

20. Jonas, 1979a, pp. 10, 52, 215, and 412 (note 26).

21. One last reference has (coincidentally?) been spared: Jonas, 1979a, p. 52. (Jonas, 1984, p. 21.)

22. Instead of a 'negative teleology', Vogel uses the symmetrically opposite concept of a 'cosmic deontology' (Jonas, 1996, p. 39). He thereby implicitly admits that normally, every deontology must be regarded as 'acosmic'; but he does not elaborate on this.

23. See, for instance, the critique of Richard J. Bernstein (1995, note 16), and of Karl-Otto Apel (1986, note 17).

24. This was one of Gilbert Hottois' main points of critique of (1995, note, 19).

25. Elsewhere Jonas makes mention of a theory of the stages of evolution, which would suggest humanity's drive for knowledge and morality to be the outcome of 'the secret final tendency of natural development' (Jonas, 1985, p. 85).

26. In fact, in 'The Burden and Blessing of Mortality', Jonas does attempt to demonstrate the objective value of life including death. Just as in *The Imperative of Responsibility*, however, he has no more to offer than an 'evidential intuition' (Jonas, 1984, p. 80).

27. The complete critique contains many different points, ranging from feasibility to desirability and respectability.

28. Hösle, 'Ontologie und Ethik bei Hans Jonas', in Böhler, ed., 1994, p. 125.

29. For an overview, see Van der Pot, 1985, 4. Hauptteil, 3. Hauptabschnitt.

30. In almost every interview contained in the special volume 'Dem bösen Ende Näher' (Jonas, 1991).

31. Jonas, 1974, pp. 132–140; Jonas, 1985, pp. 219–242. The German edition is enlarged with two interesting postscripts, in which Jonas admits his loss and puts the blame on 'the technological dictate' (Jonas, 1985, p. 239).

32. In Jonas, 1979b; in German: Jonas, 1985, ch. 1.

33. See 'Die Frage nach der Technik' in Heidegger, 1954.

34. Hösle, 'Ontologie und Ethik bei Hans Jonas', in Böhler, ed., 1994, p. 116.

35. Hans Jonas, 'Fatalismus wäre Todsünde', in Böhler, ed., 1994, pp. 455–456.

36. I want to refer to the undisputed descriptional qualities of (social) constructivism here, not to its moral recommendations (such as Bruno Latour's 'Parliament of Things' – Latour, 1987), which are not at all clear. See also Bijker, Hughes & Pinch, eds., 1987; and Latour, 1987.

37. Strachan Donnelley discusses Jonas's contribution to animal ethics in the context of an overview of the field (1995, pp. 21–29).

38. In the English text Jonas uses also the word 'custodian' (Jonas, 1984, p. 130); in German the term 'Treuhänder'.

39. Hans Jonas in conversation with Walther Ch. Zimmerli, 'Vom Profit zur Ethik und zurück. Technik-Verantwortung im Unternehmen', in Böhler, ed., 1994, pp. 224–243.

REFERENCES

Achterhuis, H. (Ed.) (1992). *De Maat van de Techniek: zes philosophen over Techniek, Günther Anders, Jacques Ellul, Arnold Gehlen, Martin Heidegger, Hans Jonas en Lewis Mumford.* Baarn, The Netherlands: Ambo.

Apel, K.-O. (1986). Verantwortung heute – nur noch Prinzip der Bewahrung und Selbstbeschränkung oder immer noch der Befreiung und Verwirklichung von Humanität? In: T. Meyer & S. Miller (Eds), *Zukunftsethik und Industriegesellschaft.* (pp. 15–40). München, Germany: J. Schweitzer.

Bernstein, R. J. (1995). Rethinking Responsibility. *The Hastings Center Report, 25*(7), 18–19.

Bijker, W. E., Hughes, T. P., & Pinch, T. J. (Eds) (1987). *The Social Construction of Technological Systems: New Directions in the Sociology and History of Technology.* Cambridge, MA: MIT Press.

Böhler, D. (Ed.) (1994). *Ethik für die Zukunft: Im Diskurs mit Hans Jonas.* (pp. 105–125). München, Germany: C. H. Beck.

Donnelley, S. (1995). Bioethical Troubles: Animal Individuals and Human Organisms. *Hastings Center Report, 25*(7), 21–29.

Engelhardt, H. T. Jr., & Callahan, D. (Eds) (1977). *Knowledge, Value, and Belief.* Hastings-on-Hudson, NY: The Hastings Center.

Heidegger, M. (1954). *Vorträge und Aufsätze.* Pfullingen, Germany: Neske.

Hottois, G. (1995). *Entre symboles et technosciences: un itinéraire philosophique.* Seyssel, France: Champ Vallon.

Jonas, H. (1954). *Gnosis und Spätantiker Geist.* vol. 1: *Die mythologische Gnosis; mit einer Einleitung zur Geschichte und Methodologie der Forschung*; vol. 2: *Von der Mythologie zur mystischen Philosophie.* Göttingen, Germany: Vandenhoeck and Ruprecht.

Jonas, H. (1963). *The Gnostic Religion: The Message of the Alien God and the Beginnings of Christianity.* Boston: Beacon Press.

Jonas, H. (1973a). *Organismus und Freiheit: Ansätze zu einer Philosophischen Biologie.* Göttingen, Germany: Vandenhoeck and Ruprecht.

Jonas, H. (1973b). Technology and Responsibility: Reflections on the New Tasks of Ethics. *Social Research, 40*(1), 31–54.

Jonas, H. (1974). *Philosophical Essays: from Ancient Creed to Technological Man.* Englewood Cliffs, NJ: Prentice Hall.

Jonas, H. (1976). Responsibility Today: The Ethics of an Endangered Future. *Social Research, 43*(1), 77–97.

Jonas, H. (1979a). *Das Prinzip Verantwortung: Versuch eine Ethik für die Technologische Zivilisation.* Frankfurt am Main, Germany: Insel.

Jonas, H. (1979b). Toward a Philosophy of Technology: Knowledge, Power, and the Biological Revolution. *Hastings Center Report, 9*(1), 34–43.

Jonas, H. (1981). Reflections on Technology, Progress and Utopia. *Social Research, 48*(3), 411–455.

Jonas, H. (1982). *The Phenomenon of Life: Toward a Philosophical Biology*. Chicago: University of Chicago Press. Originally published, 1966.

Jonas, H. (1984). *The Imperative of Responsibility: In Search of an Ethics for the Technological Age*. Trans. by H. Jonas with D. Herr. Chicago: University of Chicago Press.

Jonas, H. (1985). *Technik, Medizin und Ethik: Zur Praxis des Prinzips Verantwortung*. Frankfurt am Main, Germany: Suhrkamp. (This volume has only limited overlap with Philosophical Essays.)

Jonas, H. (1991). *Dem bösen Ende näher: Gespräche über das Verhältnis des Menschen zur Natur*. Frankfurt am Main, Germany: Suhrkamp.

Jonas, H. (1992). *Philosophische Untersuchungen und metaphysische Vermutungen*. Frankfurt am Main, Germany: Insel.

Jonas, H. (1996). In: L. Vogel (Ed.), *Mortality and Morality; A Search for the Good after Auschwitz*. Evanston, IL: Northwestern University Press. (This volume has considerable overlap with Philosophische Untersuchungen und metaphysische Vermutungen and might therefore be regarded as its English translation.)

Latour, B. (1987). *Science in Action: How to Follow Scientists and Engineers through Society*. Milton Keynes, England: Open University Press.

van der Pot, J. H. J. (1985). *Die Bewertung des technischen Fortschritts: eine systhematische Übersicht den Theorien*. 2 vols. Assen, The Netherlands: Van Gorcum.

Wendnagel, J. (1990). *Ethische Neubesinnung als Ausweg aus der Weltkrise? Ein Gespräch mit dem 'Prinzip Verantwortung' von Hans Jonas*. Würzburg, Germany: Königshausen und Neumann.

THE THING ABOUT TECHNOLOGY: TOWARD A PHENOMENOLOGY OF TECHNOLOGICAL ARTIFACTS

Peter-Paul Verbeek

"Things thing".[1] When philosophers want to have a good laugh, they take Martin Heidegger from the shelf, preferably for a phrase like this, in which he transforms a noun into a verb. This chapter, however, intends to show that the idea that "things thing" is of utmost importance for the philosophy of technology – albeit in a somewhat different way than Heidegger intended. I will discuss the importance of thinking about technology in terms of technological objects, and the ways in which this could be done. The idea that "things thing" is crucial for this, since it implies that what things "do" cannot be reduced to non-thingly aspects, but should be understood from "the things themselves".

The importance of understanding technology in terms of technological objects will be demonstrated in two steps. Firstly, by showing the consequences of paying *too little* attention to things. Ironically enough, this can be done by analyzing the philosophy of the same Heidegger who insisted that "things thing". Within Heidegger's philosophy of technology, as will be shown, a way of thinking about technology is predominant that can be characterized as "transcendentalism". Things (technological artifacts) are reduced to the conditions of their possibility and consequently disappear from the scene. Having demonstrated the consequences of this neglect of things, secondly, the outlines of a way of thinking about technology will be articulated that is better

Metaphysics, Epistemology, and Technology, Volume, 19, pages 281–299.
2000 by Elsevier Science Inc.
ISBN: 0–7623–0681–5

able to understand the role of technological artifacts in our culture. The main question will then be: How should we think about things?

HEIDEGGER'S PHILOSOPHY OF TECHNOLOGY

The importance of analyzing technology in terms of technological *objects* can be illustrated best by showing the consequences of a lack of attention for them. I will do this in the next section, by criticizing Heidegger's transcendentalism. Here I will first briefly summarize his most relevant lines of thought.

According to Heidegger, technology should not simply be seen as a means toward an end or as a human activity. Technology can be understood this way, but behind characteristics such as these, something more fundamental hides itself. In order to understand technology, we should not look among "levers, pistons, and scaffolds": technology is not "technological or machine-like".[2] What, then, is technology? Heidegger gives a rather cryptic answer. Technology, according to him, is a way of "revealing" or "un-covering" (*entbergen*). *Entbergen* is one of the words Heidegger created himself in order to be able to think what, according to him, is not thought anymore. It is his translation of the Greek *alètheuein*, which means "un-hiding" or "dis-covering": bringing from hidden-ness into un-hiddenness, un-covering what was covered. From this verb the noun *alètheia* is derived, which is usually translated as "truth". Heidegger, however, claims to have found a better translation in "uncoveredness" or "disclosedness" (*Unverborgenheit*).

We thought we were going to think about technology, but suddenly we find ourselves thinking about "truth". What does Heidegger mean if he says technology is a way of "revealing"? This question takes us right into the heart of phenomenology. What "reality" is for people, according to phenomenology, is never reality "itself" but always *their* reality. "That" to which people are related, becomes reality only *within* the relationship people have with "it". Reality, consequently, always bears the marks of this relationship. Being involved with reality implies interpreting it. People cannot get in touch with reality "itself". That is why "that" and "it" were put between quotation marks: one cannot speak about "it", because that already implies being involved with it and therefore "revealing" or disclosing it. Precisely because one cannot get directly in touch with "it", Heidegger calls it "the hidden". Within the relation people have with it, a reality comes into being: one possible reality becomes the "true" reality (*alètheia*). The "hidden" becomes "unhidden" or "revealed". "Helping to become unhidden" is exactly what *entbergen* is for Heidegger: "revealing".

If Heidegger states that technology is "a way of revealing", he therefore means that technology implies a specific relation between people and reality. In this relation a specific interpretation holds about what counts as "real". Technology is a fundamental pattern in our way of interpreting reality, one could say. Characteristic for the technological way of "revealing" is, according to Heidegger, that it approaches reality as raw material, available for manipulation. If something has been produced in a technological way, the final product is experienced as something that derives its reality entirely from its maker. In its coming about no element of transcendence is experienced anymore, as was the case for the ancient Greeks. For them, according to Heidegger, "making" meant "letting come into being": the role of the maker was seen as "helping to come into being", not "creating". For modern technology, however, "making" means "ordering" or "summoning" (*bestellen*), and what has been "ordered" (*bestellt*) has the character of "standing-reserve" (*Bestand*): a stock of materials, available for manipulation.

In order to get a thorough understanding of technology, however, we have to rise one more level of abstraction, according to Heidegger. For if "revealing" can be understood as "helping to come into being", this implies that the way of revealing is *predetermined* by the way in which "being" is understood. What does this mean? According to Heidegger, in every era there exists a dominant way of understanding "being", an understanding of what it means to "be". Disclosing reality necessarily takes place *within* such an interpretation of being. The nature of people's relation with reality, therefore, is determined behind their backs. People always already find themselves within a frame of interpretation before they can interpret reality at all. This means that we do not, according to Heidegger, disclose reality as standing-reserve because we want to, but simply because this is the character reality has for us within the current meaning of "being".

By connecting technology to "being", we arrived at the heart of Heidegger's philosophy. Being, according to Heidegger, must be understood in a historical way. It is not fixed; it does not have an "essence". In different eras, "being" has different meanings. Our current technological era should be seen as a phase in this history of being. The specific character of being that is dominant in our time, Heidegger calls "enframing" (*Ge-Stell*). This "enframing" gives our approach of reality the form of the "will to power" and reality itself the character of raw material, ready for human manipulation. For Heidegger, "enframing" is the "essence" (*wesen*) of technology – *wesen* being a verb in German, not a noun, indicating that technology must not be understood in terms of "beings", but of "being".

HEIDEGGER'S TRANSCENDENTALISM

Heidegger's philosophy of technology is not exactly undisputed. Three qualifications regularly return in the criticism: it is considered nostalgic, massive, and abstract. Because of the limited space, I will leave the reproach of nostalgia out of consideration and focus on Heidegger's supposed abstractness and massivity. It is precisely by discussing these forms of criticism that the importance of a "turn toward things" can be shown, as will soon become clear.

A good representative of the criticism of Heidegger's abstractness and massivity is Andrew Feenberg. In his article "From Essentialism to Constructivism",[3] Feenberg first praises Heidegger for his attempt to understanding technology as more than a collection of means toward ends, and for his openness to the fact that we are in a certain sense determined by our own technologies. But, subsequently, Feenberg takes a much more critical stance. According to him, Heidegger's philosophy of technology is too abstract, too remote from concrete technological practice. Feenberg shares the vision that "(t)he issue is not that machines are evil nor that they have taken over, but that in constantly choosing to use them over every other alternative, we make many other unwitting choices" (forthcoming, paragraph 13). But he considers Heidegger's hope for a "vaguely evoked spiritual renewal" "too abstract to inform a new technical practice" (forthcoming, paragraph 11): "Unfortunately, Heidegger's argument is developed at such a high level of abstraction he literally cannot discriminate between electricity and atom bombs, agricultural techniques and the Holocaust. All are merely different expressions of the identical enframing, which we are called upon to transcend through the recovery of a deeper relation to being" (forthcoming, paragraph 16). Heidegger's elaboration of this level of abstraction makes his philosophy *massive* as well: by reducing modern technology to a stage in the history of being, there cannot be any room for a *new* technology. Technology simply has the character of "enframing" (*Ge-Stell*). Not technology itself, but the technological way of thinking will change in a new stage of the history of being – a stage we can only await in openness.

I share Feenberg's criticism, but think it needs more development than Feenberg provides. Feenberg's statement that Heidegger is too abstract and massive to inform a new technological practice is correct, but in itself insufficient to criticize him adequately. Feenberg passes over the fact that for Heidegger technology is not ontic (concerning beings) in nature but ontological (concerning being): it is a matter of *being* instead of *machines*. Any Heideggerian will tell him this, insisting that any attempt to develop a better technology is itself an expression of the will to power and therefore only affirms the enframing it wanted to overcome. From Heidegger's way of thinking, the search for a better technology comes down to the attempt of Baron von Münchhausen to drag himself out of the swamps with his own hair. The only way out is "letting things be": the will not to will. Heidegger does not want an alternative technological practice. His perspective is not ontic but ontological. This means Feenberg either has to show that only an ontic approach of technology can be adequate or that something is wrong with Heidegger's elaboration of the ontological level with respect to technology. An adequate criticism of Heidegger cannot rest simply with expressing dissatisfaction with the consequences of his approach. It should be directed at the approach itself.

Heidegger's approach to technology cannot simply be called "abstract" because he interprets technology as a way of disclosing reality instead of understanding it as technological artifacts. Heidegger is "abstract" because of the way in which he analyzes the relation between technology and disclosing reality. As we saw, Heidegger gives this analysis a very specific twist: the way in which reality is disclosed by means of technology is not determined by the technologies in question, but by "being". Heidegger considers the contribution of technology to the disclosing of reality to be ontological in nature, instead of ontic. Technological "revealing", after all, always rests upon a prevailing form of "unconcealment", according to Heidegger, which is "enframing". Heidegger does not understand the "technological disclosing of reality" as accomplished by technological objects, therefore, but as predetermined by "being". Technology, in this vision, does not offer a way of disclosing reality, but is a manifestation of it.

The consequence of Heidegger's way of analyzing "technological revealing" is that concrete technologies disappear from the scene. The way in which technologies disclose reality always comes down to stating the obvious. Before technological artifacts could play a role in our access to reality, the way in which reality can show itself is already determined. Within Heidegger's approach of technology it is not even possible to ask for the specific contribution of technologies to the way in which reality is revealed. One can only ask for the ruling understanding of "reality", of which technologies are expressions instead of sources. The problem of this approach is not so much its "abstractness", as Feenberg stated, but its exclusively ontological understanding of technology. However clarifying and adequate it is as well, by making his ontological approach absolute Heidegger loses contact with technology itself: the technological artifacts in our everyday life.

Or am I wrong? After all, Heidegger illustrates his argument with many examples. And those examples by definition concern concrete technological artifacts. There are many possible ways of criticizing Heidegger, but it cannot be held against him that he did not pay enough attention to things. What counts, however, is *the way in which those artifacts are analyzed*. And this is where Heidegger's ontological approach shows its limitations. Heidegger appears not to be able to get a hold on the role *concrete technologies* play in our culture and in our ways of disclosing reality.

The best way to show this is by discussing a famous example from Heidegger himself: a hydroelectric plant in the Rhine. Heidegger writes:

> The hydroelectric plant is set into the current of the Rhine. It summons the Rhine to supply its water pressure, which sets the turbines turning. This turning drives the machines whose transmission supplies the electric current, for which the power station and its electricity network are set up to convey electricity. In the context of the interconnected outcomes of the ordering of electrical energy the Rhine itself appears as something that was ordered. The hydroelectric plant is not built into the Rhine like the old wooden bridge that has joined bank with bank for hundreds of years. Rather the stream is interrupted because the power plant was built into it. What the river is now as a stream, namely, a supplier of water pressure, it derives from the essence of the power station . . . . But, it will be countered, the Rhine is still a river in the landscape. That may be the case, but how? Only as an object that can be ordered for a visit by groups of tourists, summoned there by the holiday industry.[4]

In this example Heidegger seems to give an analysis of the way in which a hydroelectric plant reveals the Rhine. He actually seems to analyze technological revealing here in connection with a concrete technological artifact. Because a hydroelectric plant is built into it, the Rhine can only appear as a means toward the end of generating electric current; this is what the example seems to say. But does Heidegger really give an ontic analysis here instead of an ontological one? Had Heidegger just stated that the Rhine "in the context of" this technology can only appear as a supplier of energy, there would not have been any reason to be suspicious. But Heidegger states it slightly differently, and the difference says it all. He calls the technology in question "the interconnected outcomes of the ordering of electrical energy" (*ineinandergreifenden Folgen der Bestellung elektrischer Energie*).

The word "outcome" implies that the Rhine is not revealed as standing-reserve because a hydroelectric plant was built into it, but that it is *because of the revealing of the Rhine as standing-reserve that a plant was built in it*. According to Heidegger, the plant is not a source of the way the Rhine is revealed, but a consequence of it. There is no other way of revealing the Rhine than as standing-reserve. When enframing (*Ge-Stell*) offers the prevailing way of unconcealment, there is only one way of revealing, and that is "ordering" or "summoning". That the Rhine is revealed as a standing-reserve has not so much to do with the technologies that disclose it this way, as with the character of "being" in our technological era: "enframing". It is because the Rhine is revealed as standing-reserve, that something like a hydroelectric plant can be built, not the other way around.

Do we have to do with an accident here, or is this example representative of Heidegger's approach of technology? On the authority of Günter Seubold, who made an extensive study of the role of technology in Heidegger's oeuvre, I dare to hold the latter. Seubold explains that there is a "conditional relation" (*Bedingungsverhältnis*) between technological revealing and concrete technologies. This is not only true for "The Question Concerning Technology", but, e.g. also for "The Age of World as Picture", where Heidegger states: "Machine technology remains the hitherto most visible offshoot of the essence of modern technology and this essence is identical to that of modern metaphysics".[5] Seubold concludes (without criticizing it, to be clear): "[technological devices] can only be produced when the technological way of revealing has already become predominant" (Seubold, 1986, p. 195).

The example of the hydroelectric plant is not an idiosyncratic one, therefore, but appears to be representative for Heidegger's approach. Concrete technology for Heidegger is secondary to the "essence" of technology, and only from this "essence" can technology be understood. Heidegger considers the ontic level to be derivative with respect to the ontological level. If he speaks about "technology", he does not have technological devices in mind, but "enframing". This by definition makes him "abstract" in the sense that he abstracts from concrete technology.

The style of thinking that causes this abstractness may be termed "transcendentalism".[6] By "transcendentalism" I mean an absolutization of the transcendental style of understanding phenomena, namely, in terms of their conditions of possibility. Because Heidegger understands technology in an ontological way – i.e. because he sees technology as an expression of the prevailing interpretation of "being" – he reduces technology to a condition of its possibility. The manifestation of being as "enframing" *precedes* the technological artifacts that play a role in our everyday lives. This condition of possibility is subsequently made absolute: if Heidegger speaks about "technology", he does not mean concrete technology, but "enframing". Because of this style of thought, Heidegger can discern only those aspects of technology that can be understood as expressions of such conditions of its possibility.[7] If technological artifacts are being understood only in terms of the prevailing way of disclosing from which they originate, resting upon a reigning understanding of "being", these artifacts themselves cannot be seen as the source of other ways of revealing reality anymore. According to Heidegger, "being" develops behind people's backs. This view results in the often-criticized "massivity" of Heidegger's philosophy of technology. We can only prepare for a new "sending" of "being", not accomplish it, since that would only confirm technology – in the ontological sense, that is. For Heidegger, the technological way of revealing has little to do with technologies in the ontic sense.

Heidegger made visible an important aspect of modern technology. He showed that it takes a specific interpretation of reality to be able to intervene in it in a technological way. But there is much more to be discerned about technology. The role technology plays in our culture cannot be understood from this "backward-thinking" perspective only. Heidegger's vision can be countered with, e.g. the argument that technologies do not necessarily limit people's interpretation of reality to raw material for human manipulation. Modern communication technologies allow us to be engaged with people and happenings outside of our nearest neighborhood. This engagement usually does not have the character of a "will to power". A telephone does not reveal a friend as a standing-reserve but as a person, albeit in a different way than face-to-face communication does. Transportation technologies such as the car and the airplane can only be produced by manipulating reality, but from a car or a plane a forest can appear as much more than a stock of firewood or potential furniture. I realize these counterexamples are not shocking, but it is precisely their triviality that demonstrates the blind spot in Heidegger's approach. By taking as a starting point the a prioristic construction that there always is only one prevailing way of unconcealment, which in our times has the character of "enframing", Heidegger lost contact with concrete technology.

Philosophy of technology should take concrete technologies more seriously than Heidegger did. A philosophical analysis of the role of technology in our culture should not only try to reduce ("going backward") technology to interpretation, but should analyze ("moving forward") in what ways technological artifacts help to shape the relation between people and reality. In an approach like this, technology is not only understood as a reduced way of revealing, but as a *mediation* of people's relation with reality, one that opens specific ways of access to it. The nature of these "ways of access" should be analyzed. Contrary to what Heidegger advises us, we therefore *do* have to look among "levers, pistons, and scaffolds", because technology *does* have to be understood as "something technological, something machine-like." Philosophy of technology should analyze technology as technological *artifacts* instead of the way of thinking from which they originate. Like texts in postmodern philosophy were released from their authors, analyses of artifacts should not be limited to their origins.

In what follows, I would therefore like to reverse the "direction of thinking", and sketch the outlines of a program for "thinking forward" about technology instead of "backward". Technological artifacts do not necessarily have to be analyzed with respect to their origins. They actually do something in people's relation with their world. Precisely this is what Heidegger's often ridiculed statement that "things thing" can express: what a thing does, cannot be reduced to non-thingly aspects. Things should be analyzed as *things*.

TO THE THINGS THEMSELVES

How to analyze things as things? I will sketch my proposal for a philosophy of technology "from the perspective of things" against the same background as Heidegger: phenomenology. Despite all the criticism classical phenomenology has had to endure, I adhere to the term "phenomenology". Because of the way phenomenology has developed since Husserl founded it, it sometimes raises the suspicion of essentialism and romanticism. Those characteristics, however, are not necessarily connected with the aim that underlies all major phenomenological studies: understanding the relationship between people and their world. Phenomenology should not be understood as a "method" for understanding the world, a method that would allow philosophers to describe what reality "really" is. It should be seen as a way of understanding the relationship between people and their world. A phenomenology of technology, therefore, should perform an analysis of the role technology plays in this relationship.

What does this "understanding the relationship between people and their world" look like, then? The most important concept by means of which phenomenology tries to understand this relationship is *intentionality.* This concept helped to resolve the epistemological impasse that had arisen at the end of the nineteenth century between idealism and realism. For a phenomenologist, both idealism ("knowledge is a product of our consciousness") and realism ("knowledge is a depiction of reality in our minds") rest upon a misunderstanding. Consciousness and reality cannot be separated in such a radical way. A crucial feature of consciousness is, after all, that it is always directed at or involved with reality. Consciousness-in-itself does not exist. Consciousness is always consciousness-of-something. It does not make sense to speak about consciousness apart from the world at which it is directed, nor to speak about the world apart from consciousness that is directed to it. This "being directed at" is what is called "intentionality" within phenomenology.

Phenomenology did not cling to the epistemological preoccupation with consciousness. Husserl's followers gave it a much broader interpretation, and, by the end of his life, Husserl himself did the same. Husserl's analysis of the relation between "consciousness" and "reality" was expanded – mainly by Heidegger and his followers – to analyses of the relationships between "people" and the "world" they live in. Heidegger did not want to analyze "intentionality" but "being-in-the-world",[8] and replaced the term "consciousness" with "care" (*Sorge*). In doing so, he "existentialized" phenomenology. Instead of performing cerebral analyses of the transcendental structures of human consciousness, phenomenology directed itself at human beings of flesh and blood and at the full spectrum of ways in which they can be involved with their world.

By way of Heidegger's philosophy, phenomenology thus allied itself to existential philosophy. Existential philosophy, initiated by Kierkegaard, consisted of an attempt at understanding the relationship between people and their world as well. Its major aim was to get a closer understanding of the specific human way of "being". A crucial role here was played by the view that human existence cannot simply be seen as "being" but should be seen as "being there". People have a relation to their own being. They know that they exist and therefore their existence is not simply given, but has to be shaped by themselves. People are free. They are not bound by the determinants of their environment but have the ability to choose and are therefore responsible for their own being. People can, of course, only shape their being within the context of their world, which offers them possibilities and limitations. One can only realize one's existence by acting in the world with which one is involved.

Gradually, two perspectives arose within phenomenology: *hermeneutic* phenomenology, which analyzed the interpretive ways of access people have to their world, and *existential* phenomenology, which analyzed humans acting in and being involved with their world. These two perspectives logically result from the basic idea of phenomenology: people and their world cannot be understood separately, but only in their interrelatedness, which takes place over the "bridge" of intentionality. The existential perspective approached this interrelatedness from the "human side", by analyzing the way in which people can realize their existence within their relation with their world. The hermeneutic perspective approached it from the side of "the world", by analyzing the way in which a world is constituted and gets its meaning for people. The two sides cannot be separated, because they are what they are only within their interrelatedness. People are what they are only within their relation with the world, and the world is what it is only within the relation people have with it.

On the basis of both of these phenomenological perspectives, philosophies of technology developed. Philosophers such as Ortega y Gasset and Karl Jaspers stressed the role of technology in the existential aspect of people's relation with the world. Their question was: What role does technology play in the way in which people realize their existence? Heidegger's philosophy of technology, as we have seen, stressed the hermeneutic aspect of people's relation with their world, since it saw technology as a way of disclosing, a way of access to reality.

The "turn toward things" we are looking for within phenomenological philosophy of technology, then, will have to take place in both branches of phenomenology. An analysis has to be made of the role technological artifacts play in the ways in which people realize their existence in the world, on the one hand, and have access to their world, on the other.

TECHNOLOGICAL INTENTIONALITY

What does such a phenomenological analysis look like? Surprisingly enough, it is again Heidegger's philosophy that offers an interesting starting point. Heidegger's philosophy of technology, as briefly sketched above, is usually seen as the philosophy of technology of Heidegger, but actually it is his second. Already in *Sein und Zeit*, his first important work, Heidegger explicitly paid attention to technology. The analyses that can be found there are in surprising contrast with Heidegger's later philosophy of technology. Don Ihde has already called attention to this.[9] My interpretation of Heidegger's early philosophy of technology is slightly different from his, however, as will become clear.

Sein und Zeit is clearly one of the most important works in which the hermeneutic and existential "expansion" of phenomenology took place. For Husserl, as we saw, phenomenology consisted in a (transcendental) clarification of the relation between consciousness and reality. Heidegger broadens the scope to the relation between humanity and its world. Heidegger is not only interested in knowledge but in the full spectrum of possible relations people can have with their world. Knowledge is only one of these. Heidegger calls the relation between the human being and the world: "being-in-the-world". He analyzes this being-in-the-world from the perspective of the everyday involvement of people with other people and the things around them. It is only on the basis of such an everyday way of being involved with the world, that knowledge of the world is possible, according to Heidegger.

In the analysis Heidegger makes in *Sein und Zeit* of the everyday relation between people and their world, tools play a very important role (Heidegger, 1986, paragraph 15–16). According to Heidegger, the handling of tools is the pre-eminent way of entering a relationship with the world. He points out that the Greek word for "things" was "pragmata". That word expresses exactly what

Heidegger wants to make clear: things should be understood at the level of *praxis*, not at the level of theoretical analysis. The primordial way of having a relationship with things is not by critically examining them, but by being praxically involved with them. The things that play a role in this praxis, Heidegger calls "tools" (*Zeug*).

Heidegger's next question is: What makes a tool a tool? From a phenomenological point of view, this question comes down to: What place does a tool have in being-in-the-world? What makes a tool a tool, according to Heidegger, is that it makes possible a praxis: a tool is always something-to-____. It is useful; it can be used for something. Like consciousness and perception, tools cannot be understood in and of themselves but only from the relation people have with them. Just as consciousness is always consciousness of something, a tool is always a tool-in-order-to-____. The appropriate way of handling tools is not a theoretical examination, but using them and entering the praxis they make possible. The way in which a tool is present then, Heidegger calls "readiness-to-hand" (*Zuhandenheid*). When it is critically examined – when it breaks down, for instance – the tool is "present at hand" (*Vorhanden*). It does not allow for involvement through itself anymore, but itself asks for attention.

The concept of readiness-to-hand is of utmost importance for a phenomenology of things. What characterizes a use object that is ready-to-hand is that it withdraws, as it were, from people's relation with their world.[10] After all, only when people's attention is not directed toward the tool they use but to the work they do with it, is the tool present as a tool. When writing, people are not involved with their keyboards but with the text they write by means of them. Still, the fact that the ready-to-hand artifact withdraws from our attention does not imply that it plays no role in the praxis that arises around it. To the contrary: the context of the tool receives its meaning precisely from the use people make of it. "The forest is a place for getting wood, the mountain a stone quarry, the stream water power, the wind 'wind in the sails'," Heidegger writes (1986, p. 70). The way in which use objects can reveal their environment is not necessarily limited to raw material for production, however: "A covered railway platform takes the possibility of bad weather into account, and street lights the darkness" (1986, p. 71). This interpretation of technology is at right angles to Heidegger's later vision. According to early Heidegger, technology does not obstruct our access to reality, but actually enables it.[11]

This analysis can be a starting point for the "turn toward things" within phenomenological philosophy of technology. Artifacts can be ready-to-hand: they can withdraw from people's focus of attention and facilitate people's involvement with reality, which takes place through them. The artifact enters the intentional relation between people and the world, as it were. From their withdrawn position, however, artifacts-in-use do give shape to that relation. This is the second step we have to take. Technological artifacts can be understood as mediators: when used, they mediate the involvement of people with their world. Mediating should be understood in an active sense here. Artifacts are not simply media or intermediaries, they are mediators: they actually do something with the interrelation between people and their world.

Following Don Ihde, this could be called the non-neutrality of human-technology relations. Technologies are no neutral instruments, but give shape to people's relation with the world. In his analysis of this mediating role of things, however, Ihde focuses too much on its perceptual aspects. However illuminating his work is – he actually goes farther than Heidegger by showing that readiness-to-hand is one of many more possible human-artifact relations – Ihde's work on human-artifact relationships understands the relation between people and world too narrowly as a perceptual one. By approaching the mediating role of artifacts merely in a perceptual way, important aspects disappear from sight.

A good example of the mediating role of artifacts outside the perceptual realm is an example I came across when reading the newspaper some time ago. In a community somewhere in Rumania, the shanks of the rakes of the Parks Department had been shortened considerably. Those rakes, namely, had facilitated a practice that had to be ended: the employees of the Parks Department lazed about on them excessively. Shortening the shanks would make the rakes less inviting for laziness. Had the rakes only been neutral means toward the end of raking, the municipality would not have had to intervene. It is precisely because the rakes *en passant* made possible a quite different practice – which was not intended explicitly but came about in use – that intervention was apparently needed. The rake mediates the relation between the workers and the parks: it is not a simple means, but plays an active role in the way that relation is shaped.

This ability of things to give shape to the relation between people and their world can be indicated with another term Ihde coined: "technological intentionality". This term indicates at what level technological artifacts play a role: at the level of the intentional relation between people and their world. It is in mediating this intentional relation that technological intentionalities can occur. Connecting the idea of intentionality with technological artifacts should not be interpreted as an expression of realism, however. Just as the intentional relation between people and their world is not a characteristic of either people or the world, technological intentionalities should not be located in the artifacts themselves but in the relation people have with them. Phenomenology, after all, is based upon the idea that it is impossible to speak about the world separately from the people that have a relation with it and vice versa. Technological intentionalities occur within the relation between people and world that is mediated by technological artifacts. Reversely, artifacts per implication can only get their identity within this relation. The rake mentioned above was designed for raking, but within the context in which it was used it got the identity of an object for leaning on. Technologies are multistable, as Ihde calls it. They only have a "relative essence", because they are what they are only within the context in which they are used. Artifacts-in-themselves cannot be technologies: technologies are always technologies-in-order-to-____, and this "in order to" indicates the inextricable connection between technologies and their use context.

EXISTENTIAL AND HERMENEUTIC MEDIATION

As was pointed out above, two perspectives can be discerned within phenomenology: an existential and a hermeneutic perspective. Both perspectives offer ways of understanding the mediating role of technology. The "turn toward things" should, therefore, take place in both perspectives. This can be accomplished by elaborating Ihde's term "technological intentionality" in an existential and a hermeneutic direction. From a hermeneutic-phenomenological point of view, technological artifacts should be analyzed with respect to their mediation of people's access to reality. From an existential-phenomenological point of view, the analysis should concern the mediation by artifacts of people's actions: the ways in which people realize their existence in the world.

For the hermeneutic perspective, Ihde's analysis of human-technology-relations offers a good starting point. Ihde discerns several ways in which technologies can play a role in people's perceptual relation with the world. Artifacts can (1) mediate this relation, of which he discerns two variants, (2) be themselves the terminus of our experience (within the "alterity relation", in which the object is present as quasi-other), or (3) play a role only in the background, by creating a context for our experience (within the so-called "background relation"). Ihde, however, does not always connect these human-artifact relations with the hermeneutic question what they imply for the interpretive relation of people with the world. If he makes this connection at all, he mainly does this with respect to the hermeneutic role of perceptual technologies in the coming about of scientific knowledge. Technologies, however, do not only play a role in science, but in our everyday life as well. Even more: this everydayness is exactly the domain phenomenology considers to be of greatest importance. Questions should be answered such as: What do technologies like cars and airplanes do with our relation with the world? In what ways can people be present for us when our contact with them is mediated by technologies such as email or the telephone? An analysis of questions like these would make possible an "expanded hermeneutics", to use again a term of Ihde's. Hermeneutics would then really be expanded from a strongly linguistically orientated discipline to a discipline that acknowledges that things play a role in human interpretation as well. Such a "materially turned" hermeneutics would make it possible to give the hermeneutic question within philosophy of technology – What role does technology play in people's interpretive relation with the world? – a new route for finding answers, a route that avoids the transcendentalism we meet in Heidegger's hermeneutic philosophy of technology.

Existentially, an analysis should be made of the ways in which technological artifacts mediate people's existence. What do technologies such as television do with the way we spend our time and spread our attention? What do technologies such as the car and the airplane do with the way we give shape to our social lives? At the level of the smallest unit of analysis, there has already been a lot of work done within Science and Technology Studies to develop concepts for understanding the influence artifacts can have on the actions of individual actors. It is mainly Madeleine Akrich and Bruno Latour who have done this, by developing their concept of "scripts" (Akrich, 1992; Latour, 1992). A script should be understood as a kind of implicit and built-in manual: the script of an artifact is the way in which it organizes its environment. Technologies can, for instance, presuppose certain forms of knowledge or physical abilities, because of which they exclude certain categories of people from using them. Technologies can provoke certain kinds of behavior, as with a car that will not start when the driver is not wearing his or her safety belt.

This concept of "scripts", however, does not entirely escape the pitfall of transcendentalism that was found in Heidegger's philosophy of technology. Akrich describes scripts as "the product of (the) work" of "*inscribing* (a) vision of . . . the world in the technical content of the new object" (p. 208). Latour considers scripts to be the result of "delegation to non-humans" (p. 231). Here, again, the direction of analysis is backward instead of forward. What things "do" is understood primarily in terms of what has been put into them by people. This way of analyzing things runs the risk of overlooking what things can do without having been asked for it. Revolving doors, for instance, were designed to let people enter a building while keeping the draft outside. However, as many wheelchair users know, they do not only keep out the cold air. Things should not be reduced to their origins. Just as more can be read in a text than was written into it by the author, things can do more than their designers intended.

The concept of technological intentionality offers a good alternative to the term "script", because it does not suffer from transcendentalist connotations.

To the micro-perspective of Latour and Akrich a more cultural-philosophical perspective may be added. Albert Borgmann's theory of technology offers important possibilities for this. Borgmann tries to understand what role technological devices play in the ways in which we give shape to our lives. According to Borgmann, technological devices, in contrast to pretechnological "things", ask for a specific way of dealing with themselves. Devices should be understood as machineries that deliver commodities. A device is (1) a physical entity that (2) does something for you. Technological devices keep their machinery in the background as much as possible. Their commodities are put in the foreground: these should be delivered quickly, easily, safely and ubiquitously.[12] This implies that the machinery of devices should ask for as little involvement as possible.

Non-device-like "things", in Borgmann's sense, however, do not separate machinery from commodity. They only do what they do by engaging people with themselves. A hearth asks for a lot of involvement: wood has to be chopped, the hearth has to be cleaned, etc. A central heating system only asks to be turned on, after which its commodity – warmth — can be consumed. Borgmann's analysis can be seen as a deepening of Heidegger's notion of readiness-to-hand. It shows that artifacts can be ready-to-hand in two distinct ways: in an engaging way and in a consumption-provoking way. Technological artifacts can either ask for involvement in their functioning, and thus invite an engaging way of action and doing things, or for mere consumption of their commodities.

Analyses such as these can help to develop a new elaboration of the existential-philosophical perspective within philosophy of technology. They are able to locate and analyze the role technological artifacts play in the ways in which people live their lives. The focus of such an existential perspective does not primarily concern traditional notions such as the "self" or "authenticity", but people's sense of identity and the ways in which they live their lives, in relation to the material context of it.

CONCLUSION

A philosophy of technological culture must take the material culture of technology into account. Classical phenomenological philosophy of technology, especially Heidegger's, has always tried to understand technology as a way of thinking, thus treating technology as a branch of philosophy. Technologies were reduced to their conditions of possibility – a way of thinking that can be indicated by the term "transcendentalism". This way of thinking about technology did not succeed in getting an adequate picture of the role of technology in our culture. What is needed is not only an analysis of the origins of things but an analysis of what things actually do: the ways in which they help to give shape to the relation between people and their world. Technological artifacts play a mediating role in this relation, both with respect to the ways in which people have access to their world (hermeneutically) and the ways in which they can give shape to their existence within that world (existentially). This mediating role of things can become visible only if we do not reduce things to non-thingly aspects.[13] We have to think from the perspective of "things themselves." Was this not the intention with which phenomenology once started: "to the things themselves"?[14]

NOTES

1. Martin Heidegger, "Das Ding", in *Vorträge und Aufsätze* (1954), p. 172. All translations in this article are mine.
2. Heidegger, 1962, "Die Frage nach der Technik", pp. 20, 23.
3. Feenberg (forthcoming). All citations will list the paragraph number in the web document.
4. Heidegger, 1962, "Die Frage nach der Technik", pp. 15–16.
5. Heidegger, 1950, "Die Zeit des Weltbildes", p. 69.
6. Others have used the term "transcendentalism" as well to characterize Heidegger's philosophy of technology. Within the Dutch speaking countries, as far as I know, see the work of Egbert Schuurman and Maarten Coolen. In *Techniek en Toekomst* (1972), Schuurman uses the term "transcendentalists" for philosophers who "call attention to the transcendental orientation of all experience and therefore of technical matters of facts as well," and who "notice the relative autonomy of technology and its tendency toward absoluteness" (p. 60 – my translation). He considers Heidegger to be one of these philosophers. In *De machine voorbij* (1992), Maarten Coolen calls "transcendental" all approaches to technology that understand it from "the attitude that is required for people in order to have technical instruments at their disposal." Heidegger is "the great example" of this approach, because he understands technology "as an essential component of a metaphysical attitude toward being that is omnipresent in our culture." He considers Heidegger's approach to be even "transcendentalist" instead of just "transcendental" because "it sometimes looks like Heidegger denies the possibility that people can ever escape their imprisonment in the attitude of modern technology" (pp. 107–112 – my translation). My reasons relating Heidegger's philosophy of technology to the style of thinking typical of transcendental philosophy are the same as Coolen's. My reason for speaking of transcendentalism is slightly different: Heidegger makes technology's conditions of possibility absolute and puts them in the place of technology itself.
7. This transcendentalist way of thinking about technology is not unique to Heidegger, by the way. On the other side of the specter of reflection on technology, social scientists are pretty good at it as well. Under the denomination of the "Social

Construction of Technology", technologies were and are still being studied as the outcomes of social interaction. Here, technology is reduced to non-technological entities as well. It was only when the things *themselves* were seen as actors, or "actants" (Latour, 1992; Akrich, 1992), that a perspective developed that actually tried to think from technological artifacts themselves. This perspective tries to understand what things *do* instead of what their *origins* are – although Akrich and Latour appear to have trouble suppressing their sociological predisposition, as will become clear at the end of this article.

8. Cf. Ihde, 1998, pp. 18–19.

9. Cf. Ihde, 1990, pp. 31–34.

10. Cf. Ihde, 1990, p. 32.

11. This turn in Heidegger's philosophy did not appear out of the blue. An analysis of works succeeding *Sein und Zeit*, such as "Der Ursprung des Kunstwerkes", "Die Frage nach der Technik", *Gelassenheit*, "Bauen Wohnen Denken", and "Das Ding" with respect to the way in which "things" play a role shows that Heidegger gradually "turned" himself away from things. His perspective becomes more and more transcendentalist, in the sense that he reduces things more and more to what lies behind them. In a study I am writing about the turn toward things in philosophy of technology, I discuss this development in Heidegger's thinking.

12. Cf. Borgmann, 1984, p. 41.

13. This idea can be made fruitful for Industrial Design as well. Within Industrial Design, the role things play is mainly seen in terms of the functions they fulfill or the meanings they have. The notion of "technological intentionality" can be clarifying here as well, because it is able to make visible what things do as things: it shows an extra dimension of the products that are designed. See for this: Verbeek & Kockelkoren, 1998.

14. Husserl, 1980, "Wir wollen auf die Šachen selbst' zurückgehen," p. 6.

REFERENCES

Akrich, M. (1992). The De-scription of Technical Objects. In: Bijker & Law (Eds), pp. 205–224.

Bijker, W. E., & Law, J. (Eds), (1992). *Shaping Technology/Building Society: Studies in Sociotechnical Change*. Cambridge, MA: MIT Press.

Borgmann, A. (1984). *Technology and the Character of Contemporary Life: A Philosophical Inquiry*. Chicago: University of Chicago Press.

Coolen, T. M. (1992). *De machine voorbij: Over het zelfbegrip van de mens in het tijdperk van de informatietechniek*. Meppel, The Netherlands: Boom.

Feenberg, A. (forthcoming). From Essentialism to Constructivism: Philosophy of Technology at the Crossroads. In: E. Higgs, D. Strong & A. Light (Eds), *Philosophy in the Service of Things: Devices, Focal Practices, and the Quality of Life*. Chicago: University of Chicago Press. Also published on the Internet at http://www.rohan.sdsu.edu/faculty/feenberg/talk4.html.

Heidegger, M. (1986). *Sein und Zeit*. Tübingen, Germany: Max Niemeyer Verlag. Originally published 1926.

Heidegger, M. (1950). *Holzwege*. Frankfurt am Main, Germany: Vittorio Klostermann.

Heidegger, M. (1954). *Vorträge und Aufsätze*. Pfüllingen, Germany: Neske.

Heidegger, M. (1962). *Die Technik und die Kehre*. Pfüllingen, Germany: Neske.

Husserl, E. (1980). *Logische Untersuchungen.* Tübingen, Germany: Niemeyer. Originally published, 1913.

Ihde, D. (1990). *Technology and the Lifeworld: From Garden to Earth.* Bloomington, IN: Indiana University Press.

Ihde, D. (1989). *Expanding Hermeneutics: Visualism in Science.* Evanston, IL: Northwestern University Press.

Latour, B. (1992). Where Are The Missing Masses? In: Bijker & Law (Eds), pp. 225–258.

Schuurman, E. (1972). *Techniek en Toekomst.* Assen, The Netherlands: Van Gorcum.

Seubold, G. (1986). *Heidegger's Analyse der Neuzeitlichen Technik.* Freiburg, Germany: Karl Alber.

Verbeek, P.-P., & Kockelkoren, P. (1998). The Things That Matter. *Design Issues, 14*(3), 28–42.

PART II

REVIEW ESSAYS ON METAPHYSICAL AND EPISTEMOLOGICAL THEMES

THE FUTURE'S SO BRIGHT WE'VE GOT TO WEAR SHADES

Albert Borgmann

Holding On to Reality: The Nature of Information at the Turn of the Millennium. Chicago: University of Chicago Press, 1999. Pp. 274.

Dan Schiller

Digital Capitalism: Networking the Global Market System. Cambridge, MA: MIT Press, 1999. Pp. xvii + 294.

The juxtaposition of two articles in a recent issue of the *New York Times* nicely illustrates the Janus-faced aspect of the new electronic information technologies, particularly the internet and the world-wide web. Titled 'History Is Made by an Old-Fashioned chief.gov', one news story describes President Bill Clinton's participation in the first live internet chat ever held by a U.S. president. This historic event exemplifies the optimism with which many regard the internet – a hope that citizen interaction and participation in democratic government will be enhanced by the exciting advances in information technology. The other article of interest, 'Show Business Embraces Web, But Cautiously', recounts recent efforts by the big Hollywood studios to use the world-wide web to sell entertainment, both as a medium for the delivery of motion pictures and in creating new kinds of programs specifically designed for online viewing. The studios, too, hope for citizen participation, though of a different sort: Warner Brothers, for example, is developing the *God & Devil Show*, in which cartoon representations of God and Satan interview a guest (living or dead) followed by a viewer vote to determine whether the guest is to

Metaphysics, Epistemology, and Technology, Volume, 19, pages 303–314.
2000 by Elsevier Science Inc.
ISBN: 0-7623-0681-5

be sent to heaven or hell. This provides an apt illustration of the pessimism with which many regard the internet and the web, as the clamor of the marketplace further alienates the largely non-voting citizenry and the internet box in every home is used not for the upload of democratic participatory activity but instead for the download and passive consumption of ever more mindless drivel (the guest in the prototype of the *God & Devil Show* isn't Nietzsche, Heidegger, or Thomas Jefferson, but John Wayne). While it's too soon to say whether the predictions of the pessimists or optimists are more accurate, the relative placement of the articles in the *Times* is not encouraging: President Clinton's virtual town hall meeting appeared on page 20, while Warner Brothers made the front page – above the fold.

Both books under review fall into the pessimistic camp, though for different reasons. Schiller argues that despite widespread hope that the internet will usher in an era of robust direct democracy and friction-free capitalism (to use Bill Gates' phrase), what is actually occurring is a vast increase of the breadth and depth of the marketplace and the commoditization of almost everything, including education – in short, "networks are directly generalizing the social and cultural range of the capitalist economy as never before" (p. xiv). Far from being a liberating medium, "the Internet comprises nothing less than the central production and control apparatus of an increasingly supranational market system" (p. xiv).

Borgmann's pessimism is rooted in what he sees as the pernicious nature of 'technological information' itself. Though he too is concerned with the very real possibility that new information technologies will only deepen the gap between the haves and have-nots, his focus in on a more profound question, to wit, "whether the recent and imminent flood of information is good for anybody, rich or poor" (p. 4). He argues that unless we seek to balance technological information with 'natural' and 'cultural' information, humanity faces moral atrophy and a grim future indeed.

These are both excellent books and well worth reading. Though I'm not completely convinced with the details of their analyses, I do think that they point us in the correct direction. Schiller provides an interesting and accurate history of the development of information technology since the, 1950s, with an emphasis not so much on the technology but on the political machinations surrounding the emergence of the internet as we know it today. Borgmann offers a very interesting analysis of the nature of information itself (and does so in a prose style that is a pleasure to read). His aim is to provide "both a theory and an ethics of information – a theory to illuminate the structure of information, and an ethics to get the moral of its development" (p. 6). Both authors also address the changes that new information technologies are wreaking on higher education, and here I wholeheartedly agree with them that the prognosis is not good.

Everyone engaged in higher education will be interested in Schiller's chapter on 'Networking the Higher-Learning Industry', in which he gives a historical account of the development of North American colleges and universities. Schiller traces the social and institutional changes occurring before our eyes as digital capitalism advances into education, transforming higher education into the 'higher-learning industry' as it places "some of the most sensitive processes of social learning at the mercy of a proprietary market logic" (p. xiv). Universities are forced to ponder "What business are we really in?" and "What attractive and important set of services does our institution provide that people cannot obtain elsewhere better, faster, or cheaper?" (Keller, 1983, p. 121) – as if these questions even make sense in the context of higher education. (Unfortunately, the administrators of many colleges and universities have seriously considered answers to these questions, apparently not noticing that simply by accepting these questions as worthy of response, they have adopted a seriously distorted view of the social role of institutions of higher learning.)

Borgmann offers a much shorter (but very insightful) section on the impact of new information technologies on higher education (roughly, pp. 203–212). He gives a nice characterization of the program and promises proposed by distance (or distributed, or hyper-) learning advocates: "it is the extension of the promise of liberation and enrichment into a realm where authority, the confinements of space and time, and the burdens of one's shortcomings still lie heavily on the student" (p. 204). Once education is moved out of the stuffy old classroom and into cyberspace, of course, these shackles will disappear, as each student becomes "the sovereign who can choose the material, the method of presentation, and the time and place of studying" (p. 205). Additionally, "learners not only change from passive recipients to active choosers of information, they are also freed from the injuries that prejudices of gender, race, class, or physical appearance might otherwise have inflicted on them" (p. 205). So much for the promises.

As Borgmann eloquently puts it, in the explosion of information technology and the rush to apply it to higher education, "the line between genuine liberation and indulgent disburdenment [is] thoughtlessly crossed" (p. 206). While we all like to think that learning can always be a matter of joyful and spontaneous discovery, upon reflection, it's easy to recall college courses which were unpleasant at the time they were taken and only revealed their usefulness (either spiritually or practically) later. (I hear this from students taking introductory philosophy courses all the time. While some students are immediately excited by the philosophical way of looking at things, others complain that "thinking about this stuff makes my brain hurt," and only later in the semester do they admit that thinking philosophically opens up new intellectual vistas which they would otherwise never experience. Of course, some don't have this realization until much later, if at all. Such students, however, should be regarded as pathological singular points — not the norm around which to devise a mode of learning!

Borgmann's and Schiller's analyses of the current state of higher education intersect, as the former notes that the joyful spontaneity model of learning promoted by distributed learning advocates is very similar to a consumption model, with the student as consumer. "Supported by the machinery of technology, consumption is the unencumbered enjoyment of whatever one pleases," Borgmann writes. "The pleasures of consumption require no effort and hence no discipline" (p. 207). While few distributed learning advocates would claim that learning over the world-wide web requires no effort at all on the part of students, Borgmann says, "they fail to see that the discipline needed to sustain effort in turn needs the support of the timing, spacing, and socializing that have been part of human nature ever since it has evolved in a world of natural information" (p. 207). I think this is a very telling criticism of the asynchronous distributed learning model. But what of the advantages of classroom anonymity and the protection it imparts from class, race, or gender discrimination? Borgmann offers another very insightful observation:

> [I]t would be a Pyrrhic victory to deal with social prejudices by concealing them through a technological fix. People possessed by bias have most often learned to overcome it when they and their supposed inferiors were engaged face-to-face in a common and demanding task. Virtual colleges would help to perpetuate a problem that actual colleges have begun to solve (p. 208).

One might object here that Borgmann's view is too sanguine – how can actual colleges be said to have begun to solve the problem of racism, for example, when studies indicate that racial attitudes on campuses, which had been improving, began to worsen in the late 1980s? Many college campuses (my own, for example) are very homogeneous in terms of race (and, in many cases, class as well). I've taught classes in which the range of races represented runs the entire gamut from midwestern American caucasian all the way to northeastern American caucasian – not much variety there! Perhaps, however, these students could be immersed in a virtual environment in which they could interact with virtual students of differing racial and cultural backgrounds, all engaged in a common effort to complete some kind of demanding virtual task. (Such a system is currently under development at Carnegie-Mellon University. In the prototype demonstrated at the 1999 Computing and Philosophy Conference, an actual student is assigned the task of interviewing virtual people from various socio-economic groups for a campus TV news story about racial discrimination on campus, and must interact with a virtual editor and virtual co-worker who is also researching the story, and ultimately meet a deadline for completing the project. The interviews must be edited to fit the time allocated to the student's news story, and so the student is faced with choices of which subjects to include or exclude and in what order they'll be presented.) In this way, electronic information technology may actually contribute at lease one technological fix-type response to problems of social prejudice.

As the title *Holding On to Reality* suggests, Borgmann would reject such a scenario. As he puts it, an "illusory escape into cyberspace does not leave reality unharmed. At least for a time, virtuality can spread a fog of virtual confusion and blur the shape of things and events with glamour and triviality" (p. 191). Meaningful interaction with diverse human beings is reduced to a glitzy computer game. In order to understand Borgmann's contention that electronic information technology can actually harm reality, it's necessary to take a look at the theory of information he lays out.

Before defining 'information', Borgmann marks a distinction between two common ways the word is used. *Structural information* is the order inherent in physical objects; it's what is in the universe other than simply matter and energy (i.e. the way matter and energy are ordered and structured can be called 'information'). This isn't a useful notion for Borgmann, however, because under it, everything is information and nothing is not. If we take structural information, but only pay attention to what's been selected by cognition (taken in a broad sense) – that is, if we attend only to the information that is "taken up by a plant, an animal, a computer perhaps, and most especially by a human being" (p. 12) – what we are focusing on can be called *cognitive information*. (This is a pretty rough distinction, but suitable for Borgmann's purposes. The kind of information his book is about is cognitive information, and the reader gains a better grasp of what exactly this is as Borgmann's theory develops.)

Once we focus on cognitive information, we can be more precise about the notion of information by noting that it is a five-place relation: INTELLIGENCE provided, a PERSON is informed by a SIGN about some THING within a certain CONTEXT (p. 22). Beyond the basic sign/thing signified relation, CONTEXT is necessary in order to distinguish whether a particular physical object is to be regarded as a sign or a thing. (Consider a large metal stop sign. One can put it on a stand and use it as a coffee table, in which case it is a thing but not a sign; mounted on a post by the road, however, it's a sign.) INTELLIGENCE is necessary to figure out what the sign is about. Take, for example, hobo signs from the 1930s. Seeing a cat chalked on a fence, one might think it's (a) a child's drawing of a cat, (b) a hobo sign (though not understand what it referred to), or (c) a signal that a woman who's an easy touch lives nearby. Finally, a PERSON is necessary as the recipient of information. Despite Borgmann's use of a plant, animal, or (possibly) computer in originally describing cognitive information, he makes the move to restricting it to human persons early in the text without argument. In some ways, it makes his book more of a phenomenology of information instead of the abstract analysis of the nature of information its subtitle implies. Because of this, the conclusions Borgmann reaches later in the book may not be as general as he seems to think.

Given this notion of information, Borgmann distinguishes three kinds. First, there is information *about* reality (e.g. a *report* on the condition of your front lawn). Second, there is information *for* reality (e.g. a *recipe* for making an apple pie). Finally, there is information *as* reality (e.g. a *recording* of Bach sonata No. 1 in G minor, BWV 1001, which is not a report about the sonata or a recipe for performing the sonata, but is, in some sense, the music itself). These three distinctions enable us to look out at the world and classify three different categories of information.

First, there is *natural information*, which is the pristine form of information about reality. Human beings evolved in the world (the 'ancestoral environment') and with an ability to read its signs; these 'natural signs' indicate what Borgmann calls 'significant structure'. As discussed earlier, there is way too much structural information out in the world, and natural signs take us from mere structure to significant structure (p. 28). Though usually this is taken to show a flow of meaning in one direction from subject to object as the process of cognition divides the world into meaningful chunks, Borgmann takes a cue from Native Americans, for whom, he says, meaning flowed in both directions. As he sees it, "In the ancestral environment, the message of a sign is sent by a thing rather than selected by a person, though the recipient needs the capacity to gather the message from the sign" (p. 29).

Borgmann doesn't deny that meaningful structures are formed by humans, but he thinks that "as a guiding metaphor for the rise of meaning, the picture of material being shaped is misleading. It is more an issue of foreground and background, emergence and eloquence" (p. 29). Although Borgmann's notion of the eloquence of things may initially appear implausible, it is important to recall the first term in the five-place relation which comprises information, viz., intelligence. In other words, we should not expect natural signs simply to shout out their meaning to us without any resort to inference on our part. With this in mind, it is easier to understand the examples of natural signs which Borgmann offers, most of which are landmarks, e.g. "Snow-capped Lolo Peak was the sign that pointed the Salish toward the salmon on the other side of the

Bitterroot divide" (p. 25). Further, a natural sign doesn't have to be 'pure' but can be conventional. This happens when "its message exceeds what can be gleaned from its surroundings" (p. 30). One example of this kind of sign is a cairn, "a pile of rocks that by its density and steepness is set apart from its surroundings and conveys its meaning through a convention without which it means nothing" (p. 30).

Borgmann's introduction of convention might lead one to think that what he wants us to focus on is the *information about reality* aspect of these particular kinds of signs. In that sense, 'natural information' is information about nature, and that's what natural signs give us. Borgmann, however, uses the phrase 'natural information' to speak of signs that are already 'out there' in nature, and I find this notion very problematic. It's true that the meaning of Lolo Peak is determined by its surroundings — but first one must explore its surroundings to discover what it means! It only becomes a non-trivial sign *after* a person chooses to regard it as a sign. The first Salish explorer could have climbed Lolo Peak and built a huge pile of rocks on its summit to distinguish it as the one pointing to good salmon fishing, but why bother? The mountain itself, with all its physical particularities, serves as a sign. But I would claim that this isn't because it is naturally eloquent, but because it has been endowed with meaning by the explorer.

Borgmann disagrees vehemently with this and says explicitly that "there is no such thing as the original bestowal of meaning on a meaningless sign" (p. 23). In part, this is because Borgmann wants to ground conventional signs in such a way that there's no "mystery of original meaning" (see Haugeland, 1985, for a statement of this problem), and having natural signs around does the job. But the primary importance of natural signs is that they eliminate ambiguity by pinning down an objective reality about which they are informative. Borgmann illustrates how this works during a discussion of Adelbert Ames and his study of the perception of optical illusions. Consider the 'Ames room', in which one looks into a room through a peephole, seeing a very tall person in one corner and a dwarf in another. In fact, the two people in the room are of ordinary size, but the room is carefully distorted so that it appears to be in normal perspective when perceived through the peephole, giving rise to the illusion. So in looking through a peephole into a room, one might be looking at two ordinary people or a giant and a dwarf — who is to say?

As Borgmann explains it, reaction to Ames' work among social theorists falls into two major groups, conservative and progressive. The conservative take is that "Ames seems to have undermined the notion of an objective reality as a court of last appeal in cases of confusion and disagreement" (p. 12), while the progressive interpretation is that Ames has freed human beings from "rigid and authoritarian dictates of how everyone must construct his or her world from a reality that of itself allows for indefinitely many constructions" (p. 13). Borgmann disagrees with both interpretations, pointing out that while a random person looking through the peephole may regard his or her perception as ambiguous (Am I looking at two ordinary people in an Ames room, or a giant and a dwarf in an ordinary room?), the perception is not ambiguous for someone who looks through the peephole with the knowledge that it is an Ames room being observed. As Borgmann puts it, "both the ambiguity of reality and our freedom to construct it this way or that evaporate when we place things in context" (p. 13). An uncertain observer can resolve the ambiguity by entering the room, inspecting its four walls from the outside, or drilling another peephole placed to take advantage of stereopsis. By extending the context of observation, it is possible to resolve ambiguity. One could question whether this larger context is itself unambiguous, and that is where natural signs come in: natural signs give us an ultimately unambiguous context.

The second category of information recognized by Borgmann is *cultural information*, which is information *for* reality, in the sense of providing instructions for shaping reality. Borgmann contrasts it with natural information as follows:

> Natural information emerges of itself, intimates rather than conveys its message, and disappears. Cultural information, to the contrary, is wrested and abstracted from reality, carries a definite content, and assumes an enduring shape (p. 59).

Along with cultural information comes the idea of realizing information, that is, the process of converting the instructions of a design into something real. Borgmann devotes three very interesting chapters to the discussion of three characteristic ways of realizing information: reading, playing, and building. For our purposes here, however, the importance of realizing information is its association with ambiguity:

> No design can specify its realization fully. To convey exactly as much information as the thing realized, a design would have to exhibit just as many features as the thing. But then it would be a duplicate of, rather than information for, a thing. All cultural signs, or symbols as we may call them, underdetermine their realization and are in this sense necessarily ambiguous (p. 113).

Borgmann's final category of information is *technological information*, which is characterized by digital rigor, massive logic and data structures, and rapid processing; as Borgmann points out, "none of the 'pretechnological' kinds of information combined all three features" (p. 167). The primary characteristic of technological information is that it is information as reality. This idea is prevalent in contemporary culture. If one is the owner of a CD recording of

Bach's Cantata no. 10, "you no longer say that you have information about or for Cantata no. 10; you *have* the cantata; what your CD player, amplifier, and speakers produce is not something that is about or for Bach's music. It is the music itself" (p. 181). Of course, this is only true if you also have a CD player, amplifier, and speakers – otherwise the CD is unrealizable. But if you do have these auxiliary devices, the CD can be considered self-realizing. Unlike a musical score, which an accomplished musician might be able to realize (hear) in her head as she reads it (by applying her knowledge of musical notation and experience of musical instruments), one simply needs to flip on the stereo and toss in the compact disc to hear the music – no prior training or practice is required. As Borgmann puts it, "Technology as a way of taking up with reality has put the power of technological information in the service of radical disburdenment" (p. 183).

This is nice for those of us who would like to hear Bach's Cantata no. 10 but who don't have the time, interest, or (let's face it) talent for learning how to realize it from a score, or who don't have the financial means to hear it performed by a musical group (whether by buying a ticket, or hiring singers and players of instruments). This is the upside of radical disburdenment – we can hear beautifully executed renditions of musical compositions at a cost in time, effort, and money that almost any citizen can afford. The downside of this radical disburdenment comes when we don't stop at musical realization, but go for the whole enchilada – virtual reality. As Borgmann sees it,

> The computer, when it harbors virtual reality, is no longer a machine that helps us cope with the world by making a beneficial difference in reality; it makes all the difference and liberates us from actual reality. Of the five terms of information where INTELLIGENCE provided, a PERSON is informed by a SIGN about some THING within a certain CONTEXT, intelligence, things, and context evaporate and leave a person with self-sufficient and peculiarly ambiguous signs (p. 183).

The peculiar ambiguity of technological information becomes evident when we consider how ambiguity is dealt with in cultural information (i.e. information for reality). An ink smudge within the musical staves on the score for a sonata might indicate any of several different notes (or could even be a failed erasure). Its ambiguity can be resolved through realization: play the passage in each of the ways the smudge can be interpreted and use the result (the heard note in context) to determine what the smudge means. In brief, ambiguity is resolved through engagement with an existing reality, and its resolution leads to clarity. Borgmann points out that

> in virtual reality too, resolution is high and engagement is intense. Vividness and interactivity are the terms of art that define these features. But it is characteristic of virtual reality that as resolution and engagement grow, so does ambiguity (p. 186).

Virtual reality is totally ambiguous because it conveys no information about actual reality – it presents itself to us as a self-sufficient context. Unlike an actual Ames room, where one can withdraw from the peephole and examine the room in its broader context to discover its true nature, in a virtual reality presentation of an Ames room, the broader context is supplied by the virtual reality, which could itself have Ames-like characteristics. There is no way to resolve the ambiguity of one's perceptions, no final court of appeal.

Borgmann notes that on one reading, virtual ambiguity is attractive because it allows one to live in different alternative worlds. But after giving a convincing analysis of some current implementations of virtual realities (MUDs, or Multi-User Domains), Borgmann concludes that "virtual ambiguity, when accepted for what it is, renders virtual reality trivial, and, when pressed for its promise of engagement, evaporates" (p. 191). Ultimately, Borgmann points out,

> the human body with all its heaviness and frailty marks the origin of the coordinate space we inhabit. Just as in taking the measure of the universe this original point of our existence is unsurpassable, so in venturing beyond reality the standpoint of our body remains the inescapable pivot (p. 191).

So is virtual reality simply a harmless diversion that promises more than it can deliver? Borgmann sees a deeper problem: "virtuality can spread a fog of virtual confusion and blur the shape of things and events with glamour and triviality" (p. 191). This is not simply a reaction to the way fact and fiction are blurred together – e.g. on world-wide web pages, as "speculation and rumor shade over into factual claims" (p. 192) – but to the havoc Borgmann sees virtual reality wreaking on *eros* and *thanatos*. "Virtual reality has extricated sex from the depths of real life and made it available as a diversion that would be harmless if it were not for the disabilities and displacements it abets in real life" (p. 191). For example, the exploration of Fred's sexuality in cyberspace as, e.g. a bisexual polygamist with a foot fetish is not simply a harmless amusement, but spills over into real life as Fred's evolving on-line sexuality changes the actual relationship Fred has with his mate.

But is that necessarily bad? Maybe the 'real' nature of Fred's sexuality is that of a bisexual foot fetishist, and the actual relationship Fred has with his mate is not one in which Fred can fully realize himself. So while Fred's virtual explorations of his sexuality have repercussions in actual reality, this is all to the good. Borgmann would disagree. Actual reality provides various checks and balances which the ambiguity of cyberspace lacks. Really being a polygamist entails offering a lot of emotional and financial support, of which the sexual relationship is just a small part; being an actual bisexual entails forging a relationship with (at least) two partners, deciding how these partners will be related through you, and taking precautions against sexually transmitted diseases (among other things). All these non-sexual features of human relationships – the constraints of actual reality – feed back into the sexual relationship. So there are two problems here: first, the reckless decisions made in cyberspace can have deleterious effects on one's relationships in actual space and, second, lacking the complete web of consequences provided by actual reality, the explorations in virtual reality are a pale shadow of the real thing, and hence, don't really count as explorations after all.

While this may sound alarmist, conservative, or un-hip in the erotic realm, Borgmann brings this point home with respect to *thanatos* in his discussion of an AIDS sufferer's on-line sharing of his experience of dying.

> Actually to share a person's mortal illness is to feed, clean, and change that person, to suffer the person's bursts of anger and flights of hallucination . . . it is to sleep irregularly and poorly and to feel confined and at times resentful. With all that it can be an occasion of grace and gratitude. In any event, it is quite different from checking your e-mail when you are good and ready (p. 191).

Given this, one can see the source of Borgmann's pessimism for the prospects of electronic information technology providing a technological fix for, e.g. problems of social prejudice. Participation in a virtually diverse environment both trivializes and glamorizes the problems of living in a diverse society. Having a virtual friend of another race does not deeply affect one's relationships with bigoted (actual) friends – although it can allow one to feel as if he or she has actually done something noble toward living in a diverse society.

Ultimately, then, the future of electronic information technologies does not appear to be a promising one. Borgmann argues on theoretical grounds that they lead only to trivialization and glamorization of reality (which, after all, is where we must live). Schiller would point out that these are precisely those features which are attractive to capitalism – glamour sells products, and trivialization keeps one's focus on consumption. Yet the future itself need not be unpromising. Both authors feel that once the true nature of the new electronic information technologies is recognized, we may begin to take steps to counteract their perniciousness. For Borgmann, this means that we must not allow ourselves to be overwhelmed by the glamour of technological information: "we have to become again readers of texts and tellers of stories"; we must not allow technological information completely to supplant natural and cultural information (p. 231). From Schiller's perspective, we must recognize that "digital capitalism has strengthened, rather than banished, the age-old scourges of the market system: inequality and domination" (p. 209). Instead of relying on technology itself to free us, we must take active political steps to ensure that its development is in the direction of liberation rather than domination.

BRIAN ROSMAITA
Kent State University

REFERENCES

Haugeland, J. (1985). *Artificial Intelligence: The Very Idea*. Cambridge, MA: MIT Press.

Keller, G. (1983). *Academic Strategy: The Management Revolution in American Higher Education*. Baltimore: Johns Hopkins University Press.

Lacey, M. (1999). History Is Made by an Old-Fashioned chief.gov. *New York Times*, 9 November, late ed., A20.

Pollack, A. (1999). Show Business Embraces Web, But Cautiously. *New York Times*, 9 November, 1999, late ed., A1.

DIGITAL REALITY CHECK: ARE COMPUTERS TRANSFORMING PHILOSOPHY?

Terrell Ward Bynum and James H. Moor

The Digital Phoenix: How Computers Are Changing Philosophy.
Oxford: Blackwell,, 1998. Pp. 412.

An air of celebration runs through *The Digital Phoenix.* Computers are changing the way things are done in almost every area of life, and philosophy shares with other academic disciplines many of the new techniques and practices the technology provides. All the topics addressed by the informative essays collected in Part 2, for example, are relevant to other professionals working in the humanities: web-based teaching, the use of multimedia in teaching and research, the role of 'list servs' and electronic 'bulletin boards' in professional and pedagogical communication, etc.

By contrast, Part 1 probes a unique relationship between computers and philosophy. In 'Global Information Ethics', Terrell Bynum provides a good overview of how a brand new field in the burgeoning area of 'applied ethics' has opened up around computers. And James Fetzer's 'Reflections on the Program Verification Debate' contains an enlightening account of the backlash encountered when a philosopher tried to point out conceptual ambiguities in the way practicing computer scientists expressed the aims of a particular project. Yet these two examples do not suggest the radical nature of the thesis promoted by the vast majority of essays in *The Digital Phoenix.*

As the editors put it, "a new philosophical paradigm" is emerging "like a phoenix" from the ashes of the old (p. 1). The impact of the computer, in other

Metaphysics, Epistemology, and Technology, Volume, 19, pages 315–322.

ISBN: 0–7623–0681–5

words, does not consist in the creation of new sub-disciplines within philosophy (such as 'computer ethics' and 'philosophy of computer science'), but rather in an essential transformation of the way philosophy is done. Perhaps the most ambitious attempt to foresee the future of philosophy is Eric Steinhart's 'Digital Metaphysics', which argues that "reality is ultimately a massively parallel collection of universal metaphysical (non-classical) computing machines" (p. 126). Of course, conceiving the digital computer as the physical means for representing a monadic metaphysics is reviving an old dream of Leibniz's. For such a metaphysics might, in principle, be able to generate a set of algorithms by which humanity could calculate rational solutions to all life's complicated problems. Steinhart's essay, however, is highly speculative, and only experts in the philosophy of physics will be in a position to critically assess the degree to which his argument fits recent scientific developments. Paul Thagard's 'Computation and the Philosophy of Science' provides a better introduction to *The Digital Phoenix*, because the book claims to be a 'sampler' of the 'rich diversity' of approaches to the relationship between computers and philosophy (p. 4), but Part 1 (taking up more than three quarters of the book) actually concentrates on a diverse sample from one important approach. In a concise explanation of that approach, Thagard identifies the cogency of the book's radical thesis with the success of a specific practice – computational modeling.

According to Thagard, the philosophy of science is central to a "picture of philosophy . . . primarily concerned with producing and evaluating *theories*" since "the most impressive knowledge gained by human beings comes through . . . science" (p. 48). And his description of the practice of computational modeling in this area suggests how a long-standing problem might be overcome. Science, that is, has impressed philosophers largely because of its perceived rationality. Yet a theory of rationality has traditionally been expected to specify rules or algorithms that guarantee solutions to problems in a finite sequence of steps. The complexity of scientific enquiry, therefore, has resisted the demands of full-fledged rationality that seems inevitably tied to the realm of formal logic. After all, when rationality operates within specific domains, criteria of relevance are required for data inclusion and then for mediating possible conflicts between domains. This creates a kind of variation on the Platonic 'third man' scenario, destroying any hope of delineating a necessary connection between problem and solution. Into this context, however, Thagard introduces a rich *working* theory of rationality based on one crucial claim: "we can understand human thinking by postulating mental representations akin to computational data structures and mental procedures akin to algorithms"

(p. 49). Given the hypothesis of computationalism, the rationality actually operating in creative biologists or physicists can be experimentally modeled.

Thagard emphasizes that in modeling scientific enquiry "algorithms . . . cannot guarantee that optimal discoveries will be made from input provided," but rather the models "employ *heuristics*, approximate methods for attempting to cut through data complexity and find patterns" (p. 50). Lindley Darden, inspired by the possibility of devising heuristic rules for identifying the most reasonable scientific explanations among a set of candidates, describes one experiment in computational philosophy of science in 'Anomaly-Driven Theory Redesign'. It is performed with the TRANSGENE system, designed to provide a 'functional representation' of "the components of the theory of the gene" (p. 64) and aimed at the specific goal of investigating the "reasoning in the redesign of . . . [the] scientific theory to improve it to remove an anomaly" (p. 73). TRANSGENE treats the theory as a "device with the functions of prediction and explanation" (p. 65), and hence it fulfills both the backward and forward-looking demands set down by Thagard, namely, to describe "the structure of knowledge and the inferential procedures by which knowledge can be increased" (p. 48). Experiments in other areas expand the range of the computational project. For example, John Pollock's 'Procedural Epistemology' describes OSCAR, an AI system designed to embody rational cognition; Selmer Bringsjord, in 'Philosophy and 'Super' Computation', even claims that the "seminal OSCAR system is intended to be a full-fledged person" (p. 236). And Mark Bedau's essay in the philosophy of biology, 'Philosophical Content and Method of Artificial Life', refers to the BOIDS and TIERRA programs that model the processes characteristic of (non-human) life-systems.

These accounts are enlivened by excitement regarding the way technical innovation fuels increasing theoretical sophistication. Bedau's fascinating discussion, for instance, notes that traditionally "AI models are top-down-specified serial systems," whereas "most natural systems exhibiting complex autonomous behavior seem to be parallel, distributed networks of low-level communicating àgents' " (p. 136). Hence he uses the connectionism influential in contemporary philosophy as a context for understanding relevant models, even as he specifies its practical limitations. Indeed, Bedau expresses a core value of the computational project: "the discipline imposed by expressing a model in feasible computer code entails a salutary precision and clarity and insures that all hypothesized mechanisms could actually work in the real world" (p. 137). Pollock adds that such discipline will "have a very therapeutic effect on a profession that is overly fond of handwaving," and Darden dismisses "an armchair philosopher speculating" at a time when "AI systems allow philosophers to experiment with knowledge and reasoning" (p. 63). In sum, although Thagard summarizes what is to be "gained by adding computational modeling to the philosophical tool kit" (p. 56), more consistent with the experimental spirit he and others describe is the position that philosophers who are not up to speed with the relevant technology can "fairly be accused of professional incompetence" (p. 55). This judgement is inseparable from the radical thesis of *The Digital Phoenix*.

The computational project is certainly not restricted to philosophy of science or epistemology. In Theodore Scaltas' 'Representation of Philosophical Argumentation', for example, the ARCHELOGOS system works on the analysis of historical texts, and in 'How Computers Extend Artificial Morality', Peter Danielson argues for the efficacy of treating morality as a 'cognitive device'. Is it time, therefore, for philosophers to change their approach to the discipline in which they have been trained, retool, and seek out appropriate philosophical laboratories? Such revolutionary euphoria is absent in Henry Kyburg's 'Epistemology and Computing'. This cautious essay does acknowledge computationalism's intellectual discipline, and Kyburg underlines one of its undeniable values: "it was difficult and time consuming for Carnap to work out the detailed model theory corresponding to very simple languages; a philosopher doing similar work now could build a computer program to automatically characterize models of quite complicated fragments of first order languages" (p. 45). Still, if Kyburg admires the project of instantiating a general theory of rationality in the OSCAR system, for example, he insists that it must be possible to test OSCAR's rationality against competing systems. And in this 'great rationality contest', decisions regarding both the relevance of the test *data* as well as the *criteria of success* in performing with the data, are not only exceedingly 'thorny'; they depend upon 'traditional philosophical' thinking (p. 44).

Kyburg opens up a limited, internal criticism of one experiment, but its significance is magnified in Hubert Dreyfus' broad critique of the computational project in 'Response to my Critics'. According to Dreyfus, research in GOFAI (good old-fashioned AI) "assumes that intelligence is based on making inferences and asks how this is done"; the central issue in modeling intelligence, in other words, is the procedural matter of "how to make the right sort of inference" in order to solve problems (p. 207). This requires a set of specific and general 'relevance axioms' of the sort designed by Douglas Lenat for his CYC system. Dreyfus argues, however, that no GOFAI system operating in a non-formalizable domain can be successful because "what counts as relevant depends on the current context . . . but how we classify the current context itself depends on the relevant information" (p. 209). He is much more sympathetic to the conception of intelligence underlying connectionism. For the architecture of simulated neural networks entails that systems "learn from examples and respond in similar ways to similar cases without . . . any rules or principles which serve as a theoretical representation of the structure of the domain in which they work" (p. 205). If, however, an expert system is able to engage in increasingly sophisticated pattern discrimination without having to analyze those patterns "into context-free features" (p. 205), might Dreyfus' deep skepticism regarding computationalism be unwarranted?

James Moor believes so. In 'Assessing Artificial Intelligence and its Critics', he argues that DEEP BLUE's victory over Gary Kasparov falsifies Dreyfus' prediction that no computer could ever compete with human expertise in a non-formalizable domain. In his absorbing and accessible account of the various improvements made to the expert chess system, moreover, Moor exploits Dreyfus' ambiguous analysis of the significance of different computational architectures. Particularly interesting is his description of how DEEP BLUE calculates its successful moves: it uses "heuristics to prune the search tree, to allot certain time to different branches of search and to create the proper mix of depth and breadth of search" (p. 221). DEEP BLUE can match the judgement of a human player precisely because chess-specific criteria of relevance control its number-crunching. Moor concludes that skeptics must simply be emotionally bothered by the idea "that the activities of computers are so obviously a matter of calculation and not judgement" (p. 221) – and hence by the materialistic underpinning of computationalism. After all, he does distinguish the calculative heuristics of computers from the 'holistic intuition' of human judgement for the purposes of argument. But eventually he makes explicit that the hypothesis to which he is committed has no place for the concept of judgement because "at bottom human insights and intuitions arise from calculations as well, calculations of the human brain" (p. 221).

From this perspective, ironically enough, what stands out in Dreyfus' critique is his constructive insistence that human intelligence be understood in terms of 'embodiment'. Our powers of *reason*, in other words, are thoroughly contextualized by a 'pre-understanding' of the world rooted in the way the human *body* interacts with things. 'Perception and manipulation' are the activities by which humans get a 'maximal grip' on a situation (p. 200). Even though criteria for the meaningfulness of data are bound up with the deeply social character of our intelligence, the latter, therefore, is itself shaped by a bodily life that is the ultimate source of cognitive relevance. There is nothing highly original in Dreyfus' view. It draws on philosophers such as Merleau-Ponty who are mentioned nowhere else in the book. And it fits nicely with the concept of 'tacit knowledge' in a marginalized philosopher of science such as Michael Polanyi, for example, as well as Charles Taylor's view that expert judgement is best clarified by skilled physical performances shaped by a specific socio-somatic background. Nevertheless, the sole dissenting voice against the radical thesis of *The Digital Phoenix* forces the crucial question: not whether human intelligence can be adequately understood in terms of the body, but whether "the astonishing phenomenon of the body," as Nietzsche once put it, can be adequately understood in terms of calculation. By answering affirmatively, "calculation not judgement" must be read as a slogan of (what John Searle calls) 'strong AI' and hence a very ambiguous materialism.

In 'Qualitative Experience in Machines', for instance, William Lycan argues for the relative unimportance of the *medium* of intelligence. The contingent fact that humans are embodied in a specific way, he says, scarcely precludes the possibility that our intelligence could "run on considerably different hardware" (p. 173). This is consistent with Moor's assumption that "chess is the paradigm of an intellectual activity" (p. 215). Yet Moore himself points out that a central problem for the computational project has been the absence of a "rich theory of general intelligence" to "guide the design" of a computer system able to operate successfully in a wider domain (p. 228). Chess-playing lacks the socio-somatic texture of the broad spectrum of our intelligent activities (from ballroom dancing to political debate). Given the central problem, therefore, surely chess is an exceedingly poor choice of a paradigm. Indeed, the choice is justified only if the goal is to engineer a disembodied inference machine, namely, a computer that can functionally represent the sort of intelligence a computer is most capable of representing. Bedau's essay on artificial life, concentrating on systems with technical specifications designed to represent the less cerebral end of the spectrum of somatic activity, seems tailor-made to break out of this vicious circle. Instead, the latter morphs into an equivocation. For the BOIDS program represents behavioral regularities according to rules, but Bedau repeatedly refers to BOID behavior as 'following' or being 'determined by' rules. This creates the misleading impression that the same rules represented in computer simulations are *causally efficacious* in the behavior of actual, biological life forms.

Paul Churchland's discussion "of how social and moral knowledge . . . might actually be embodied in the brains of living biological creatures," in 'The Neural Representation of the Social World', undercuts the idea that even the cognitive capacities of humans are caused by "internalized rules-to-be-followed" (p. 162). At first glance this is strange, because he agrees with other contributors that the impact of "cognitive neuroscience and connectionist artificial intelligence" on philosophy "is destined to be revolutionary" (p. 169). But his choice of an exemplary model – the EMPATH neural-network designed to discriminate emotional states on human faces – is the first indication that

Churchland's argument is going in a different direction. For example, he says that mastering the social complexity of teenage life is "a cognitive achievement at least equal to earning a degree in physics" (p. 159), and his praise for George Eliot's or Henry James' ability to capture "the intricate structure of human social space" (p. 158) would not be out of place in a Martha Nussbaum essay. Churchland's concluding defense of an ethical theory centered on 'practical wisdom' stands as a blunt rebuke to Peter Danielson's aptly named 'artificial morality' research project which produces a working computational model only by chopping off anything that does not fit his procrustean theory. The more general point is that the revolutionary destiny of the computational project will require background theories that do justice to the profound richness of the phenomena being modeled.

How persuasive is this brilliant essay? Consider the striking statement that "Aristotle's perspective and the neural network perspective here converge" (p. 167). An Aristotelian ethical theory could easily be embraced by Dreyfus, but skeptics will still point out that the convergence of this theory and neural networks takes place only in the promise of an EMPATH-like system. What is really being expressed is a hope or prediction that a more sophisticated neural net will eventually be capable of instantiating the requisite ethical theory. With Churchland, however, this promise is reminiscent of Thomasina, the visionary in Tom Stoppard's play *Arcadia*, who playfully poses a question in 1806 that appears to make no sense before the invention of the computer can bring together her love of the irregularity of picturesque art with the formal elegance of mathematics in chaos theory: "if there is an equation for a bell, there must be an equation for a bluebell, and if a bluebell why not a rose?" William Lycan, on the contrary, continues to insist that those who reject "calculation not judgement" must be "human chauvinists" (p. 184) bothered by materialism, rather than possible advocates of somatic complexity. He uses his familiar Henry and Henrietta *thought* experiments. Yet they have been imaginatively superceded by the example of *Star Trek's* Data, and Dreyfus explicitly affirms that "there will someday be androids like Data" (p. 210). The relevant issue is the skeptic's claim that when Data arrives he will be based on some fundamentally new scientific discovery – not GOFAI or neural networks.

By juxtaposing humankind with "some futuristic, science-fiction technology" (p. 172), Lycan does provoke intriguing questions of human identity left untouched by other contributors. Of course, virtual reality technology has provided philosophers with a new – and, arguably, an experimental – means of exploring these questions. *The Digital Phoenix*, however, does not address how this pivotal form of computing is transforming the human sense of self (the influential figure of Sherry Turkle is mentioned only in passing, in Bynum's overview of computer ethics). For its radical thesis requires a broad survey of computationalism which inevitably restricts the breadth of Part 1's sample of other approaches to relevant issues. Indirectly, moreover, that focus does a disservice to Part 2. Several essays are only intended to be short, informative descriptions of how philosophers are using computers in the course of their professional activities. But Robert Cavalier's 'Multimedia and Research in Philosophy' and John Dorbolo's 'Philosophy Teaching on the WWW' contain insightful analyses of pedagogical issues that would benefit greatly from richer theoretical surroundings. With reference to a hypertext, for instance, interacting with a computer is more like entering into another world than communicating with another person. Contrary to the assumptions of good, old-fashioned AI, therefore, perhaps a hypertext is best conceived on a cyberspace rather than mechanical self-model, as Benjamin Woolley argued in *Virtual Worlds* (Oxford: Blackwell, 1992). Although it is natural to pose this question given Part 1's focus, this is not perceived to be the place to engage in substantive philosophizing. The irony is that the topics of Part 2 will probably interest a wider range of professional philosophers.

RODERICK NICHOLLS
University College of Cape Breton, Canada

PHILOSOPHY OF MIND AND TECHNOLOGY?

David Chalmers

The Conscious Mind: In Search of a Fundamental Theory. New York, NY: Oxford University Press,, 1996. Pp. vi + 414.

Daniel Dennett

Brainchildren: Essays on Designing Minds. Cambridge, MA: MIT Press, 1998. Pp. vi + 418.

Technology has often served not only as a metaphor but often as a literal model for what was otherwise inadequately understood. Plato compared the application of a Form, an abstract, preexisting, and perfect pattern, to the work of a craftsman who crafts an object from a plan. Leibnitz thought incorporeal minds and bodies both behaved like clocks predestined to harmoniously unwind. In the last century, as Norbert Wiener points out, a living organism was thought of as "a heat engine, burning glucose or glycogen or starch, fats, and proteins, into carbon dioxide, water, and urea" (Wiener, 1948, p. 53). Today functionalists in the philosophy of the mind propose the electronic computer as not only a model of the mind, but one that holds the most promising vision yet of how the mind might be material. The history of the functionalist view is richly described in David Bolter's, 1984 book *Turing's Man.*

Two recent, clear, and engaging books – *The Conscious Mind* by David Chalmers and *Brainchildren* by Daniel Dennett – agree in one respect but disagree in another concerning the prospects of a functionalist account of the mind. Both agree that cognitive or psychological abilities can be understood in

Metaphysics, Epistemology, and Technology, Volume, 19, pages 323–334.
2000 by Elsevier Science Inc.
ISBN: 0-7623-0681-5

functionalist terms, but Chalmers divides the mind into a phenomenal part and a psychological or cognitive part. He then proposes that

> [t]he division of mental properties into phenomenal and psychological properties has the effect of dividing the mind-body problem into two: an easy part and a hard part. The psychological aspects of mind pose many technical problems for cognitive science, and a number of interesting puzzles for physical analysis, but they pose no deep metaphysical enigmas. . . . [There are problems about "learning, memory and language" (p. 4) concerning the psychological aspects of mind and] these problems are all serious, but they have the character of puzzles rather than mysteries. . . . The phenomenal aspects of mind are a different matter. Here, the mind-body problem is as baffling as it ever was. . . . The hardest part of the mind-body problem is the question: how could a physical system give rise to conscious experience? (pp. 24–25).

The suggestion that cognitive and linguistic ability pose "no deep metaphysical enigmas" is challenged, notably but not only, by arguments in John Searle's *Minds, Brains and Science* (1982) and Saul Kripke's *Wittgenstein on Rules and Private Language* (1982). In this review I will nevertheless concentrate on what Chalmers dubs the 'hard problem'.

In regard to the hard problem, Chalmers and Dennett occupy opposite poles of one of the most interesting conflicts in philosophy of the mind: the prospect of fitting consciousness into a physical world. Thomas Nagel has said "without consciousness the mind-body problem would be much less interesting. With consciousness it seems hopeless" (Nagel, 1979, p. 166). In *The Conscious Mind*, Chalmers argues vigorously that the qualitative contents of consciousness, qualia, cannot be understood to be part of the physical world, and for a dualist property view of the mental. According to that view, human beings have both physical and nonphysical mental properties. Dennett's book *Brainchildren* is a collection of essays on various topics from consciousness and animal minds to artificial intelligence. The essays take up threads of views Dennett has argued in numerous books and papers for thirty years. In particular, and in contrast to Chalmers' book, Dennett argues that consciousness can be understood as part of a physical world, that it can be understood within the vocabulary of artificial intelligence, and that qualia do not exist. It is this conflict between Chalmers and Dennett on which I will focus.

The difference in question between Chalmers and Dennett can be seen – or so I will suggest – in connection with paragraph 304 from Wittgenstein's *Philosophical Investigations* (1958). Wittgenstein's interlocutor, distressed by admonitions against privacy, protests, "But you will surely admit that there is a difference between pain-behavior accompanied by pain and pain-behavior without any pain." Wittgenstein responds, saying, "Admit it? What greater difference could there be?" But again the interlocutor protests, "And yet you again and again reach the conclusion that the sensation itself is a nothing." Wittgenstein responds,

> Not at all. It is not a *something*, but not a *nothing* either! The conclusion was only that a nothing would serve just as well as a something about which nothing could be said. We have only rejected the grammar which tries to force itself on us here.

Briefly and at the end, I suggest that Dennett in recent work denies not only that the pain itself is a something but also denies that it is not a nothing. Chalmers, by contrast, insists that the pain itself is a something. Wittgenstein's position is a reasonable compromise between Chalmers and Dennett.

Most would agree with John Searle when he says, "everyone knows that a feeling of pain is one thing and the behavior associated with the pain is another" (Searle, 1997, p. 137). Yet Searle thinks some philosophers overlook the difference and need to be reminded of it:

> I think we all really have conscious states. To remind everyone of this fact I [ask] my readers to perform the small experiment of pinching the left forearm with the right hand to produce a small pain. The pain has a certain sort of qualitative feeling to it, and such qualitative feelings are typical of the sorts of conscious events that form the content of our waking and dreaming lives. To make explicit the differences between conscious events and, for example, mountains and molecules, I [say] consciousness has a first person or subjective ontology (Searle, 1997, p. 120).

Searle's remark is made in a review of Daniel Dennett's, 1991 book *Consciousness Explained*. According to Searle, Dennett denies the existence of consciousness and so he needs to be reminded of the effect of a pinch. The pain, the effect of the pinch, is thought by some philosophers, Chalmers included, to have an intrinsic character, a character logically independent of its relations with physical states which it causes or which cause it, to which a subject has privileged, direct and largely incorrigible access and which eludes a subject's ability to fully communicate to others. And this is exactly what Dennett denies.

In *Brainchildren* Dennett asks,

> Don't our internal discriminative states *also* have some special 'intrinsic' properties, the subjective, private, ineffable properties that constitute *the way things look to us* (sound to us, smell to us, etc)? No. The dispositional properties of those discriminative states already suffice to explain all the effects: the effects on both peripheral behavior (saying 'red!', stepping on the brake, etc.) and 'internal' behavior (judging 'Red!', seeing something as red, reacting with uneasiness or displeasure if, say, red things upset one). Any additional 'qualitative' properties or qualia would thus have no positive role to play in any explanations, nor are they somehow vouchsafed to us 'directly' in intuition. Qualitative properties that are intrinsically conscious are a myth, an artifact of misguided theorizing, not anything given pretheoretically (pp. 143–144).

By contrast, in the *Conscious Mind*, Chalmers argues for the very view that Dennett here rejects. He argues that consciousness has irreducible qualitative properties, qualities that cannot be reduced to anything physical. He repeatedly draws attention to the qualitative character of conscious states, like Searle's pain that results from pinching the forearm, which any conscious creature can know wholly but only by attention to its own case. Chalmers says, "there is nothing we know about more directly than consciousness" (p. 3). "[I]ntuitively, our access to consciousness is not mediated at all. [Rather], conscious experience lies at the center of our epistemic universe; we have access to it *directly*" (p. 196).

Of course, others do not have access to me nor do I have access to others in the way that each of us has access to ourselves. Nonetheless, thinks Chalmers, phenomenal contents and their physical counterparts correlate. Indeed, the former supervene naturally on the latter. "[C]onsciousness arises from functional organization but is not a functional state" (p. 249). He asserts the coherence between physical states and conscious states as a psychophysical law (pp. 242–246). Yet

> the basic evidence comes from familiar cases: ultimately, for me, in my own case. The apparent correlations between awareness [Chalmers' name for the physical, functional counterpart of consciousness] and consciousness in my own case are so detailed and remarkable that there must be something more than a mere chance regularity. There must be some underlying law (p. 243).

And what there is, is not merely whatever causes behavior of a particular sort. To define, say, a pain as "whatever is going on in me when I behave in such and such a manner" would be a functional definition. Chalmers finds such definitions wanting.

> When we talk of a green sensation, this talk is not equivalent simply to talk of 'a state that is caused by grass, trees and so on'. We are talking about the *phenomenal quality* that generally occurs when a state is caused by grass and trees. If there is a causal analysis in the vicinity, it is something like 'the kind of *phenomenal* state that is caused by grass, trees and so on'. The phenomenal element in the analysis prevents an analysis in purely functional terms (p. 23).

Chalmers offers several familiar arguments that the mental does not logically supervene on, and so cannot be reduced to, anything physical. He argues first that zombies are logically, if not empirically, possible. "I confess that the logical possibility of zombies seems . . . obvious to me" – as obvious as it is obviously logically possible that there be a mile high unicycle and possible since "it seems quite clear that there is no hidden contradiction lurking in the description" (p. 96). Second, it is logically possible that from one person to another the color spectrum is inverted. "[O]ne can coherently imagine a physically identical world in which conscious experiences are *inverted* or . . . imagine a being physically identical to me but with inverted conscious experience" (p. 100). His third argument is that consciousness and the physical world are epistemically asymmetrical. "My knowledge of consciousness, in the first instance comes from my own case, not from any external observation" (p. 102). This makes consciousness unlike the rest of the universe. Unlike anything else, the qualitative character of consciousness is accessible only from a first person point of view. "Our grounds for belief in consciousness derive solely from our own experience of it" (p. 101). Fourth, one could know everything that there was to know about the brain, the nervous system, and physics of light, yet not having ever experienced colors, still not know what it is like to see red. And so also, knowledge of physical facts about the brain and nervous system is not logically sufficient to know about the qualitative character of one's own inner conscious states. "For consciousness to be entailed by a set of physical facts, one would need some kind of analysis of the notion of consciousness – the kind of analysis whose satisfaction physical facts could imply – and there is no such analysis to be had" (p. 104). Thus finally, there is no analysis of expressions like 'red sensation' or 'pain' in functional, causal, or physical terms. Therefore, consciousness does not logically supervene on the physical.

Although there is no logical supervenience, Chalmers holds that consciousness does naturally supervene on physical states functionally described. Despite the logical or conceptual possibilities that the foregoing involves, Chalmers insists that there cannot be a strong disassociation between experience and cognition but, rather, there is coherence between consciousness and cognition. As already noted, "consciousness arises in virtue of the functional organization of the brain" (p. 247). The natural, empirical impossibility of disassociation between experience and cognition he argues with thought experiments concerning fading, absent, and dancing qualia. Imagine that there is no coherence between consciousness and cognition. Now imagine someone whose neural cells are gradually replaced by electronic circuits yet the difference — a difference which results from the absence of coherence between consciousness and cognition – goes unnoticed, and so judgments remain unaffected. In this case we would have to say that I am systematically wrong about everything I am experiencing. I say that I am having a brilliant red and orange experience but they are, say, a rather dull pink. Thus I am totally out of touch with my conscious experience and incapable of getting in touch.

> This seems quite implausible. Here we have a being whose rational processes are functioning and who is in fact *conscious*, but who is utterly wrong about his own conscious

> experiences . . . . We have little reason to believe that consciousness is such an ill-behaved phenomenon, and good reason to believe otherwise (p. 257).

Now suppose a system – call it 'Cog' – functionally isomorphic with, say, myself. If we can also suppose that it is absent qualia, then while there is something that it is like for me to be me, there is nothing that it is like for Cog to be Cog. Chalmers undertakes to undermine this imagined case by supposing the gradual replacement of my neural circuits with chips made of what Cog's chips are made of. Thus, at some point I would have to lose all consciousness while functioning normally or my consciousness would have to gradually fade while I continue to function normally. It is implausible that consciousness be completely lost, that there might be *suddenly disappearing qualia*. "If this were possible, there would be brute discontinuities in the laws of nature unlike those we find anywhere else" (p. 255). But, fading qualia are implausible since as they faded and so became qualitatively different, I would remain functionally unchanged. I would thus be wrong about everything I was experiencing. I would say that the light is red, that it looks red to me, and I would stop when in fact it is red, but I nonetheless experience the light to be, say, green. This Chalmers rejects. "One of the most salient empirical facts about consciousness seems to be that when a conscious being with the appropriate conceptual sophistication has experiences, it is capable of forming judgments about those experiences" (p. 257).

A similar conclusion is drawn concerning what Chalmers calls dancing qualia. Imagine my neural cells replaced by circuits making me physically like a robotic twin, yet the original neural cells are not removed. Rather, they are hooked up so as to allow switching back and forth between cells and circuits. All the time I am functionally invariant. Now if we were to say that because of the different wiring there is a change in qualia, then again I do not notice. "My experiences are switching [say] from red to blue but I do not notice any change . . ." (p. 268). As in the case of fading qualia, there is simply no room for such a change to take place, "unless it is in an accompanying Cartesian disembodied mind" (p. 269) – that is, a mind whose qualities do not cohere with physical, functional organization.

These thought experiments argue for psycho-physical coherence. But how do our functionally constituted concepts get hooked up with phenomenal contents? How do a red sensation and the judgment 'red sensation' get together? Functionally construed, concepts are physical structures that take a physical system from informational inputs to behavioral outputs. But Chalmers property dualist position renders phenomenal contents causally inert. "[C]ontemporary science tells us that the physical world is more or less causally closed" (p. 125). "[I]t seems that consciousness is *explanatorily irrelevant* to our claims and judgments about consciousness. This result I call *the paradox of phenomenal judgment*" (p. 177). Thus phenomenal contents, qualia, do not causally explain judgment.

Neither is phenomenal content constituted functionally by judgment. Thinking and judging employ concepts, and phenomenal or qualitative content was to be nonconceptual – that is, not a function of judgment. "There is something more to phenomenal belief than a phenomenal judgment" (p. 206). But also, for Chalmers, there seems more to phenomenal experience than belief. Experience does not even require concepts.

> [W]e have experiences that we do not have beliefs about, and so do not know about. Further, one might have an experience without conceptualizing the experience in any way. To have an experience, and consequently to be acquainted with the experience, is to stand in a relationship to it more primitive than belief: it provides evidence for our beliefs, but it does not itself constitute belief (p. 197).

One wonders how experience might provide evidence yet be unconceptualized. Once the gap is opened, the dualism of concept and phenomenal content it is not easily closed. Yet Chalmers tries. He offers it as conceptual that acquaintance is epistemic, thus cognitive. But how is acquaintance to be understood? Do we impose concepts on the 'passing show' as the classical rationalist might think, or does the 'passing show' impress ideas on us, as the empiricist would propose? Chalmers answers this question saying that "the content of our concepts and beliefs is constituted by something over and above our physical and functional structure . . . the content of phenomenal belief is constituted by phenomenology itself" (p. 207). "Our core epistemic situation already includes our conscious experience" (p. 195). But he does not tell us what the constitutive relation involves. *A* is constitutive of *B* if *A* is basic, elemental, institutive for *B* or a part of *B*. It now seems that the gap is closed by compromising, if not undoing, the distinction between the phenomenal and the psychological.

Chalmers, of course, wants to close the gap so as to guarantee the certainty of phenomenal belief (about this certainty, see pp. 192–200), and so the psycho-physical coherence argued for by the thought experiments above. He asks, "what is it that justifies our beliefs about our experiences, if it is not a causal link to those experiences, and if it is not the mechanisms by which the beliefs are formed? I think the answer to this is clear: it is *having* the experiences that justifies the beliefs" (p. 196). But there is some awkwardness in the suggestion that justification results from having, and thus that having direct, unmediated access to an experience, is immediate evidence for the existence of the experience. The alleged access and resulting certainty ought to render vapid the suggestion that experience is evidence for belief which it justifies. Justification is needed only when there is a gap of sorts between the evidence and what it is evidence for. The evidence in this case is necessarily evidence; so, ordinarily it makes no sense to suppose that one might be wrong.

Chalmers' views – that the character of phenomenal states is intrinsic, ineffable, accessible only from a first-person point of view, that first person claims about phenomenal states are incorrigible, and that they are logically independent of any physical states – lead to what Wittgenstein called a "private language." Considering the possibility of an inverted color spectrum between Mary and her inverted twin, Chalmers says,

> note that when Mary has a red experience for the first time, she learns something *different* from what is learned by her inverted twin, who has green experiences where Mary has red. Mary learns that [things both she and her twin call] red things cause experiences like *this*, whereas her counterpart learns that they cause experiences like that (p. 206).

Thus for neither Mary nor her twin is the content of the concept 'red experience' exhausted by the relational aspects of the concept. For example, that it is caused by looking at fire engines, or that they both concur that the color of the fire engine is called 'red'. Thus Mary and her inverted twin, or me and my inverted twin, each have an inner 'beetle'. On this point Chalmers is worth quoting at length:

> My qualitative concept 'R' plays little direct role in communicative practices. In that way, it resembles Wittgenstein's 'beetle in a box'. My inverted counterpart has a different concept 'G' in the vicinity, but others understand him to be saying the same things that they understand me to be saying, assuming that their own situations remain constant across the two scenarios. This reflects the 'ineffability' [of phenomenal states]: despite the rich intrinsic nature of red sensations, there is little I can say to communicate this difference apart from pinning it down via various relational properties, and assuming that others have the same associated experiences. That is, it appears to be the relational concept of 'red experience' that carries the communicative burden. This ineffability can be seen as providing indirect support for the explanatory irrelevance of experience: the fact that there is little one can say that captures the intrinsic quality of the experience meshes with the fact that the quality plays no direct role in governing cognitive processes . . . the content of our concepts and beliefs is constituted by something over and above our physical and functional structure, so that no reductive account of the belief contents will succeed. My inverted twin and I might be physically identical, but our corresponding qualitative concepts are distinct. . . . Here . . . is a case where the content of a phenomenal belief is constituted by phenomenology itself (pp. 206–207).

In contrast to Chalmers, and as his remarks quoted at the opening of this review make clear, Dennett thinks there are no qualia. He is not, of course, suggesting that we forego saying things like "That stout has a bitter aftertaste," "I can't get that Coltrane melody out of my head," or "He carries on as if he is in pain but he is not in pain." What Dennett objects to is the notion that our ontology must include any more than what our best physical theory requires to explain the mentioned verbal behavior. Chalmers points out that the physical world is causally closed, and so phenomenal states are causally inert. Dennett concludes that there is therefore no explanatory use for them and they ought to be banished from our ontology.

We can characterize the difference between Chalmers and Dennett as follows. Whereas Chalmers would perform *modus ponens*, Dennett would prefer *modus tollens* on a conditional such as the following: if a functional isomorph without conscious experience is conceiveable, then consciousness is something in addition, and no causal, functional vocabulary is sufficient to capture what the world (really!) contains when one is in a conscious state.

Chalmers proposes that qualia are private objects. Wittgenstein says "[a]lways get rid of the private object in this way: assume that it constantly changes, but that you do not notice the change because your memory constantly deceives you" (Wittgenstein, 1958, p. 207). Dennett constructs thought experiments – he calls them intuition pumps – that aid imagination here. Their imaginative thought experiments make both Chalmers and Dennett entertaining as well as provocative reading. Dennett's thought experiments are designed to argue that there is no inner 'beetle' to which we have direct access. Suppose information about colors, so that this information can be machine processed, is encoded in numbers. But now in addition, suppose that evil computer scientists come along and insert into the system a routine that alters the numbers in such a way as to result in color inversion. This routine is inserted after 'sensor' stimulation but before the effects of perception cascade throughout the system. It is within this gap that, were there such things, qualia would be located. Thus, say, the machine identifies as blue what we identify as yellow. Moreover, we can suppose that the machine's memory can be inverted in corresponding ways, thus insuring that the machine behaves as previously. In particular, the appropriately altered color information would be remembered to be what we all call 'yellow'; it would be associated with memories of glorious sunrises, etc.

Now we can ask:

> Are the qualia still inverted? Well the numbers are .... But lacking the capacity for such direct inspection, the robots would be unable to detect any difference in their own color behavior or 'experience', and [the] robot would be oblivious to the switch from its earlier states ... unless there were some leftover links ... which it could use for leverage – for identifying, to itself, the way things used to be (p. 151).

The inversion fantasy "depends crucially on there being a central place from which all 'subjective' color-effects flow" (p. 152). If there is no such clearinghouse – and there is not, according to Dennett – "the whole idea of qualia inversion becomes undefinable for the system. And that is precisely the situation we now know to exist in human perception" (p. 152).

Dennett offers a Multiple Drafts model of consciousness according to which there is no Cartesian Theater and no Master Judge. There is not "a place where it all comes together" (p. 132), nor is there some final incorrigible referee of the character of conscious states. "The creation of conscious experience is not a batch process but a continuous process" (p. 134). There are lots of states competing to enter consciousness but no one place nor moment when the threshold of consciousness is passed. There is no "Cartesian bottleneck or turnstile" (p. 132). Rather, there are a series of micro-takings or micro-judgments "creating larger and larger fragments of what we call narrative" (p. 134), a narrative of conscious life which is in constant revision. "So there is a sort of stream of consciousness, with swirls and eddies but . . . *there is no bridge over the stream*" (p. 134). There is "no fact of the matter about whether one is remembering experiences or misremembering experiences" (p. 135).

In his essay 'Quining Qualia', and in his book *Consciousness Explained*, Dennett offers imaginative variations on the foregoing intuition pump. These intuition pumps are in turn variations on the Wittgensteinian argument that there is no criterion of mistake, and so no such thing as judging right or wrong, when gawking inwardly and taking stock of our own mental innards. I consider an example from 'Quining Qualia' (Dennett, 1990). Chase and Sanborn are coffee tasters for Chase and Sanborn (CS). Chase thinks that CS tastes as it has always tasted, but although he once liked that taste he no longer does. Sanborn thinks that the taste of CS has changed and he does not like the way it now tastes. How to distinguish the two cases? On the one hand, Chase might be misremembering what CS used to taste like and may, misremembering its earlier taste, only dislike its current taste; so Chase is in Sanborn's boat. On the other hand, as Chase himself thinks, his tastes may simply have changed. Then again, Sanborn may be misremembering the taste of CS. It tastes as it has always tasted to him but that taste no longer appeals to him. Or, as Sanborn thinks, the taste may be the same but his tasters have changed and that taste no longer appeals to him.

To further complicate the story, suppose that Chase is surgically rewired so that the coffee whose taste does not appeal to him since it is, say, acidic and pungent, suddenly does appeal to him and is judged rich and full-bodied. There are two ways that this can be understood. First, the compensatory adjustments are prior to taste experience, prior to qualia; they are pre-qualia. Thus his reaction is what it was long ago. If post-qualia, then the qualia are not returned to what they were, but only his reactions to them have changed. How are these two alternatives to be settled? Introspective evidence cannot settle the matter since introspection is what is in question. But neither can any neurophysical evidence settle the difference since there are always alternative ways of interpreting the physical evidence. Physical evidence can settle the question only if there is some point in the brain that can be identified as the transform which serves as the threshold of consciousness. But there is no such transform, no Cartesian Theater, and no Master Judge. Current qualia can be interpreted as either input to a memory-accessing device or as the output of such a device. That is, this taste at time T2 may be given, and then compared with the memory of the taste at time T1. Or, *this* taste at time T2 may be the product of a comparison in the memory issuing as output the judgment that this is taste2. One can misremember the past and judge the past the same as my current qualia. Or one can judge that this current taste is rich, smooth, and full-bodied, and then remember that in the past this is what taste1 was. Dennett calls the latter an Orwellian alternative and the former a Stalinesque alternative (pp. 133–134).

Dennett concludes that there is no fact of the matter about the character of the alleged qualia. This, coupled with their causal inertness, leads him to conclude that they have no explanatory value and so do not exist. Descriptions are beside the point; they can be mistaken. There is nothing available to introspect that settles the question. Rather, we need to explain the judgment that, say, the coffee tastes the same as always. And this, according to Dennett, can be done in functionalist terms. For Dennett, the qualitative character, a qualia, is not only not a something but it is not a nothing either. Chalmers agrees that the judgment, a psychological phenomenon, can be functionally explained. But consciousness of the qualitative state is ineffable and cannot be understood in functionalist terms. A qualitative state, a qualia, for Chalmers, is a genuine something.

Wittgenstein offers a compromise. With Chalmers we can say that consciousness has a genuine qualitative character, a character that cannot be described in functionalist vocabulary nor in any other physicalist vocabulary. Yet, with Wittgenstein and Dennett, we can say that an inner state stands in need of an outer criterion. The beetle, as Wittgenstein says, is not part of the grammar. Thus, with Dennett, we can say that the inner qualitative state is not one viewed by a Master Judge in a Cartesian Theater. But this Wittgensteinian compromise does not require that the qualitative character of conscious states be denied. Dennett can seem to deny that, provoking Searle to want to pinch him. Dennett says, "I claim, then, that sensory qualities are nothing other than the dispositional properties of cerebral states to produce certain further effects in the very observers whose states they are" (p. 146). Space prohibits developing this compromise further here, but it is a compromise worth exploring in more detail.

RICHARD LIEBENDORFER
Minnesota State University, Mankato

REFERENCES

Bolter, D. (1984). *Turing's Man: Western Culture in the Computer Age*. Chapel Hill: University of North Carolina Press.

Dennett, D. (1991). *Consciousness Explained*. Boston: Little Brown.

Dennett, D. (1990). Quining Qualia. In: W. G. Lycan (Ed.), *Mind and Cognition: A Reader*. (pp. 519–547). Cambridge, MA: Blackwell.

Kripke, S. A. (1982). *Wittgenstein on Rules and Private Language: An Elementary Exposition*. Cambridge, MA: Harvard University Press.

Nagel, T. (1979). What Is It Like To Be A Bat? In: *Mortal Questions*. (pp. 165–180). Cambridge, England: Cambridge University Press.

Searle, J. (1982). *Minds, Brains and Science*. Cambridge, MA: Harvard University Press.

Searle, J. (1997). *The Mystery of Consciousness*. New York: New York Review of Books.

Wiener, N. (1948). *Cybernetics, or, Control and Communication in the Animal and the Machine*. New York: Wiley.

Wittgenstein, L. (1958). *Philosophical Investigations*. Trans. by G. E. M. Anscombe. Oxford, England: Blackwell.

LIVING WITH MACHINES

Harry Collins and Martin Kusch

The Shape of Actions: What Humans and Machines Can Do.
Cambridge, MA: MIT Press, 1998. Pp. xi + 212.

Bonnie A. Nardi and Vicki L. O'Day

Information Ecologies: Using Technology with Heart.
Cambridge, MA: MIT Press, 1999. Pp. xiv + 232.

When they work together well, how do people and technology interact? Both of the books I discuss here are centrally concerned with answering versions of this question, though quite different versions.

Bonnie Nardi & Vicki O'Day, in *Information Ecologies: Using Technology with Heart*, treat this question as a practical one about organization, and to a lesser extent about the development of technologies. Their central notion is that of an 'information ecology', defined as "a system of people, practices, values, and technologies in a particular local environment" (p. 49). The 'ecological' part of the metaphor is intended to evoke the diversity, fluidity, and fragility of these systems: we often value ecosystems for the number of niches that have developed within them, and guard against their rapid collapse. The metaphor is supposed to allow Nardi & O'Day to situate themselves between the extremes of uncritical acceptance and blanket condemnation of technology, because they can see some ecologies as more healthy than others. (They discuss relatively few works on technology, most of which are positioned at these untenable extremes).

Because people do not exist without practices, values, and technologies, by their definition almost any human environment can count as an information

Metaphysics, Epistemology, and Technology, Volume, 19, pages 335–343.

ISBN: 0-7623-0681-5

ecology. However, Nardi & O'Day's examples implicitly limit the applicability of the metaphor: they focus on information technologies, and environments such as workplaces and schools, spaces – which might be virtual spaces – where people spend large chunks of time and engage in sustained interactions with a number of other people. These are the environments to which their metaphor applies best.

Nardi & O'Day's main examples stem from five small ethnographic studies of information ecologies. Two of the five are accounts of what they call "keystone species," people – or technologies, though they do not provide any examples of technological keystones – who are essential to healthy environments. Two more of their stories are enthusiastic accounts of the introduction of particular information technologies into schools. And the last is a study of what they see as an unhealthy information ecology, involving monitoring of neurosurgical operations.

Of the showcases for the educational possibilities of computers I'll mention one: Pueblo is a text-based virtual world in which elementary school students and their teachers participate. There is a sense in which Pueblo is a city being built from the ground up by children, using infrastructure provided by Xerox PARC and researchers at Phoenix College. What is lovely about Pueblo is the way in which the children, who are mostly underprivileged inner-city residents, create rich descriptions of the environments they create, and intricate connections among their different components. That sort of richness, Nardi and O'Day imply, is absent from their more constrained schoolwork.

In their study of corporate libraries Nardi and O'Day find, perhaps unsurprisingly, that librarians are keystones. Librarians have skills in locating information efficiently, but more importantly they have skills in articulating patrons' library needs better than the patrons can themselves. Very similar to librarians, and another example of a keystone species, are what Nardi and O'Day call 'gardeners'. Gardeners enjoy working with computers and enjoy solving other people's computer problems, but do so on a more or less volunteer basis. They are essential to many information ecologies because they perform computer support services as insiders. Even when formal computer support is readily available, because they better understand the activities and people they are supporting gardeners can often do a better job than can outsiders. As such they can customize their help to local needs. Nardi and O'Day don't note this, but one might speculate that if computer technicians received the type of training that many librarians do they would probably behave more like gardeners. At the same time they might see their status level suffer if they were perceived as solving human rather than purely technical problems.

Information ecologies look something like the heterogeneous networks described by actor-network theorists like Michel Callon, John Law, and Bruno Latour. For example, the role of librarians and gardeners is to mediate between people and computers, translating the interests of the one into the terms of the other. And like actor-network theorists, Nardi and O'Day claim to be interested in the causes of stability, in what makes for a stable ecology. But their metaphor affords Nardi & O'Day no way of distinguishing between a stable ecology in which human life is impoverished and one in which it is rich, presumably their real goal. McDonald's fast food outlets are clearly very successful information ecologies: efficient, flexible, and even integrating extraordinarily diverse elements. Yet they are hardly model systems.

For Nardi & O'Day people are primary in the sense that what is really at stake in healthy ecologies is the health of the humans. This brings their use of the metaphor very close to the 'technology as tool' metaphor that they criticize: how do artifacts affect people when they are introduced into an environment? Yet the ecological metaphor doesn't really help much to tell us what to expect, or what constitutes healthy human practices. It does not tell us what we should value, and why. Although the subtitle of the book promises insight into "using technology with heart," the heart is largely an uncomfortable addition to information ecologies.

There are other limitations to Nardi & O'Day's perspective. The authors recommend thinking and acting locally: choose technologies that enhance your environment. Yet as readers of *Research in Philosophy and Technology* know, local problems do not always have local causes and local solutions. The emphasis on locality also ignores imbalances of power. It seems difficult to make technological choices in a democratic way, in which all affected people have a voice.

The strength of *Information Ecologies* is not in its theoretical perspective, though, but in its down-to-earth recommendations and its optimistic stories. If you were considering automating a feature of your workplace, or installing a new communication or database program, Nardi and O'Day are the sort of people you would want to hire as consultants. This is not completely surprising, given their backgrounds: they wrote this book while they were researchers in the high-tech industry (Apple, AT&T, Xerox PARC, SRI, and Hewlett-Packard are all acknowledged). Probably the most valuable three pages of the book consist of a categorization, with examples, of questions that one should ask before acting to change an information ecology. Among preliminaries, *analysis questions*, for example, "are concerned with motivations, opinions, and relations between things." They include such questions as: "What is the real goal?" and "Who should participate in forming and carrying out a technology plan?" Those are deepened with *visioning questions*, such as: "If we could do anything, what would be great?" and "What values . . . do we bring to this planning process?" The result of asking such questions is supposed to be a more careful decision-making, design, and implementation process, geared to maintaining and creating healthy information ecologies. So while I don't think that Nardi & O'Day's theoretical apparatus captures what it is that they might achieve with their questions, that does not take away from the value of the questions.

I turn to Harry Collins & Martin Kusch's *The Shape of Actions: What Humans and Machines Can Do*. For Collins & Kusch practical problems are entirely in the background as they address the essential differences between humans and machines, and then examine the nature of human/machine interaction. The authors, a philosophically-minded sociologist of science and a sociologically-minded philosopher respectively, create a theory to describe which types of human actions could, in principle, be automated, but they have no recommendations about which actions we might choose to automate.

I think that *The Shape of Actions* is most neatly situated in a context of the critique of behaviorist psychology, though the same debate has been played out many times: Collins & Kusch place themselves in the context of arguments about the scope of artificial intelligence, and they might have referred as easily to questions about the natural and human sciences, or questions about reductionism in the philosophy of mind. But the critique of behaviorism is probably where the issues are most clear and straightforward.

Drawing on and developing positivist thinking in the philosophy of science, behaviorists argued that minds could only be accounted for scientifically in terms of relations between observable inputs and outputs. They therefore had to be treated as black boxes whose inputs were stimuli and outputs behaviors. Behaviorism failed for a number of reasons, but probably most important was the re-articulation of cognitivist psychology, which sees humans as fundamentally intentional creatures. Stimuli and behaviors are mediated by meaningful thoughts, but there is no way of creating tight correspondences between either stimuli and thoughts, or thoughts and behaviors; those relations are situationally produced. Typing at a keyboard is easily thought of as a behavior, but understanding it requires understanding the meaning of what is being typed, as well as the meaning of various causes and contexts. Thus psychology, and the rest of the human sciences, depend upon investigations of meaning.

Actions are meaningful behaviors. In philosophy the action/behavior distinction is constructed to reflect issues of responsibility. For Collins & Kusch the distinction is constructed to reflect issues of intentionality. Intentionality, in turn, is understood in social terms: quite plausibly, they claim that meanings are essentially connected to forms of life. For Collins & Kusch actions are behaviors that make sense only in the social context in which they occur, and are thus comprehensible only to the extent that the analyst shares a form of life with the actor.

One might think that the distinction between actions and mere behaviors captures the distinction between that which cannot be automated and that which can. But many actions can be automated, if the behaviors that instantiate them are well-enough defined or if people can treat them as such. This book is a development of Collins's earlier forays into this terrain, following the general lessons of his *Artificial Experts: Social Knowledge and Intelligent Machines* (Cambridge, MA: MIT Press, 1990). Collins & Kusch are, however, attempting to create a systematic account, so they devote much of *The Shape of Actions* to making distinctions and to classifying actions and interactions. The central distinction, however, is between what they call 'polimorphic' and 'mimeomorphic' actions.

Polimorphic actions – the spelling is right, creating a pun to draw attention both to multiplicity and to their social derivation —can be instantiated by a number of possible behaviors, only some of which are appropriate in any specific context. Writing a love letter, voting with one's conscience, and being playful are some of the examples that Collins & Kusch give to represent a broad range of polimorphic actions. These are all polimorphic because the actions cannot be completed merely by copying previous instances of the behaviors that instantiate them: a love letter which is merely a copy of a previous love letter would be a failure or a deception; a joke which is told over and over again ceases to be a joke; a vote with one's conscience has to be related to reasons.

Mimeomorphic actions, by contrast, are actions that can be performed by copying some limited repertoire of behaviors over and over again. While a computer-generated love letter does not, in general, count as a love letter, a computer-generated invoice counts all too well as an invoice – in fact, in most modern cultures a computer-generated invoice generally looks more appropriate than a hand-written one. When I make a purchase on my credit card, the clerk at the store inputs the size of the purchase into an electronic cash register. That is usually the last human input into my bill until it arrives at the local post office, ready to be delivered. Invoicing can be automated because the key contents of the invoice are supposed to rigidly reflect orders, purchases, and other money owed. No decisions have to be made that have not been made already by the writers of software, the makers of hardware, and the designers of templates. The rigidity of the process essentially limits the repertoire of the invoicing machinery. This is not to say that the process is simple. Indeed, the credit card bill that arrived for me today is correctly addressed both in print and in a machine-readable code, has three pages of charges from three different countries, correctly credits my last payment, calculates interest on the basis of a daily rate, credits an airline frequent flyer account, and, because I pay my credit card bill electronically, does not include a payment envelope in the package I receive. I have no idea how many other bills this particular system generated, but it numbers at least in the hundreds of thousands. Yet in many years of scrutinizing these bills I have never found an error.

Polimorphic actions cannot be automated, whereas mimeomorphic ones can. For philosophers familiar with debates about reductionism the point that Collins & Kusch are making will look familiar: it is in most of its essentials a claim about multiple realizability. If actions are realizable by a sufficiently wide variety of behaviors, or, because appropriate behaviors are situationally defined, are unpredictably realized by a variety of behaviors, then they cannot be reduced to those behaviors. Polimorphic actions are not definable in terms of behaviors, but rather in intentional terms.

Proponents of artificial intelligence believe that all or almost all actions are mimeomorphic like normal cases of credit card invoicing, in that they can be broken down into a finite number of decision rules, none of which have any grey areas. Opponents of artificial intelligence believe that many important types of actions are polimorphic like writing love letters, and require decision-making skills that only a social actor can have. Collins & Kusch do not devote much space in *The Shape of Actions* to proving that there are genuinely polimorphic actions; they treat this as a largely conceptual point, one already well-enough made in the literature, and intuitively obvious. Nonetheless, they provide a number of complex examples intended to make that point, and to show that the category is not a straightforward one. For example, riding a bicycle in traffic is polimorphic because bicyclists have to communicate with other people on the road, which they do through subtle glances, movements, and signals. Yet riding a bicycle on a deserted road is mimeomorphic because it only requires continual rebalancing, and reacting to road conditions. In principle one could build an automatic bicycle for riding in severely constrained conditions. One could not build a mechanical bicycle courier, though, because it would be unable to understand the goals, and therefore the actions, of other occupants of the road.

With this in the background, how do humans and machines interact? Simplifying somewhat, since machines are incapable of polimorphic actions, interactions have to reserve the polimorphic portions of all actions for humans. Take the use of spell-checking program, one of Collins & Kusch's examples. Such programs do not check spelling in the sense that people do, because they merely compare the words typed with words in a dictionary. They produce both false negatives, when they do not note the difference between 'there' and 'their', and false positives, when all of the proper names, neologisms, and even slightly obscure words are highlighted. False negatives can only be corrected by a human spell-checker doing the job again. False positives are corrected by users simply ignoring them – in a trial just now I found myself positioning the cursor on the 'ignore all' button and clicking it repeatedly. So while spell-checking programs perform useful work, they do not check spelling. Checking spelling is an action done by a person, who may use a program to catch some of the more egregious or easily-overlooked errors. Of course many users of these programs do not realize that there is a distinction.

In human/machine interactions, the humans typically have to correct the output of the machine. They also have to constrain the input. Computer programs, for example, can only understand a very limited range of actions as commands. I cannot just ask my computer to 'go over' the document, or to 'check it over', as I might suggest to myself or request of a secretary. Instead I have to type a specific combination of keys or to select an option via a quite specific set of movements and clicks on my trackpad. This means that what would be a polimorphic action when interacting with humans – in this case asking for copy-editing – often becomes a somewhat different sort of mimeomorphic one when interacting with machines.

How would Collins & Kusch understand Nardi & O'Day's librarians or gardeners? Roughly, those librarians and gardeners perform mediating functions between other humans and machines. When lay users try to run a search on a large database, they typically fail to get what they want, to a greater or lesser degree. The users, operating in natural language contexts and not knowing about the rigidities of the database – and not having the time or inclination to learn about them – do not know how to think about their goals in the terms of the search engine and database. Meanwhile, of course, the computer does not have the meaningful context to be able to understand users's goals. Librarians are essential mediators, because they know how to translate user terms into computer terms, and vice versa. They allow the computer, acting mimeomorphically, to perform desired actions. In addition to shaping its inputs, they may also correct the computer's output, to again configure a search to a user's needs. While this is simply a redescription of Nardi & O'Day's account, it offers a diagnosis of the problems of human/machine interaction that librarians solve, and explains in more general terms why they are a 'keystone species'.

By itself Collins & Kusch's theoretical position does not tell us much about where one should stand with respect to different technologies, or to the politics of technology in general. The authors are studiously non-normative. In a short discussion of automation at McDonald's, for example, their goal is to contest the standard understanding of deskilling, but not to contest McDonald's practices. "Training takes fifteen minutes and full efficiency, according to McDonald's founder, is reached in half an hour." While this looks like almost complete deskilling, according to Collins & Kusch it is in fact the replacement of specialized skills by ubiquitous ones. Workers at McDonald's still have to understand the central point of the exchanges that go on, so that they can correct the errors of the machines and deal with the accents, obliquenesses, and idiosyncrasies of customers. McDonald's cannot survive without human skills, though it has developed a strategy whereby it can do extremely well if franchises are almost entirely staffed by people who have only the (tremendous) skills that most people share. Those skills are not highly valued, which is demoralizing for the employees, and McDonald's can hire at the lowest wages and accept high staff turnover. But Collins & Kusch's main point here is to show the limits of automation, to puncture the pretensions of fans of artificial intelligence, not to take a normative stand on fast food.

McDonald's is a telling case, though, not just for how it manages employees, but for how it manages customers, products, suppliers, and so on. For example, through marketing and repeat exposure, customers have been trained to accept McDonald's products for what they claim to be, to accept a machine-produced hamburger as a hamburger, and so on. With this we are quickly taken back to issues of living with technology: a constantly recurring theme in the criticism of technology is that human life is impoverished when people interact with technology where once they interacted with other people. This is typically attached to one or another version of Marxist or Weberian or Heideggerian analyses of modern rationalities.

Collins & Kusch have something to add, though they only make part of the point. McDonald's has created a set of environments in which the range of appropriate actions, behaviors, and expectations has been restricted. They have to be restricted because otherwise the mechanical parts of the environments would be unable to function: the machines cannot act polimorphically. Collins & Kusch's contribution, then, is a somewhat different explanation than the standard one of the impoverishment that takes place in some socio-technical systems.

Yet some technical innovations increase the range of appropriate actions and expectations, as Nardi & O'Day's Pueblo example shows. The children, who have few resources and opportunities, are given enormous new freedoms when they are allowed to interact in the Pueblo environment. Their actions are restricted to the virtual realm, whatever that means, but the virtual realm opens up a range of new highly valued activities that would otherwise be unavailable to the children: it allows them to create and explore structures that they consider beautiful, interesting, and valuable. Their actions are also limited by each other's actions and creations, but that is part of what gives Pueblo a feeling of reality.

All of this is too simple to count as an analysis of either McDonald's or Pueblo, but it points in the right direction. Some technologies, like McDonald's, most visibly restrict the scope for human actions, as part of a trade for efficiently running machines. Other technologies, like Pueblo, most visibly expand the scope for human actions; the machines are a support service for action. We would not want to live without either type, but we certainly cannot settle for only the former.

SERGIO SISMONDO
Queen's University, Canada

SUPERFICIAL METAPHYSICS AND TECHNOLOGY

Mark C. Taylor

Hiding. Chicago: University of Chicago Press, 1997. Pp. 354.

In the foreword to *Hiding* – a foreword entitled 'How to Read this Book: A Note to the Reader from a Concerned Friend' – Jack Miles tells us that it's a mistake to look for any arguments in this book. "To read *Hiding* by coming to grips with its central argument would be to misread it," Miles writes, "because Mark C. Taylor, its author, is not trying to come to grips with you. You can't wrestle with a man who won't wrestle back, and Taylor won't" (p. 1). "To read this book right, you have to read it wrong," he continues:

> Reading a book wrong is getting lost in what is superficial or merely charming about it – its interesting examples, its striking illustrations, its literary style, the way the book looks on the page, and so forth – and failing to come to grips with its central argument. But that's just the right way to read this book. Its surface fascinations are considerable. Enjoy them. Don't try to link them into a thesis that you can accept or rebut. Don't read *Hiding* as a book, in short. Instead, take it as a trip (p. 1).

After reading this advice, what most impressed me about Mark Taylor's new book was how many arguments I could find in its pages. This is not to suggest that the book is nothing but arguments, presented without embellishments and without any style. On the contrary, here you will find all sorts of things that you would not expect in the typical philosophy book – like pictures, for example, typography that refuses to stay inside the margins, hot pink pages, several different types of paper, a cover that you will wish you could tear off (once you realize what is hiding there), and in general a look and feel that says "experimental art book, suitable for your coffee table" instead of "treatise on

Metaphysics, Epistemology, and Technology, Volume, 19, pages 345–350.
2000 by Elsevier Science Inc.
ISBN: 0–7623–0681–5

certain issues in metaphysics, epistemology, and axiology, suitable for a dusty bookshelf." *Hiding* looks a lot more like the latest issue of *Wired* than the latest issue of the *Review of Metaphysics*. But to argue, as Miles does, that *Hiding* contains no arguments is to go too far. Instead, I would argue that the book should be read as a hypertext, one layer of which consists of traditional philosophical arguments which are meant to persuade an audience well-versed in the tradition of western philosophy that certain assumptions that pervade that tradition must now be called into question.

Above all, Taylor wants to call into question the assumed dichotomy between superficiality and depth. There is a central argument to *Hiding*, and here it is: Depth and superficiality are not in simple opposition; depth is itself another surface. Everything we experience is superficial; surfaces folded within other surfaces, hides hiding hides. But "if depth is but another surface, nothing is profound . . . *nothing is profound*. This does not mean that everything is simply superficial; to the contrary, in the absence of depth, everything becomes endlessly complex" (p. 18). To read *Hiding* without coming to grips with that central argument would definitely be to misread the book. This argument, however, is not just executed in the form of premises and conclusions arranged in standard form; it is also effected by the book's typography and graphics, by the way the text deliberately blurs the distinction between signing and designing. The surface of this book is also an experiment in the principles of superficiality for which Taylor is arguing, and these elements of the text also serve to advance that argument. In addition, the central argument of the book is advanced by Taylor's explorations of superficiality in areas traditionally ignored by academic philosophers, such as dermatology, tattoos, body piercing, fashion, Las Vegas architecture, and virtual reality machines. All of these different layers are linked together in the hypertext of *Hiding*. Surface connects with surface in a text that both presents and embodies an argument about surfaces.

Hiding is divided into five chapters, the first and last of which are the most ambitious and theoretically rich. Taylor frames Chapter 1 in terms of the question of detective stories.

> If nothing is profound, is it any longer possible to write detective stories? Who would be the detective? What would be the detected? Would there be anything left to hide? (p. 18).

In response to this question, Taylor refers us to the work of Dennis Potter and Paul Auster, whose stories are all about detectives who look for clues on the surface which reveal the hidden truth, but discover in the end that there is no hidden truth, only layer upon layer of skin, and the endless complexity that goes with it. The references to Potter's teleplay 'The Singing Detective' are especially prevalent – the title of chapter one, for example, refers to the nightclub where that story begins – and readers would be well-advised to find a copy of this program (it ran in six installments on the BBC in 1986 and on PBS in 1988) and watch it. Much of chapter one will be incoherent unless you understand Taylor's use of this detective story, and this is a crucial part of the text, because the question that Taylor poses about the possibility of detective stories is also a question about the possibility of philosophy and criticism in general. Taylor's discussion of detective stories is important because it establishes the nature of his own project in the rest of the book — the way that he will go about analyzing surfaces that have no depth, and yet are endlessly complex.

In the remainder of the book, Taylor adopts the role of the superficial detective that he has discovered in Potter and Auster. He analyzes surfaces to appreciate their multi-layered complexity instead of attempting to find clues there that point to a deeper meaning that is hidden beneath the surface. In Chapter 1 the surfaces that Taylor investigates include the science of dermatology, phrenology as it was practiced in the nineteenth century, and championed by – among others – Hegel, psychoanalysis, the skin disease psoriasis, and composite portraiture. Along the way, he also considers one period from the surface of his own life, when his father and mother died within the space of a few years. These superficial investigations continue in Chapters 2–5 with extended studies of tattooing, body art, fashion, advertising, pyramids, and Las Vegas architecture.

I suspect that the last chapter will be the most interesting part of the book for readers of this journal, since it is here that Taylor's investigation focuses on electronic technology. In this chapter Taylor attempts to come to terms with virtual reality not as a 'specific technology' but rather as "an effective figure of the postmodern condition" (p. 301). "As modernity gives way to post-modernity, reality is virtualized and virtuality is realized," he writes (p. 268). We now find ourselves living in a world where "[f]rom media, entertainment, and telecommunications to artificial life, neuroscience, and electronic money, virtual technologies are transforming what was once called 'reality' " (p. 268).

Poststructural critical theory has not been equal to the task of understanding this change in our cultural condition, Taylor argues, because it has generally remained stuck on the level of criticizing any form of totalization. "While this sustained critique has been enormously important," Taylor maintains, "little is to be gained by ceaselessly repeating gestures of resistance that have lost their historical necessity and have become utterly familiar and predictable" (p. 272). Poststructural criticism has too often contributed to a politics of identity that is not much of an improvement over the totalizing structures that it meant to displace. And so, Taylor argues, it is time to go beyond merely attacking hegemony in the name of difference.

> While hegemonic systems surely have not been eliminated, it is our differences that increasingly are tearing us apart. Social and cultural survival depends upon our ability to conceive and realize complex structures that can function holistically but not totalistically (p. 272).

In Chapter 5, Taylor argues that the networks and webs that constitute our emerging virtual reality represent structures which are holistic and yet not totalizing, and thereby respond to this social and cultural need that poststructuralism has failed to address.

To make a case for this conclusion Taylor's first task is to trace out a genealogy of virtual reality that reconnects our current condition with its technological, philosophical, and social roots. In terms of the technological genealogy of virtual reality, Taylor discusses the key innovations in hardware and software that have taken place in the last fifty years, such as Ivan Sutherland's 1965 paper 'The Ultimate Display', and his creation of the program Sketchpad, which fundamentally changed the paradigm of computing by allowing user and machine to interact through the display screen. Taylor's social genealogy of virtual reality focuses on changes in society and culture such as the extensive use of glass in architecture that began in the early nineteenth century, the invention of the department store that closely followed, and the evolution of cinema.

In the philosophical genealogy of virtual reality, everything begins with Kant. The first *Critique* establishes a separation between experience and the thing-in-itself, knower and known, real and ideal. The third *Critique* attempts to overcome these oppositions by means of the inner teleology or "purposiveness without a purpose" that Kant finds in beautiful works of art and in nature. But because we can never know reality in itself, we can never be sure that the idea of inner teleology is objectively true. It is, at best, a regulative ideal which helps us to make sense of our experience; we can't be certain that internal teleology is actually constitutive of reality. So Kant's third *Critique* leaves the real and the ideal in opposition.

Hegel attempted to reconcile this opposition by making Kant's idea of inner teleology not just a logical idea, but an ontological idea that gets worked out in history. Inner teleology for Hegel is not just a regulative idea, it is a constitutive Idea. It structures both thought and reality. But this means that the thing-in-itself, which Kant put out of our reach, has now evaporated. In Hegel's system, there is nothing outside the concept. It is both comprehensive and comprehensible. This means that the thing-in-itself has essentially become a fiction, but this realization remained to be worked out by Nietzsche.

"While Hegel historicizes, systematizes, and ontologizes Kant's aesthetic idea, Nietzsche fictionalizes, relativizes, and desystematizes the Hegelian concept" (p. 289). Nietzsche argues that reality is always relational, which means that reality is always virtual.

> Nietzsche realizes that the mind is not hardwired; perceptual and conceptual matrices are in a state of constant flux. Never existing in isolation, these matrices form something like a web or network within which 'reality' apparently is fabricated (p. 290).

The philosophical genealogy of virtual reality moves from Kant's grid of categories through Hegel's system of ontologic, finally arriving at "the open fabric of the Nietzschean web" which "creates a fragmentary field in which aleatory associations generate unexpected patterns that disorganize as much as they organize" (p. 290).

Taylor argues that as a result of these three processes of virtualization, we now find ourselves in a condition where traditional oppositions which have served as compass points for our experience – such as mind/body, human/machine, and natural/artificial — have begun to slip.

> What once seemed to be hard and fast oppositions now appear to be interfaces in which neither term remains the same. Virtual reality involves neither the synthesis of opposites nor the suppression of one term by the other, but gives rise to a different order of reality that eludes traditional classificatory structures. As the material becomes immaterial and the immaterial materializes, binary structures like human/machine, mind/body, and natural/artificial are thoroughly transformed (p. 303).

In the remainder of Chapter 5, Taylor investigates these transformations through the phenomena of artificial life, DNA computers, theories of the brain as a virtual machine, and the body as an information system. All of these interface the biological and the mechanical in a way that blurs the binary oppositions that have distanced humans from the technological world that surrounds them. And all of these, he further argues, are examples of networks that are holistic but not totalizing:

> Can the difference between the material and the immaterial any longer be clearly defined? Is there any reality that is not virtual and any virtuality that is not real? What is surface? What is depth? These distinctions do not merely disappear in a synthetic unity but are reconfigured as endless interfaces, which, though undeniably superficial, are nonetheless profound. Interfaces hide nothing but other interfaces (p. 324).

Obviously, there is a lot more hiding in *Hiding* than I have covered in my superficial review of this superficial book. I have focused on Chapter 5 because of the particular relevance of virtual reality networks to philosophy of technology, but readers will find in Chapters 2 through 4 equally interesting treatments of topics as diverse as Alexander Graham Bell's affection for pyramids and Doctor Masaichi Fukushi's Medical Pathology Museum at Tokyo University. *Hiding* is a fascinating example of a philosophy book that transgresses the traditional boundaries of philosophy books. Anyone who is interested in the interface between philosophy and technology or in the possibility of expanding the horizons of poststructuralist criticism will not be disappointed by this text.

STUART DALTON
University of Hartford

PART III

REVIEWS: METAPHYSICAL AND EPISTIMOLOGICAL ISSUES

TECHNOLOGY AND SCIENCE: TRADING ZONES, DISCONTINUITY, AND INSTRUMENTALIZATION

Peter Galison

Image and Logic: A Material Culture of Microphysics. Chicago: University of Chicago Press, 1997. Pp. xxv + 955.

Peter Galison's *Image and Logic: A Material Culture of Microphysics* is a series of case studies, spanning the twentieth century, of the development of detectors in particle physics. Because of the depth of the individual chapters (each of the seven empirical chapters is about one hundred pages in length), together the case studies amount to a history of detectors and their builders. The 'image' and 'logic' of the title refer to two traditions in microphysics, one centered on singular telling images, and one on statistically manipulated counts of events. Each of these has its own instruments, which have come together in recent years.

In its focus on scientific instruments, *Image and Logic* is not a typical history of either science or technology, and certainly not a typical book likely to be of interest to philosophers. On the science side, theories and experiments are more standard subjects of interest for historians and philosophers. On the technology side, philosophers usually focus on widely-available or widely-seen artifacts, but most of the instruments Galison discusses have never been produced in large quantities; some of them, especially the large detectors used from the 1950s on, have not even been commercially produced. So Galison's attention is on the instruments' design, and their use in very circumscribed contexts. In both of these areas, familiar themes emerge: the design process is as much

Metaphysics, Epistemology, and Technology, Volume, 19, pages 353–357.

ISBN: 0-7623-0681-5

bricolage as rational planning, involving ad hoc and approximate use of physical theory; in their use, these instruments provoke labor issues, including issues to do with the stratification of labor by sex, the automation of some routine tasks, and the organization of the laboratory around the complexities and dangers of machines. As such *Image and Logic* might be valuable to philosophers of technology interested in seeing the work of one of the most able and philosophically-informed young historians of science, writing on a technological topic.

Galison's best-known contributions to the philosophy of science come in the form of two connected ideas, which structure this book. The first is the notion of a 'trading zone', an area in which scientific and/or technical practices can fruitfully interact, without requiring full assimilation. For example, between 1895 and 1911, the physicist C. T. R. Wilson developed the cloud chamber by creating opportunities for trades between atomic physics and meteorology. Wilson's research started in an attempt to understand cloud formation by re-creating clouds inside closed glass chambers. He was led to the hypothesis that clouds formed around ionized particles, and from there to questions about ionization. Because he had been educated in physics at Cambridge and had contacts in the Cavendish Laboratory there, Wilson could draw on the most up-to-date ideas in atomic physics and the study of radioactivity; he thought, for example, that clouds might form in the atmosphere because of radioactive cosmic rays. Meanwhile, his colleagues at the Cavendish used each of his developments to work on condensation physics, from a perspective completely divorced from meteorology. In the end, Wilson had a chamber in which clouds formed along the tracks of individual particles, which were photographed and studied. That cloud chamber became a central instrument in atomic physics, and the photographs it produced became vivid evidence for the reality of particles and the nature of particular interactions.

Wilson's cloud chamber is the first of Galison's cases. For a slightly different use of the 'trading zone' notion, we might look to the last. In a chapter that doesn't fully belong in this book, Galison turns from detectors to models and simulations that are important in the use of detectors and in the interpretation of results. Models and simulations become necessary because predictions cannot, in general, be derived from theories. A model is typically a structure that makes use of theoretical insights, or the basic framework of a theory, in an attempt to describe or make predictions about some class of material situations; simulations do similar work, but in situations that are complex enough to demand instantiation on a computer. Galison's example of a physical model is one created by Richard Feynman and Richard Field to describe 'jets' generated by hadron and lepton collisions. Feynman and Field employ 'plausible physical assumptions' without committing themselves to the truth or the details of the underlying theory. They can therefore say, "[W]e think of our jet model, not as an interesting theory to be checked by experiment, but rather as a possibly reliable guide as to what general properties might be expected experimentally" (p. 649). In so doing Feynman and Field are engaging in an activity that doesn't fit neatly into pre-established categories; they are engaged in making the goods of different groups available and valuable to others, but by changing the nature of the goods. Similarly, in Galison's examination of Monte Carlo simulations, the objects are neither theory nor experiment. Because simulations are symbolic, they cannot belong to the realm of experiment, but because they produce unpredicted data, and because the details of their inner workings are opaque, they cannot belong to the realm of theory either. This is an uncomfortable situation for simulators:

> Caught between a machine life and a symbol life, computer programmers in physics risked becoming pariahs in the sense that many university physics departments found them neither fish nor fowl: they could successfully apply neither for theory positions nor for experimental jobs. At the same time the simulators were also irreplaceable intermediaries between experiment and theory (p. 732).

Galison is not alone in such observations about models and simulations: most of the philosophers, historians, and sociologists studying them are struck by their intermediary location (see, e.g. Morgan & Morrison, 1999; Sismondo & Gissis, 1999). Models and simulations provide neat evidence of the incommensurability of scientific practices.

The notion of trading zones contributes to Galison's second well-known idea, a thesis on the disunity of science, developed in part through some fresh analysis of positivist and post-positivist philosophy of science. Figures such as Thomas Kuhn and Paul Feyerabend have insisted that observation is fully theory-dependent, allowing them to argue for the discontinuity of science through time. But Galison argues that instruments, theories, and experiments change at different times: science's periodization is intercalated. The cloud chamber was succeeded by nuclear emulsions, which were in turn succeeded by bubble chambers. But each of the later detectors was initially used for the same types of experiments and observations, and fit into the same theoretical context as its immediate predecessors had. And when theories changed, there was no immediate change in either experiments or instruments. Discontinuity in one realm, then, is at least generally bounded by continuity in others. On this picture science gains strength from the fact that its key components do not change in tandem. Like Wittgenstein's thread – "the strength of the thread does not reside in the fact that some one fibre runs through its whole length, but in the overlapping of many fibres" – science maintains stability by being disunified.

A disunified science requires trading zones so that its many fibres are in fact twisted around each other. Even while disunified, science hangs together and has a certain stability. Experimenters, engineers, and theorists interact well enough so that each can do work that bears on the work of the others. Collaborations can be successful even if the cultures and practices that are brought together don't agree on problems or definitions. Situations can emerge in which simplified languages – pidgins – allow parties to trade goods and services without concern for maintaining the integrity of practices.

What comes across in *Image and Logic* is that trading zones are ubiquitous. It certainly seems as though almost every interesting piece of work takes place because of trades, and not because of purely discipline-bound moves. Every collaboration and every borrowing of a concept or tool involves subtle adjustments in which practices are made to intersect, even while they do not line up against each other. If this is right then we should see almost all scientific practices as slightly incommensurable with each other. This is, I would argue, a straightforward consequence of practice-centered pictures of science. Making practices central displays incommensurability because *full* meanings of particular concepts or actions must be understood with reference to practices. But practices, which are bounded at points of discontinuity, are simply not the sort of things that can be straightforwardly commensurable. But because of its near-universal applicability, the 'trading zone' metaphor misses its mark: there aren't discrete zones in which trades take place. As Galison uses it, the metaphor is pre-modern, but science might be an activity so rich in exchanges that its economy is in this respect modern, even though there are no general standardized currencies in which to measure trades.

Once we see Galison's picture of intercalated periodization, the reasons for his focusing on some of the instruments of particle physics become more clear. Instruments are essential to the continuity of science, bridging changes in theory and experimentation. As such, Galison wants to claim that they are a key third component of scientific practice, standardly ignored by philosophers, historians, and sociologists.

Galison insists that a modern scientific instrument is not, as Bachelard claims, simply a 'reified theorem'. At the end of the book it is hard to believe that anybody could ever have thought that it was. As I've mentioned, detectors have not changed with changes in theory, nor have they followed predictably from changes in theory. More important, the histories of these detectors show that theory is well-removed from their design and construction. For example, the making of nuclear emulsions was clearly infused with craft knowledge.

Most of the experts were located in the large photographic film companies Kodak and Ilford, and the details of their processes were and are industrial secrets, but Pierre Demers, a chemist at the University of Montreal, was willing to reveal more:

> [the precipitated emulsion] comes down as long noodles or shreds which are washed thoroughly in cold running water .... A good rule is as follows. A noodle is melted between closed hands. Then the hands are rubbed, half opened, and are smelled. A very small trace of alcohol is easily detected. When the smell of alcohol disappears, washing is usually halfway or two thirds through (p. 192). Demers also describes the source of gelatin used for one step in the process: the bones of calves are better than those of pigs, because calves eat wild mustard and other plants containing key sulfur compounds.

Each of Galison's chapters contains such stories of the craft knowledge required to build detectors, or the serendipity involved in their design. For example, in, 1956 the Japanese physicists Shuji Fukui and Sigenori Miyamoto, after learning about the chance creation of an inexpensive and relatively low-tech 'hodoscopic chamber' by two Italian physicists (using abandoned U.S. Army radio equipment), set out to make a large-scale version of their own:

> Stuck in the laboratory without any technical assistance, they foresaw nothing but tedium in the prospect of wrapping 5,000 little glass bulbs in black paper. To lighten their burden, the physicists piled a few bulbs, unwrapped, with gas of varying pressures on top on one another. They turned the pulser on and the lights off. Out of the darkness came a remarkable image that Conversi and Gozzini could not have seen because of the dark paper: inside at least some of the tubes, *tracks* were forming (p. 476).

Although this is an important book, much of its detail will be of interest primarily to historians of physics. It represents ten years of research and writing – a number of the chapters are revised versions of published articles – by a *very* energetic and prolific historian; clearly Galison was loath to leave too much of that ten years' research out of the book. As a result, few people will read the entire work. It is nonetheless well worth dipping into, for its theoretical insights and for its many valuable examples.

SERGIO SISMONDO
Queen's University, Canada

REFERENCES

Morgan, M. S., & Morrison, M. (Eds) (1999). *Models as Mediators: Perspectives on natural and Social Science*. Cambridge, New York: Cambridge University Press.

Sismondo, S., & Gissis, S. (Eds) (1999). Modeling and Simulation. *Science in Context, 12*(2), pp. 247–376.

THE UNKNOWABLE BEING OF LIGHTNESS

David Park

The Fire with the Eye: A Historical Essay on the Nature and Meaning of Light. Princeton, NJ: Princeton University Press, 1997. Pp. xiii + 377.

Park's history of humanity's thought regarding the nature of light and associated matters strikes me as outstanding for three reasons. First, it is highly readable. Second, in addressing apparently metaphorical properties of light, it helps one to appreciate *how different* people's views about light have been. Finally, Park's willingness to provide a perspective based on current optical theory is invaluable, not only for providing another glimpse of how views of light have altered across history, but also for motivating the dynamics of historical change. Inevitably, for an account of this breadth, there are problems of detail, some of which I take up in what follows. Yet the book is marred, in my view, by a sloppiness (both intellectual and typographic) greater than necessary, even given the breadth just mentioned. On the whole, though, I found Park's work an enjoyably provocative introduction to the history of optics. Before attempting to expand on the three points above (especially the second and third), which I think raise interesting issues about how to do history of science, some remarks on just what history Park provides us seem in order.

Park divides his account into three parts of rather unequal length: the ancient, medieval, and modern. In surveying the history that he offers, I shall follow this division.

Three of the book's ten chapters are dedicated to 'ancient' optics. For Park this covers the period from about 600 BCE to roughly 1200 CE, and the region in and around the Mediterranean Basin. This part of the book commences with

Metaphysics, Epistemology, and Technology, Volume, 19, pages 359–363.

ISBN: 0-7623-0681-5

a timeline listing the major figures covered as well as a selection of important (non-optical) historical events. I found this an excellent tool in tracking the subsequent discussion, and was saddened that the timeline did not cover all the figures discussed in the text (a map would have been handy too). Chapter 1 provides the philosophic, scientific, and theological background against which theories of light and vision were developed in the Hellenic and Hellenistic periods. Park gives us the obligatory lines on Thales, but concentrates on the pre-socratics with something to say about vision (Empedocles, Leucippus, and Democritus), and on Plato and Aristotle. On the theological side, we are treated to the views of Mani and the Church Fathers on the spiritual aspects of light. Chapter 2 sets out three ancient theories of vision: the atomist's *eidola*, and Empedocles' and Aristotle's visual rays. Chapter 3 tracks theories of vision through the mathematical work of Euclid (which is illuminated with a series of helpful diagrams, as are many other otherwise confusing points in the text), the experiments of Ptolemy, and the theories of Philoponus, to the critical developments of the Arabs Alkindi and Alhazen.

Faced with the impossibility of saying something sensible about Plato or Aristotle in a couple of pages, I have only two quibbles with this portion of Park's work. First, it would have been good to do more to overcome the Euro-centric nature of this work. Chinese work on optics is surely worthy of mention, and Alhazen's contribution to optics is worthy of more space than Park provides. A brief browsing of Selin's massive work should make this clear – H. Selin, ed., *Encyclopedia of the History of Science, Technology, and Medicine in Non-Western Cultures* (Dordrecht, Netherlands: Kluwer, 1997), see especially pp. 793–799. My second quibble is more substantial. One of Park's virtues is to get the reader to take alternative points of view seriously. His inclusion of the thought of Mani and the Church Fathers (pp. 23–32) as being about the literal properties of light is a fine illustration of this virtue. Why then does he undermine this enterprise by suggesting, on no evidence that is clear to me, that the thought of Thales (p. 48) and Plato (e.g. p. 13) is to be regarded as mythic or metaphorical? Surely consistency demands treating Plato as seriously as Mani, unless strong evidence to the contrary can be found.

The medieval portion of Park's work is the shortest, being composed of just two chapters. It covers the period from about 1200 CE to 1600 CE, and is firmly European in focus, somewhat more forgivably than in the ancient section of the text. Chapter 4 covers the views on light of the High Middle Ages. We are given brief accounts of the spiritual aspects of light, which Park regards as still important at this time, and of the gradual reawakening of a more 'scientific' approach to light and visual phenomena in the work of such figures as Roger Bacon, Robert Grosseteste, and Albertus Magnus. All of this is interspersed with Park's usual asides on relevant aspects of the cultural background, such as the growth and structure of medieval universities. In chapter 5, Park starts us off with the massive work of Witelo (circa 1280), which he sees as the summation of all the optical knowledge then available in Europe. The rest of the chapter is devoted to describing the practical, and to some degree theoretical, challenges to this view that eventually led to Kepler and the 'new optics'. These challenges include the need for a good theory of reflection, some account of the new optical instruments, eye glasses, and the telescope, and perhaps most importantly, the development of a mathematical theory of perspective by artists such as Brunelleschi and Alberti in fifteenth-century Italy. A particular strength of this chapter is Park's ability to make it clear that a conceptual revolution was needed to meet these challenges; optical instruments needed to be seen as aids, not impediments, to vision (p. 125), and the theory of perspective was accepted as a way of 'capturing' reality (p. 127).

Three points are troubling about this portion of the book. First, a minor point: Why does the medieval period finish with Galileo? It seems the answer to this must lie in Kepler's theoretical contributions, which are regarded as largely correct to this day. But this is to impose current judgments of intellectual value on history, a weakness that Park had been successfully avoiding to this point. More seriously, Park's quick summary of medieval thought seems to cross the line between summary and 'gloss'. Granted not every claim in such a broad-ranging work can receive the evidential support or conditionalization that it deserves. However, when Park offers us, without further comment, claims such as, "beauty was not something the people in the middle ages thought much about" (p. 96), or "In the beginning [scholasticism] was all right, but in the end it failed to satisfy a thirst for new knowledge" (p. 105), he oversteps the limits of reasonable summary. Park's final sin in this part of the book is one of omission. To this point he has been interested in following the more spiritual aspects of light, since he sees them as literal rather than metaphorical properties. This interest is a great strength of the book, particularly in the way that it reminds the reader that there has been more than one view across history on what the properties of light are that a good theory must account for. Park's 'sin' is that this approach to light drops out of the story at this point, *without* explanation. That the spiritual aspects of light become metaphoric in the modern period is not the most controversial of claims. The why and how of it, however, surely deserve at least a little discussion, especially given Park's previous interest in these spiritual aspects.

The final portion of Park's work, the last five chapters, covers the period from 1600 to the present. It is, despite taking up half the book, somewhat more sketchy than either of the earlier parts. This is, however, more than understandable, given the vast amount of material it needs to cover. Still, some of the omissions, for example Bishop Berkeley (p. xii), seem a trifle peculiar. Chapter 6 discusses the mathematization of optics by Kepler, Snel, Descartes, and Fermat. I found the simple experiments that Park suggests (e.g. p. 164) to be an exciting and useful addition to the text. Chapter 7 provides an account of the discovery of the properties of light, for example its speed and color, that took place in the seventeenth and early eighteenth centuries. Pride of place goes, naturally enough, to Newton and Huygens, but Park still finds space for the lesser figures and background that make his history so enjoyable. Chapter 8 covers the theoretical rebirth of the wave theory of light in the later eighteenth century, due primarily to the work of Young and Fresnel. It is a compliment to Park that he also covers the work of Goethe and Euler, which many a whiggish historian has passed by. Chapter 8 also covers the development of optical tools such as the telescope, microscope, and spectroscope during the 1700s. Once we get to Chapter 9, Park is reduced to giving us the highlights of the history of optics. There are, however, *real* highlights to be had. In this chapter we get an account of the growth of the theory of electromagnetism, via Faraday, Maxwell, and Hertz, which is then tracked onward to the search for ether and the work of Michaelson-Morley and Einstein. Park ends the chapter with a return to the issue of vision, and the work of Helmholtz, which has languished for some time. Chapter 10 gives a quick glimpse of quantum mechanical thought about light, an even briefer review of current optical tools, including the laser, and then returns to 'philosophic' issues about the nature of light. It is at this point that Park concedes the "unknowable being of lightness." The book concludes with a useful glossary, endnotes, and bibliography, the utility of which would be enhanced by the removal of an unusual number of copy-editing errors.

Park's work as a synthesizer and condenser is outstanding in this portion of the book. Having more rather than less material to work with seems to bring out his best. Some of the problems from the medieval portion of the work remain, though; Park tends, in places, toward the romantic. One example is the way Helmholtz is given sole responsibility for the rise of German science in the nineteenth century (p. 304). There are also touches of glossing, such as the claim that the most important problem facing science at the end of the nineteenth century was the nature of the medium supporting electromagnetic radiation (p. 269). This can hardly be considered a balanced view, given the status of thermodynamics at that time. Finally, Park remains quite whiggish in places, for example in his account of the Michaelson-Morley experiment, which he describes as the Lord talking to his children (p. 295). This is odd, as Park effectively counters these same tendencies in many other places.

Other than these minor points, I would quibble with Park only over his account of Descartes, which seems to me unusually harsh, and his discussion of Newton's views of light. Newton may well have been both a particle theorist at some points and a wave theorist at others, or just confused, but Park's discussion (pp. 206–207) merely confuses the reader without clarifying Newton.

I want to return at this point to an issue I mentioned in my opening remarks, namely, the question of how the history of science ought to be done. In Park's view, "the history of thought resembles the astronomer's expanding universe. As our mental cosmos widens, ideas once closely related spread out . . ." (p. xi). Park lives up to this powerful vision in two ways. First, he incorporates spiritual as well as physical aspects of light into his story, two ideas that, on his metaphor, are initially close but later separate. Second, he reminds us, periodically, of our current perspective on the issues at hand, in order to make clear just how greatly our mental cosmos has expanded. Without endorsing Park's expanding world view, it is still possible to acknowledge his methods as powerful tools for both displaying and grasping difference. We cannot share Ptolemy's world view, but we can appreciate both what it was and that it differs from ours. Park's work is, I believe, an excellent example of how to achieve this position with respect to other world views. The book would benefit from a somewhat more cautious and consistent approach, but these remarks should not detract from Park's achievement. He has given us a readable and stimulating historical introduction to central issues concerning the nature of light.

STEPHEN CROWLEY
Indiana University

PHILOSOPHY AND TECHNOLOGY?

Joseph C. Pitt

Thinking About Technology: Foundations of the Philosophy of Technology. New York: Seven Bridges Press, 2000. Pp. xiii + 146.

Thinking about technology began to appear widely in academic presses shortly after World War II, when governmental craving for Big Science produced great technological feats, such as the Manhattan Project and the atomic bomb. With this craving came also a realization that there were many political and moral issues to worry about. Do the ethical differences between using fission technology to build nuclear weapons for dropping on civilians and to run power plants to generate electricity mean that technology is neutral? What are the dangers with pollution and hazardous waste once nuclear power is widely adopted? Is technology as pervasive as Jacques Ellul claimed in 1954? Can technology, as Langdon Winner wrote, become 'autonomous'? Must we worry about technology, as Hans Jonas suggests, in order to throw up a 'heuristics of fear' about its use? Can the risks of technological application be reasonably assessed, as Kristin Shrader-Frechette maintains? Such questions have framed and deeply influenced the discussion of technology in the circles of philosophers and sociologists, including the early pioneers, from Buckminster Fuller, Ivan Illich, and Joseph Agassi, to later commentators such as Albert Borgmann and Don Ihde. Joseph Pitt's contribution to this ongoing discussion is welcome, even though he seems to limit himself to two major concerns. To do so he has obviously had to ignore much of the existing philosophy and technology literature. A gesture of acknowledgement would have been much appreciated by those laboring for years in the field, unless what is at issue for Pitt is a reclaiming of the field from the sociologists of knowledge and other critics back to the bastion of proper philosophical analysis.

Metaphysics, Epistemology, and Technology, Volume, 19, pages 365–367.

ISBN: 0–7623–0681–5

Pitt's two major concerns are: first, how to define technology, and second, how to respond to the warnings of Winner about the alleged autonomy of technology. (Heidegger's ideas find themselves somehow into the discussion, but it is not clear how they relate to the concern, for example, with an explanatory model for technology.) Pitt's treatment is always lucid and straightforward, analytical and concise. However, I will also point out some shortcomings.

Let's begin with his attempt at defining technology, an attempt he admits is difficult because "nowhere can I locate *Technology* pure and simple" (p. x). He offers the definition of technology as "humanity at work" (p. xi) and insists that technology be distinguished from science (p. 1), only to admit that there is a 'symbiotic' relationship between science and technology (p. 6). He proceeds to suggest that technology "must be seen in broader terms than have been used previously" (p. 10), as if this hasn't been the case for the past fifty years.

Pitt rightly claims that to understand science and technology we should pay some attention to the epistemological frameworks used by both, but here is the rub. For him, it seems, epistemology is defined in standard 'analytic' terms, whether of the Harvard or Pittsburgh kinds – or, as some like to call them, old-fashioned Logical Positivists. So an attempt by epistemologically-minded critics, from Agassi (the Popperian rebel) to the French iconoclasts Lyotard and Latour (the philosophically-inclined socio-anthropologists) is not even considered. What about the Edinburgh gang? How do they all fit into Pitt's idea of epistemology, not to mention philosophy? Are their early contributions not relevant because they diverged from the standard philosophical tradition that took hold of American universities after Carnap?

Let me explain why these questions are crucial for the philosophy of technology. As far as 'philosophies' (or schools of thought) go, philosophy of technology is a relatively young field, a field that attracted thinkers who fell in-between intellectual stools. (Carl Mitcham and Paul Durbin come to mind, as well as the phenomenologist Borgmann, and others) Some were philosophers who were nudged by Heidegger to think beyond the confines of Continental thought and apply American concerns to their study; others were mavericks of their day, such as Buckminster Fuller, Ivan Illich, C. P. Snow, and Norbert Wiener. The newly-founded field attracted them because they could make a difference in framing questions and in applying the best intellectual tools at their disposal, without worrying about orthodoxy. Something similar may also go for more recent contributions that emphasize the special characteristics of postmodern technoscience, which again, are completely ignored by Pitt. The reason for bringing up these omissions is that many of the authors and approaches mentioned have important implications for some of Pitt's statements. For instance, Pitt says that "it naturally follows that technology is epistemically prior to science" (p. 24). But many of those who analyze the phenomenon of technoscience argue that there is no way to so easily classify and categorize. Put differently, there is no 'priority' if there is 'symbiotic' relationship between science and technology — which is perhaps why 'technoscience' isn't such a bad term.

As for Pitt's concern with Winner's approach to technology, there are a number of responses worth mentioning – although let me confess that I'm not interested in adjudicating between their respective claims. Technoscience is not confined to theoretical or practical analyses of the kind offered in chapters 5 and 6, but is rather so pervasive that it may defy, as Pitt admits repeatedly, any single model. What Winner was worried about, to be fair to him, is to ensure human control over the development and dissemination of technological methods and instruments so that the excuse "technology is autonomous" would not be left unchallenged.

In this context, I would have liked to hear more from Pitt about health care research and provision, either in clinical settings or in terms of the Human Genome Program. Does it matter if technoscience is understood in theoretical or practical terms? Does this affect the questions of power relations, as Marxists remind us? What about governmental interference in technoscientific research, in terms of monitoring, funding, and licensing? How is the public invited to participate in these discussions? What about the so-called Information Age and all the social, political, moral, and economic problems emanating from the use of computers and the Internet? What about private and public methods of surveillance linked to new technologies?

I raise such questions not in order to assess Pitt's present contribution to the field, but in order to express my personal gratitude that someone trained in the analytic tradition is willing to plunge into this theoretical and practical mess, and find some provisional answers to complex questions. I hope his essay becomes an invitation to added participation in this conversation, a conversation open to anyone with interest in preserving humanity and seeing it at work, as Pitt so eloquently reminds us, without compromising human aspirations.

RAPHAEL SASSOWER
University of Colorado, Colorado Springs

EPISTEMOLOGY AND TECHNOLOGY?

Robert F. Port and Timothy van Gelder

Mind as Motion: Explorations in the Dynamics of Cognition.
Cambridge, MA: MIT Press, 1998. Pp. x + 587.

This book is a collection of papers that explore the dynamical aspects of mind. Contemporary cognitive science usually looks at the mind as a computational device, and more traditional views have tended to look at the mind as a place where thought occurs, much as space was thought of as a container for motion. The central theme of this book is that mind itself is a dynamical entity best understood through the theory of dynamical systems which applies widely to physical, chemical, and biological phenomena. Since the mind is natural, and based in these types of systems, a satisfactory dynamical theory of mind would permit more direct integration with the biological components of the mental.

The majority of chapters deal with behavior and neural dynamics, but some deal with language and cognition. On the whole, the book presents no integrated theory, but just scratches the surface of a dramatically new approach to mind.

The introduction by the editors gives a nice overview, more ambitious than most of the chapters. They emphasize the mathematical and physical force aspects of dynamical systems (eliminating the problem of how the mind causes things, which plagues more traditional approaches); however the role of force disappears after its introduction, and appears in only two of the empirical papers. This is a major problem with the present state of the development of the dynamical approach: without force, it can at best be descriptive and not explanatory. Another lapse is the lack of discussion of function, which is surely important to understanding intentional activity. Again, only two of the chapters mention function explicitly. Some discussion of open systems would have been

Metaphysics, Epistemology, and Technology, Volume, 19, pages 369–370.

ISBN: 0–7623–0681–5

helpful, since they are a very important subclass of dynamical systems, as has been known since von Bertalanffy introduced *General Systems Theory*. It is very likely that the mind cannot be understood except as an open system (Humberto Maturana and Jerry Fodor notwithstanding).

The second chapter, by Alec Norton, is an introduction to dynamical systems. It is fairly elementary, but highly formal, and makes no mention of force or function. I would recommend much earlier sources on General System Theory, since they are both more sophisticated, and have a better sense of what they are doing – e.g. Ludwig von Bertalanffy, *General System Theory* (New York: George Braziller, 1968). Highly formal dynamical treatments of mental properties aren't much better than computational approaches as explanations or as tools to work with. Norton presents dynamical systems theory as a branch of formal mathematics. As such it has little to do with force or causation. Otherwise, the tutorial is good.

The next chapter, by Esther Thelen, discusses the idea of embodied cognition. Thelen seems to miss one of the most interesting aspects of open systems when she claims that "[d]evelopment has no independent dynamics" (p. 78). Open systems can be self organizing, and to some extent can develop independently of genetics and environment. A great opportunity has been overlooked here. Thelen uses the word 'spontaneous' a number of times, but never defines or explains it. At least she considers force, though she restricts herself to behavior.

Later chapters range across issues from decision making, through sensorimotor skills, to aspects of language. The final chapter by Marc Giunti considers dynamical models of cognition. Again, there is no mention of forces, which tends to trivialize the claims, and makes the argument that the mind is a dynamical system easier than it should be. Nonetheless, he addresses some of the central problems of a dynamical theory of cognition, and lays out some paths for future research.

The book as a whole shows that there are many mental activities that are not easy to understand from the computational point of view, and are more tractable to a dynamical approach. At the same time, the material covered just scratches the surface, and leaves many questions unanswered, presenting no coherent new paradigm. Perhaps that is all we can expect at this early stage of development of the dynamical approach, but I wish more of the authors had steeped themselves in previous work in Systems Theory, which encountered and solved some of the problems three decades ago.

JOHN COLLIER
University of Newcastle, Australia

TECHNOLOGY AND ONTOLOGY?

Brian Cantwell Smith

On the Origin of Objects. Boston: MIT Press, 1996. Pp. xii + 420

Brian Cantwell Smith, a computer science professor with broad interdisciplinary credentials, turned to metaphysics when he became convinced that "developing an adequate theory of computation requires developing a full theory of ontology" (p. 27). Smith's ambitious study seeks nothing less than a new foundation for ontology. This review focuses on three questions. Can ontology do what Smith asks of it? What does Smith provide? How is Smith's work related to other ontologies?

Smith notes correctly that significant practical and legal issues are sometimes expressed in ontological terms:

> To decide whether a given system is or is not an implementation of a standard program (Unix, say) or whether or not one company's chip violates a competitor's copyright (as for example whether AMD violated Intel's copyright on the i486), requires more sophisticated individuation criteria, and more subtle understandings of issues of reduction and supervenience (e.g. do identity conditions cross implementation or abstraction boundaries?) than are supplied by any known ontological theory (p. 47).

But unlike Smith, I doubt that ontological theory has anything to offer lawyers and practitioners concerned with copyrights and implementations. As complex artifacts, software programs and computer chips are analogous to Hobbes's ship of Theseus. Is it the same ship after all the boards have been replaced? Is Mail VI a new program or simply a modification of Mail V? I take these to be practical questions that can be settled in different ways depending on simplicity, goals, and convenience. Lawyers draw sharp lines in gray areas. While Smith presents interesting computer-related examples of indistinct identity and individuation, I was not convinced that metaphysical study had

Metaphysics, Epistemology, and Technology, Volume, 19, pages 371–372.

ISBN: 0-7623-0681-5

anything to contribute to their clarification. Nor did the computer-related ontological puzzles appear to raise previously unappreciated philosophical puzzles.

As an ontologist, Smith aims "to steer a path between the Scylla of naive realism and the Charybdis of pure constructivism" (p. 3). This laudable goal well describes the work of many philosophers who defend a distinction between the real (e.g. electrons) and the merely conventional (e.g. constellations). As evidenced by debates between realists and anti-realists, these are turbulent waters. Smith introduces a term – 'registration' – that is intended to include perception and "all intentional actions that can legitimately be said to have some form of content" (p. 194).

> At its most basic level, the world is depicted as one of cosmic and ultimately ineffable particularity: a critically rich and all-enveloping deictic flux. Neither formally rigid nor nihilistically sloppy, the flux sustains complex processes of registration: a form of interaction, subsuming both representation and ontology, in which 's-regions' or subjects stabilize patches of the flux . . . (p. 347).

How persons, words, and objects hang together in the broadest sense is of course a central issue in philosophy. But I did not find Smith's concept of registration to be helpful in clarifying or even describing these mysteries.

Smith's call to begin with ontology seems to be undermined by his eventual position. If things are as contingent and conventional as he suggests, then stabilizations of the flux should come first – stabilizations that do not necessarily depend on theory. The lawyer worried about copyright infringement needs precedents and guidelines because the flux allows so many possibilities. There is a difference between theories about stabilizations and theories that actually bring about stabilizations, and it is not clear that Smith acknowledges this distinction.

Smith does not undertake the hard work of relating his study to well-developed alternatives. I would have appreciated comparisons with W.V.O. Quine's talk of 'posits' and his famous recipe, "to be is to be the value of a bound variable." Contemporary discussions of realism and anti-realism often highlight Michael Dummett's philosophy of language. Parallels could have been developed between Smith's claim that "individuals are epistemic, not ontological features of physics" (p. 182), and Kant's distinction between phenomena and noumena.

The material from computer science is intriguing. But it is not clear that this is truly a work of philosophy.

DREW CHRISTIE
University of New Hampshire

A HISTORICAL METAPHYSICS OF ELECTRICITY

Ronald C. Tobey

Technology as Freedom: The New Deal and the Electrical Modernization of the American Home. Berkeley, CA: University of California Press, 1996. Pp. xvii + 316.

This book on the history of technology casts revealing light on pivotal and controversial issues in the philosophy of technology. What is the prime mover in the development and spreading of technology? With what degree of explicitness are public decisions regarding the course of technology made? To what extent are moral issues tied to technological ones in public discussion? How large is the scope of individual consumer choice?

Ronald Tobey's answers to these questions are historical. They pertain to the electrification of the American home from, 1920 through 1960. It is the burden of the philosophy of technology that it has to pay attention to history. If it does not, it is likely to fall into error or, worse, vacuity. But debts to history and social science also mark the opportunity of philosophy of technology, the chance to make a significant and helpful contribution to the public conversation and to aspire to what Hilary Putnam has approvingly called "the idea of philosophy as redemptive" (Borradori, 1994, p. 59).

Tobey's book is extraordinarily helpful to philosophers of technology because it covers a crucial period in the development of technology (in fact it is one of Tobey's accomplishments to have shown that period to be such), and because it does so in a philosophically sensitive way without preempting philosophical questions. Tobey combines exacting quantitative research (presented in appendices) with narrative accounts, and he illustrates the broad

Metaphysics, Epistemology, and Technology, Volume, 19, pages 373–375.

ISBN: 0-7623-0681-5

national policies and currents with the engaging particulars of Riverside, California. His prose is clear and serviceable throughout and occasionally produces jewels of precision and illumination. The first two paragraphs of 'Public Enframing' (p. 92) provide an eloquent sketch of how we first shape our subdivisions and houses, and then our subdivisions and houses shape us.

That such shaping takes place in a public, explicit, and consequential way is the major theme of Tobey's book. We tend to think that technological progress moves like a steady rain, covering everything and insinuating itself gradually. We take it for granted that the benefits of technology trickle predictably from the upper to the lower classes in what Fred Hirsch has called "dynamic egalitarianism," where in time the comforts of the middle class will be equal to those the rich are enjoying now. And like Hirsch we are inclined to blame the 'static' (or perhaps better 'synchronic') inequalities on "the neglect of active egalitarian politics" (Hirsch, 1976, p. 166).

In fact, however, the blessings of electricity were at first confined to the upper class and would have remained there for who knows how long had it not been for vigorous liberal, if not quite egalitarian, political endeavors. The electric power companies in the, 1920s had supplied the homes of the rich with electricity but thought of the domestic market at large as barren, and concentrated on industrial customers.

It required enormous energy on the part of social reformers and liberal politicians to bring electricity to middle and lower class homes. To electrify a home, however, is to change it from the ground up, and to succeed in this endeavor is to change not only the shape of a house but its ownership. Neither landlords nor renters were inclined to commit the investment needed for an electrical home. Home ownership was a necessary element of modernization.

The case for modernization was made publicly and on moral terms. Reformers and politicians deplored the drudgery of housework (pp. 40, 51, and 61). And they thought of electricity as a vehicle of hope and happiness. Roosevelt articulated this view when he revisited Tupelo, Mississippi, after it had been connected to cheap electricity from the Tennessee Valley Authority (p. 96; see also pp. 47–48, 51, 120–121, and 208):

> What I saw on those [earlier] trips, what I saw of human beings, made the tears come to my eyes. The great outstanding thing to me for these past three days has been the change in the looks on people's faces. It has not been only a physical thing. It has not been the contrast between what was actually a scarcity of raiment or a lack of food two years ago and better clothing and more food today. Rather it is something in people's faces. I think you understand what I mean. There was not much hope in those days. And yet today I see not only hope, but I see determination and a knowledge that all is well within the country, and that we are coming back. And there is another side of it [TVA]. I have forgotten the exact figures and cannot find them here in this voluminous report at this moment, but the

> number of new refrigerators that have been put in, for example, means something beside just plain dollars and cents. It means a greater human happiness. The introduction of electric cook stoves and all the other dozens of things which, when I was in the Navy, we used to call 'gadgets', is improving human life. They are things not especially new so far as invention is concerned, but more and more are they considered necessities in our American life in every part of the country.

The reform, however, would not have got underway without the deep apprehensions that had settled on industry and commerce through the depression and without the political acumen and resolve of Franklin Delano Roosevelt. Even so, the federal government succeeded not through a wholesale program of public utilities and public housing, but by establishing benchmarks and furnishing demonstration projects through the Tennessee Valley Authority, the Federal Housing Administration, and the like. The several initiatives began to gather momentum in the mid-thirties, were greatly accelerated through the war, and gave this country its present broadly middle-class structure by the mid-nineteen fifties. These roughly two decades constituted an explosively transformative period in the history of technology.

Tobey notes in conclusion how quickly and completely these revolutions have settled into normalcy and oblivion. Most of us have been born into a world of abundant electrical light, electronic information, and the comforts of electric refrigeration, cooking, and cleaning. No one turns a light switch or uses an appliance with a feeling of relief or gratitude engendered by a recollection of darkness and drudgery any longer. Nor is there any awareness of the social and political struggles we owe our comforts to.

Thus concluding his account, Tobey shows that answers to the leading philosophical questions must be given in one way for the period from, 1920 through, 1960 and likely in quite another today. This conclusion returns us to the tasks of the philosophers of technology. They would be foolish to aim at a timeless account of the essence of philosophy. Yet they would be just as mistaken if they confined themselves to analyzing the wrinkles of a social fabric they accept as whole cloth. As Tobey has shown, there once was a searching public and moral debate about the fate of technology. However naive some of the assumptions of that debate may have been, something like it is much needed today, and philosophers should be among the first to instigate it. No matter the failings of the profession as a whole, there are by now a good many philosophers of technology who have devoted themselves to just that task.

ALBERT BORGMANN
University of Montana

REFERENCES

Borradori, G. (1994). *The American Philosopher.* Chicago: University of Chicago Press.
Hirsch, F. (1976). *Social Limits to Growth.* Cambridge, MA: Harvard University Press.

PART IV

REVIEWS: GENERAL

FROM BOON TO BANE: EVOLVING LOCAL PERCEPTIONS OF A NUCLEAR WEAPONS PRODUCTION PLANT

Len Ackland

Making a Real Killing: Rocky Flats and the Nuclear West. Albuquerque, NM: University of New Mexico Press, 1999. Pp. xi + 308.

Len Ackland sets out to discover "why the citizens of the world's greatest democracy willingly participated in building huge nuclear arsenals capable of destroying the human species" (pp. 3–4). In the process he sketches out tentative answers to two additional questions: why did these citizens eventually reverse their course? and what is the 'legacy' left behind by these decisions? Ackland's answers are based on a wealth of interviews and primary documentary sources. From these he weaves an intimate story of the life of the Rocky Flats Nuclear Weapons Plant from the points of view of workers, managers, neighbors, protestors, reporters, and a host of other mainly local actors. The story displays a quite different texture and substance from the usual political histories of the U.S. military industrial complex.

Ackland's story begins with the Church family and their ranch, backed up against the foothills between Boulder and Denver, Colorado. Much of Church Ranch was taken by the US government in 1951 for construction of the Rocky Flats facility. Ackland returns to conversations with family members periodically throughout his story, giving voice and insight into the ambivalent, changing attitudes of Rocky Flats' neighbors. The Church family, like many other neighbors of the weapons plant, alternated between anticipation of economic benefit, apprehension about spillovers from the plant's activities, and

Metaphysics, Epistemology, and Technology, Volume, 19, pages 379–383.
2000 by Elsevier Science Inc.
ISBN: 0–7623–0681–5

resentment of perceived broken government promises. Ackland also follows the actions and reminiscences of several Rocky Flats workers who are pulled in multiple directions by Cold War patriotism, the security of high-paying employment, and anxiety over safety and health. Likewise, several plant managers are recurring characters, with their concerns about meeting production deadlines and maintaining plant security weighed against health and safety worries similar to those of the workers. Beginning in the sixties antinuclear and environmental protesters are added to the mix. Throughout, the 'major players' – politicians, high-level appointees and civil servants, the local media, and the like – make cameo appearances, but the local citizens are in the spotlight of Ackland's story.

Through the eyes and voices of these citizens we come to understand the answer to the first question. Citizens accepted and participated in the nuclear weapons enterprise because they did not know and were not particularly curious about what the plant did; they were patriotic and believed their leaders when they told them the plant was necessary for national security; they saw opportunity for economic security and gain for themselves and their families and communities; and they were not aware of the potential environmental, health and safety risks, at least in the beginning. In short, the advantages and disadvantages of such an enterprise in their midst were unclear and countervailing. Ambivalence led to acceptance and inaction.

The balance began to tilt ever so slightly after Mother's Day, 1969, when a fire nearly breached the confines of one of the plutonium-laden weapons assembly buildings. The fire 'awakened' local scientists, activists, and journalists to both the local threat and the global hazard posed by the plant's activities, leading by the mid-1970s to increased media scrutiny, public meetings, and protest demonstrations. Over the next decade, local activists in the anti-nuclear, peace, and environmental movements made common cause over Rocky Flats. The movement to shutdown Rocky Flats continued to build until its climax in 1989, when nuclear weapons production at the plant was finally ended. That was the same year in which the Berlin Wall came down. The crumbling of the 'Evil Empire' eroded the Cold War ideology that had protected the plant until that time. The ideological shift became fully apparent on June 6, 1989, when 75 agents from the FBI and EPA conducted a surprise raid on the plant to obtain evidence of violations of environmental laws. Such an action would have been unthinkable just a few years earlier. The convergence of the environmental and peace movements during the seventies and eighties, together with the decay of the Soviet Union, shifted both national and local public perceptions about the desirability of living next to a nuclear neighbor.

Attention today focuses on what's been left in the wake of this episode in our military history–the 'legacy' of Rocky Flats, as Ackland terms it: How and to what extent can the 6550-acre site be cleaned up? What have been the long-term health effects on those who worked in the plant? Have surrounding properties been contaminated? What future land uses might be appropriate? Is this something that we will be able to put behind us after a few years, or will its consequences linger on for generations? What did we lose and what have we gained from Rocky Flats? Was it worth the price? And most of all, have we learned anything at all about how we govern ourselves?

Ackland's answers to these kinds of questions are abbreviated, at best. And perhaps that is understandable, since these are questions that the community is still struggling to answer for itself. Ackland's tone is rather wistful, if not downright pessimistic, when he addresses these questions. He is skeptical of the adequacy of cleanup plans, the $6.3 billion cost estimate, and the 2006 completion deadline. He is critical of the inability of the U.S. Department of Energy to answer workers' questions about delayed health effects or to provide adequate health insurance coverage.[1] He raises questions about the studies that have concluded there is no off-site contamination.[2] He does not really address the question of future land use – on- or off-site–at all, which is one of the current major focuses of Rocky Flats' neighbors. Nor does he address a second issue of major concern currently, the problem of waste transport off site.[3] He is dubious about our being able to put the Rocky Flats Era behind us, because of the long-lingering nature of nuclear contamination problems. Similarly, he suggests that costs and benefits of the endeavor do not balance out; our national-local partnership in peace-through-strength was not worth the price we will have paid by the time our descendants can close the book on the episode.

And have we learned anything through making "the worst mistakes of the twentieth century"? This question is never really answered, though one suspects Ackland thinks not. Certainly there is good reason to wonder today whether there has been any 'policy learning' at any level – local, national, or global. Locally, there are some hopeful signs. There was little or no opposition to the federal government's takeover of Rocky Flats in the 1950s. Today, the decision-making process is no longer shrouded in deep secrecy, so the public is much better informed. Environmental groups and neighboring residents, with the help of their Congressman, Mark Udall, are advancing a plan to turn the Flats into open space for wildlife preservation and recreational uses.[4] This approach is similar to that followed with the closure and cleanup of Rocky Mountain Arsenal on the northeast edge of Denver. But, motivation to turn the land to more economically 'productive' uses remains strong. As long as cleanup continues at the facility, the employment base remains; in fact, the number of jobs at Rocky Flats in its cleanup phase has actually increased substantially over employment levels during weapons production. But once decontamination is completed – or once federal decontamination dollars dry up, whether the job is completed or not – a major piece of the area's economic base will disappear. Consequently, local governments and private interests press for development of the real estate in and around the Rocky Flats facility. Labor unions and businesses cast about for job- and profit-generating industrial uses of the old plant facilities.

Nationally, our stock of nuclear weaponry has presumably been reduced, but there is no hint of ending production altogether. And globally . . . well, recent developments in nuclear weaponry in India and Pakistan are only one alarming signal. According to Jonathan Schell's recent article, there are still 31,000 nuclear weapons remaining in the world.[5] The prospects for ending reliance on nuclear weaponry is bleak. Hence, the ecological, health, and safety problems associated with their manufacture will continue for the workers and neighbors of weapons production facilities in this country and elsewhere on the globe.

KAREN B. WILEY
Colorado School of Mines

NOTES

1. Since Ackland's book was published, the US Department of Energy has released a comprehensive study on worker exposure and cancer rates at the nation's 14 nuclear weapons plants. The study concludes what has previously been denied: that workers at the plants were indeed exposed to high levels of radiation and chemicals and that these produced higher-than-normal cancer rates. Legislation has been introduced in Congress to compensate workers and their families for the consequent illnesses and deaths. (Matthew L. Wald, 'US. Acknowledges Radiation Caused Cancers in Workers', *New York Times*, January 29, 2000).

2. In, 1990, DOE provided funding to the Colorado Department of Health and Environment to study Rocky Flats' health effects on neighboring communities. The nine-year study was completed last August and concluded that cancer risks to the public in the area were low. Technical reports are available on the study's website: <http:/ /www.cdphe.state.co.us/rf>. ('Study Results Indicate Low Cancer Risks Around Rocky Flats', *The Advisor: A Publication of the Rocky Flats Citizens Advisory Board*, Winter, 1999, p. 3).

3. Mixed chemical and radioactive wastes from Rocky Flats were to be shipped to the Waste Isolation Pilot Plant in New Mexico for permanent underground storage. Although shipments were started in June, 1999, New Mexico has raised a number of barriers to continued transport in its permit for WIPP. (Cat Lazaroff, 'Millions, State's Rights, Worker Safety at Stake in Nuclear Shipment Lawsuit', *Environment News Service*, 11/5/99, <http://ens.lycos.com/ens/nov99/1999L–11–04–06.html>.

4. Rocky Flats Open Space Act, H.R.2179, 106th Congress, 1st Session.
5. Jonathan Schell, 'The Unfinished Twentieth Century: What We Have Forgotten about Nuclear Weapons', *Harper's Magazine*, January 2000, p. 43.

SHAPED BY THE ATOMIC AGE

Michele Stenehjem Gerber

On the Home Front: The Cold War Legacy of the Hanford Nuclear Site. Lincoln, NE: University of Nebraska Press, 1997. Pp. 334.

Hal K. Rothman

On Rims and Ridges: The Los Alamos Area Since 1880. Lincoln, NE: University of Nebraska Press, 1997. Pp. xiii + 384.

In *On the Home Front* and *On Rimes and Ridges*, Michele Gerber and Hal Rothman each give detailed descriptions of particularly remote parts of the United States that are suddenly transformed by a secret government project. Gerber looks at the Columbia Basin, a desolate area in Washington state frequently plagued by dust storms, that becomes the site of the Hanford Engineer Works. Rothman discusses the arid Pajarito Plateau in New Mexico that becomes the seat of part of the Manhattan project.

In *On the Home Front*, Gerber provides a detailed account of the history of the facilities that produced plutonium for the atomic bombs that ended Word War II and the missiles that shaped the Cold War. After a description of the construction of the plants and their contribution to the region's economy, Gerber describes the history of airborne, river-borne, and groundwater contaminants in three separate chapters. Her account reveals a peculiar mixture of extensive and precise environmental monitoring, with shifting norms about tolerable doses and a lack of understanding about the impact of radioactive substances on the environment.

The main weakness of the book is the absence of a chapter on the different production techniques employed at the Hanford site. Without this, it is difficult

Metaphysics, Epistemology, and Technology, Volume, 19, pages 385–387.
2000 by Elsevier Science Inc.
ISBN: 0–7623–0681–5

to put into proper perspective the long lists of disposals of radioactive substances. Nevertheless, Gerber's patient and careful description provides a wealth of material illustrating how scientists and bureaucrats treat substances the nature of which they do not fully understand. Gerber also does a good job in correlating the disposals of what she calls 'complex waste' with the growth of knowledge of the effects of radioactive materials on humans, the food chain, and the environment overall. Of particular interest are her descriptions of the erratic health advisories, such as the advice to cut down sagebrush and Russian Thistle, to use iodized salt, or to drink only Richland's milk.

In the final chapter, Gerber describes the events that follow when the Department of Energy releases nineteen thousand documents about the Hanford site. In this she devotes most attention to discussions on the future of the site and the cleanup operation started by the Department of Energy. The patriotism that accompanied the early years of the Hanford operation is now replaced by class action lawsuits and the unwillingness of other states to store part of the nuclear waste generated there.

On the Home Front was first published in, 1992. The 1997 paperback edition comes with an epilogue, in which Gerber discusses the environmental cleanup at the Hanford site during the years after publication of the hardback edition. By, 1994, approximately 18,400 persons were involved in the cleanup process, a far greater number than has ever worked at the site, and with an annual budget of $1.88 billion.

Whereas Michele Gerber glances only briefly over the history of the Columbia Basin before the Hanford Engineer Works, most of Hal Rothman's *On Rims and Ridges* deals with the barren Pajarito Plateau before it becomes the seat of Project Y, the arm of the Manhattan Project that developed the atomic bomb. Rothman describes in great detail how in less than a hundred years the Pajarito Plateau falls from one extreme into the other. Before the railroad crossed the upper Rio Grande in 1879, no rules or laws were governing the use of this land and its scant resources, and there existed only a loose sense of communal ownership. One century later, not an inch of the land remains unclaimed, and there are constant disputes regarding land ownership and usage rights.

Rothman does an excellent job in showing the different notions of property rights and land use that come to play a role on the plateau. He examines not only the different cultural notions – such as the differences between Native American, Spanish, and Northern European notions of property – but also those that follow from a changing relationship to the land. First there is the subsistence farmer, who makes the plateau his home and depends on its sustenance for that of his own. Next comes the archeologist, who makes his career dependent on taking the land away from others so as to preserve the ancient Indian artifacts. Finally there is the physicist, to whom the location is little other than a diversion (beautiful sunsets and scenic mountain trails) from work that is fully detached from the region, in that it could have been done almost anywhere. Curiously, the impact of the last group is by far the greatest.

On Rims and Ridges was also first published in 1992. The 1997 paperback edition, like Gerber's book, comes with a new epilogue. Rothman too addresses how the end of the Cold War affects the region. In this he contrasts the Los Alamos community, which is characterized by an unbridled faith that to all problems there are technical solutions, with the emergence of a New Age culture in nearby Santa Fe, which is characterized by anti-technological and anti-progress ideals. Simultaneously, Los Alamos, now no longer under the generous protectorate of the Department of Energy and the University of California, is for the first time in its history forced to compete in the world at large.

CORNELIS DE WAAL
Indiana University

OF MACARONIS, PROSTITUTES, AND EXCISE MEN

Miles Ogburn

Spaces of Modernity: London's Geographies 1680–1780. New York: Guildford Press, 1998. Pp. 340.

Typically, discussions of modernity and the city concentrate on the industrialized and consumer-oriented nineteenth-century city and the figure of the flaneur. Miles Ogborn bucks that trend by concentrating instead on London of the seventeenth and eighteenth centuries, and though he includes the modern figure of the flaneur, it is joined by many others, including the prostitute, the tax man, the policeman, the dandy, and the imperial hero. By choosing this time and place, Ogborn is able to show modernities under construction, highlighting how they are partial, contradictory, particular, and always in process – thus complicating our understanding of modernity and the city in the nineteenth and twentieth centuries as well.

Spaces of Modernity consists of a series of essays, each examining a modern practice and space tied to seventeenth- and eighteenth-century London. These are introduced by a chapter examining the multiplicity of understandings of modernity in contemporary discourse and marrying them within a plural, complex, and geographically and temporally situated model of modernity. Each case-study chapter focuses primarily on one aspect of the process of modernity: shaping the modern self, inventing the public sphere, consumption and artifice, forming the bureaucratic state, and using the systematic collection of information to change connections across time and space. Each of these aspects of modernity is examined as it is situated physically and geographically in a particular place and activity. Chapters 2 through 5 focus on different sites

Metaphysics, Epistemology, and Technology, Volume, 19, pages 389–393.

ISBN: 0-7623-0681-5

within the city of London – the Magdalen Hospital, the streets of Westminster, and Vauxhall Gardens – using them and their associated activities and discourse as a window into changing concepts of self and society. Chapters 6 and 7 focus on the spatial and social practices implicated by excise taxation and the Universal Register Office, examining how these practices remap England, the city of London, and the relationships between London, England, and the rest of the world.

Though each chapter focuses on one particular aspect of modernity, these aspects are not entirely separable from each other, and thus themes carry through from chapter to chapter, with each chapter informing the others within the mind of the reader. Ogborn does not explicitly mark out the continuities between chapters, leaving them instead, in his own words, to "sit somewhat uneasily alongside each other" (p. 231). His brief (seven-page) concluding chapter draws the case studies together as examples of "modernity's implication of institutional transformations of space and social relations with new identities," but does not explore how they elucidate each other (p. 233). A further examination of the connections and tensions between the case studies would open up more insights and questions about modernity, its spaces, and especially gender and the modern subject.

The modern self and the public sphere, the themes of the first two case-study chapters, thread particularly strongly through the book as a whole. The importance of the self and the public as themes is also highlighted by the theoretical sections at the beginning of these two chapters. Chapter 2, 'The Magdalen Hospital', begins with a discussion of historians' competing stories of the formation of the modern self, while Chapter 3, 'The Street', begins by laying out both Habermas' model of the public sphere and eighteenth-century political philosophers Shaftesbury, Mandeville, and Hume's understandings of the public and its relationship to the individual. Later chapters do not begin with this sort of philosophical and theoretical overview, but instead jump directly into the specifics of the sites and practices examined.

Ogborn discusses the modern self through the site of the Magdalen Hospital, a charitable institution founded in 1758 with the purpose of reforming prostitutes through a regime of work, proper activities, self-reflection, penitence, and isolation from the outside world. These prostitutes, victims of deception, the unpredictability of modern life, and the evils and anonymity of the city, were confined in the Magdalen to remove them from these forces. Within the Magdalen, life was fully predictable and closely surveilled by staff, other inmates, God, and ultimately themselves, as they internalized "the external eye" in the form of self-reflection and self-control. Ideally, these disciplined modern women were reabsorbed into the family and a life of subservience, but in their self-possession they also came uncomfortably close to the autonomy of the prostitute (for many the iconic modern woman).

It is somewhat ironic that, while Ogborn suggests that "the first modern individuals – with all the ambivalences that implies – *were* women," the modern individuals around whom the remainder of the book is organized are nearly all male (p. 74). Throughout, the gendering of these modern individuals is marked and remarked upon, but it is explored in detail only in the chapters dealing with prostitutes and Macaronis, figures whose female sex or perceived femininity were problematic.

In 'The Street', the disciplined modern self returns as the male pedestrian, whose civilized behavior and self-control is necessary to the proper functioning of the public sphere. In Chapter 5, 'Excise Geographies', it recurs in the form of the excise man, the tax assessor who ideally reflects the mathematical impartiality of his profession through his proper behavior, freedom from local entanglements, and total avoidance of excess. In his disciplined, self-controlled person, the excise man embodies the rational bureaucratic state.

In other chapters, the modern individual is presented as more problematic, particularly in a modern world of anonymity and artifice, in which people can appear to be what they are not. In Chapter 4, 'The Pleasure Garden', the Macaroni, the insufficiently masculine dandy whose self-presentation is overly artificial and too closely tied to consumption, is presented as a problematic modern individual, modern in his ties to fashion and the international world of goods, but improper in his lack of self-control and in his artifice. This concern with artifice is also central in Chapter 6, on the 'Universal Register Office', an agency which collected information on services and goods for sale and wanted. In his discussion of Joseph Reed's 1761 farce, 'The Register Office', Ogborn shows that the proper functioning of a register office depends on the honesty and lack of artifice of the people registering their positions and services, as well as on the honesty of the man running the office. When, as in the farce, information is not given truthfully and individuals misrepresent themselves, the office functions improperly, and often in the service of prostitution.

This brief survey of the role of the individual in Ogborn's modern geographies also highlights the importance of the public sphere, and the relationship between the public and the individual upon which it depends, throughout the book. In 'The Street', Ogborn sets out Hume's view of a civilized public sphere in which private interests are regulated by "the state, the reasoned self, and civilized, feminized sociability," arguing that this Humean view is necessary to understanding the paving of Westminster (p. 89). This view of a public sphere in which private interests are balanced by the state and the disciplined modern self is not only useful for understanding the change from haphazardly individual, to regulated, to government paving of the street, but is also important in other chapters. For example, the Universal Register Office was predicated on a public made up of disciplined individuals, without which the public service of the office cannot function, just as the system of excise taxation was predicated on excise men disciplining their individual interests in the interest of the public purse.

The role of the public is strongest in the chapters addressing physical public space: Vauxhall Gardens and the streets of pedestrians and prostitutes. 'The Pleasure Garden' is an exegesis of an incident at Vauxhall gardens in which a group of Macaronis who were improperly staring at Mrs. Hartley, an actress, were challenged by Mrs. Hartley's escort, a pugilistic parson. In this incident the Macaronis disrupted the public sphere by improper behavior and dress, and compounded this by disguising a footman as a gentleman and sending him to box the parson, thus breaking rules of honesty. The prostitutes housed at the Magdalen were problematic because they created disorder through not taking their proper public roles as mothers and through destroying marriages, putting their private desires before the public good.

These two themes – the modern individual and the underpinning of the civilized public sphere – come together in a gendered form in the body of the prostitute, with whom the book begins, and whose figure recurs elsewhere. A 'public woman', after all, is a prostitute. The staring of the Macaronis, for example, challenges Mrs. Hartley's identity as an honest woman, especially as her fame as an actress, a woman who is looked at for pay, makes her respectability more tenuous. The space of Vauxhall gardens, "a site for flirtation and assignation," constructed of self-conscious illusions, and based on consumption, serves to heighten the ambiguity introduced by their gaze. The danger of prostitution, of the combination of sex and commerce, is also central to the farce 'The Register Office', in which older lords and ladies ostensibly wishing to hire servants are actually searching for seduceable flesh, and a procuress uses the office to find fresh faces to sell.

Ogborn more explicitly ties his chapters together through the space of London and the centrality of geography to his analysis. Unfortunately, the map of London included is difficult to read, as it is a small grayish image of a very large and detailed map (a smaller edition of a 13 foot by six and one-half foot map). The sites discussed in each chapter are impossible to find for a reader who does not already know their locations. However, the physical geography of these sites is less important than their imaginary geography, which Ogborn recounts in detail. The reader understands the importance of the Magdalen Hospital's site outside the bustle of the city and yet connected to it, as well as the regulated spaces of breweries policed by excise men. In the larger-scale geographies of excise taxation and information, the importance of London as center and anchor, from whence men and information go forth and return, tying together the city, the nation, and the world, is likewise clear.

The geographies which Ogborn traces are not only the literal geographies of sites and activities, but also the imagined geographies on which ideas are based. For example, the self-reflective, penitent individual promoted by the Magdalen hospital is based on the imaginary space of the solitary cell, even though the actual space of the hospital had wards, not cells. Similarly, Shaftesbury's, Mandeville's, and Hume's understandings of the public and its relation to individuals, discussed in 'The Street', are each based on imagined urban geographies of public space and the street.

By moving smoothly from philosophical understandings of modernity, the individual, and the public sphere to geographies, stories, incidents, and objects in which these ideas can be seen at play, Ogborn tells a complex and fascinating tale. Ogborn has a wide range of analytical methods at his disposal and wields them all well, employing paintings, prints, maps, newspaper accounts, plays, and a wide variety of other texts to paint complex and lively vignettes of modernity in process.

JESSICA SEWELL
University of California, Berkeley

TYCHISM, AGAPISM, AND SYNECHISM

Charles Sanders Peirce

Chance, Love, and Logic: Philosophical Essays. Ed. Morris R. Cohen. Lincoln, NE: University of Nebraska Press, 1998. Pp. xlv + 318.

Charles Sanders Peirce (1839–1914) is probably one of the few American philosophers, in the genuine Greek sense. Whitehead compared him to Aristotle; not surprisingly, Pierce had a profound interest for pre-socratic philosophers. He worked in geodesy, astronomy, psychology, chemistry, physics, mathematics, logic, and philology. For him the goal of philosophy was to explain the universe at large, not mere sophistry. For this, even as the son of a highly respected American mathematician, Benjamin Peirce, his work was during his life not well received. Slowly, people started to discover his monumental work. First known for his pragmatism, he later on became famous for his contributions to mathematical logic and his investigations in semiotics, a field that he anticipated incredibly.

During his life Peirce published many papers and articles in journals, but only one book, on photometry. After his death, a comprehensive edition of his works was published by Harvard University Press (*Collected Papers of Charles Sanders Peirce, 1839–1914* – Hartshorne and Weiss, eds., 1974), which is presently out of print. A more ambitious project to publish all of his work chronologically is in progress. Five volumes have been released so far. (For more information about the Peirce Edition Project, see http://www.iupui.edu/~peirce. The general Peirce website is http://www.peirce.org. As a side product of this project, two books have also been edited: *The Essential Peirce*, volumes 1 and 2, Houser & Kloesel, eds., 1992.)

Metaphysics, Epistemology, and Technology, Volume, 19, pages 395–397.
2000 by Elsevier Science Inc.
ISBN: 0–7623–0681–5

The book under review is a new edition of the first posthumous anthology of Peirce's works. Edited by Morris R. Cohen in 1923, it includes two series of papers. The first is a series of six papers published in *Popular Science Monthly* during 1877–1878 under the generic name of 'Illustrations of the Logic of Science', here renamed 'Chance and Logic'. The second is the first series of Peirce's papers published by his friend Paul Carus, former student of the revolutionary German mathematician Hermann Grassmann, in his journal *The Monist*. Published during 1891–1893, these papers are generally known as Peirce's 'scientific metaphysics'; Cohen's heading is 'Love and Chance'.

These two 'books' are quite well known nowadays and can be found, for example, in volume 1 of *The Essential Peirce*. Most of their contents are also included in the Dover anthology (Buchler, ed., 1940). But as Kenneth Laine Ketner notes in his introduction to this new edition, Cohen played a fundamental role to divulgate Peirce's work and in particular these two books. So this edition does justice to Cohen's efforts. Moreover, this anthology includes a very interesting introduction by Cohen himself, as well as an essay by John Dewey, 'The Pragmatism of Peirce'. It contains also a ten page bibliography of Peirce's published writings prepared by Cohen at that time. Ketner in his introduction gives a few more references, including his book *A Comprehensive Bibliography of the Published Work of Charles Peirce* (1986).

As a proem for the two books, Cohen chooses a short, interesting text of Peirce's, 'The Rules of Philosophy', in which Peirce situates himself between Cartesianism and Scholasticism.

Cohen wanted to publish his anthology under the title *Tychism, Agapism, and Synechism*, but his wife dissuaded him. The more fashionable title *Chance, Love, and Logic* is quite ambiguous. Chance and love can be considered as translations of tychism and agapism, but how did synechism become logic, and what has logic got to do with love?

Someone who knows nothing about logic or Peirce may want to buy this book for odd reasons. A book involving logic and love will generally not involve standard logic, such as first order classical logic, or even paraconsistent logic. And, in fact, in this anthology logic is also used with a quite different meaning. This might be confusing because Peirce is considered the founder (together with Frege) of modern logic, due to his contribution to quantification theory. But the reader will not find such a contribution here, nor will he or she find other technical works of Peirce, such as his contribution to many-valued logic.

It is not difficult to accept Peirce's conception of logic as the general methodology of science, as when he says "each chief step in science has been a lesson in logic" (p. 9). But one has to understand Peirce's notion of agapism in order to assess his claims that "to be logical men should not be selfish" (p. 73).

A connection between mathematical logic and synechism, which may explain why logic has been substituted for synechism in the fashionable title, appears in 'The Law of Mind', here the third part of 'Love and chance'.

For Peirce, synechism is "the tendency to regard continuity . . . as an idea of prime importance in philosophy" (p. 203). He states that "the present is connected with the past by a series of infinitesimal steps" (p. 205), and claims that one can reason soundly about infinitesimals. He was himself convinced of this by studying the work of Cantor.

Peirce argues that what distinguishes the finite from the infinite is a special type of reasoning, De Morgan's syllogism of transposed quantity, which applies to finite quantities but not to infinite ones. He then argues that there are only two grades of magnitude of infinite collections and that they can be distinguished by mathematical induction which applies to numerable collections but not innumerable ones.

Although this analysis is interesting, and Peirce was one of the first to recognize the high value of Cantor's work, it looks defective today, not only because there are more than two kinds of infinite collections, but because mathematical induction can be generalized (cf. transfinite induction) to apply to innumerable collections, provided they are well-ordered. If one accepts the axiom of choice, Peirce's distinction is meaningless.

JEAN-YVES BEZIAU
National Laboratory for Scientific Computing, Brazil

REFERENCES

Buchler, J. (Ed.) (1940). *Philosophical Writings of Peirce: Selected Writings.* New York: Dover.

Hartshorne, C., & Weiss, P. (Eds) (1974). *Collected Papers of Charles Sanders Peirce, 1839–1914.* 8 vols. Cambridge, MA: Harvard University Press.

Houser, N., & Kloesel, C. (Eds) (1992). *The Essential Peirce: Selected Philosophical Writings.* 2 vols. Bloomington, IN: Indiana University Press.

Ketner, K. L. (Ed.) (1986). *A Comprehensive Bibliography of the Published Works of Charles Sanders Peirce with a Bibliography of Secondary Studies.* (2nd ed.), rev. Bowling Green, OH: Philosophy Documentation Center, Bowling Green State University.

PERELMAN'S INFORMATION

Michael Perelman

Class Warfare in the Information Age. New York: St. Martin's Press, 1998. Pp. 154.

This book has the virtues and faults of what it is – a left-wing corrective to information hype from Hayek to hypertext. It goes beyond moralizing to provide a genuinely broader perspective on the computer age: we are reminded, for example, not only that the relentless exultation in technology ignores the human cost in job loss and work regimentation, but also that a field hand, employing specialized information to judge when a melon is ripe for picking, is as much a 'knowledge worker' as Bill Gates. On the other hand, the book is marred by a leftist addiction to pessimism. For example, the idea that information technology will belong only to those who can afford the fancy hardware is already being undermined by the development of cheaper computers and still cheaper information appliances. It seems obvious that the corporate sector, since it wishes to reach everyone, has an interest in promoting accessibility, not restricting it.

At times Perelman seems to forget that he has promised to talk about the class system, not the market system. Many evils, particularly those resulting from restricting the free flow of information, are blamed on an excessively privatized information market. But keeping much more information in the public domain is entirely compatible with a class system, so it is hard to see why excessive privatization is an evil of the class system rather than of an under-regulated free market. In the second half of the book class almost drops out of the picture, yielding center stage to market inefficiencies whose bad effects patently cut across class lines.

Metaphysics, Epistemology, and Technology, Volume, 19, pages 399–400.

ISBN: 0-7623-0681-5

The book is of interest to philosophers primarily because its arguments pinpoint areas which cry out for philosophical analysis. Consider, for example, the hasty claim that "economics doesn't work any more" because it gets the economic status of information wrong:

> The fundamental theorem of economics concludes that goods should be priced at their marginal cost – the cost of producing one more unit of production. . . . [I]n the case of information, the cost of transmitting this information on to others is minimal. . . . In other words, the marginal cost of information is effectively zero (p. 88).

There is a gap in this reasoning: how does "the cost of producing one more unit" of information become "the cost of transmitting" it? But this gap reveals insufficient attention to the notion of information itself. One would think that the marginal cost of information was the cost of producing one more unit of information. Perelman seems to believe that I can produce one more unit just by transmitting another copy of my information-bearing sentence. But even naive philosophers would do better. They would locate the information in a (true, justified) sentence-type or a proposition, not in the sentence-tokens or the inscriptions that express the proposition. If I tell two people that birds fly, I transmit one unit of information twice. I do not produce a new unit of information, even once.

MICHAEL NEUMANN
Trent University, Canada

THE RHETORIC OF SHARING THE EARTH

Tarla Rai Peterson

Sharing the Earth: The Rhetoric of Sustainable Development. Columbia, SC: University of South Carolina Press, 1997. Pp. xii + 238.

Tarla Rai Peterson's *Sharing the Earth: The Rhetoric of Sustainable Development* is best looked at in terms of three distinct sections, each of which can be read separately. The first section consists of an historical overview of the concept of sustainable development beginning with the work of Gifford Pinchot, an American leader of the conservation movement during the early 1900s, and concluding with a discussion of the Bruntland Commission Report. The second section consists of a discussion of the work of two social theorists, American literary critic and polymath Kenneth Burke (1897–1993) and German sociologist Niklas Luhmann. The theoretical positions of Burke and Luhmann form the philosophical basis for Peterson's subsequent treatment of three case studies: The Earth Summit Meeting in Rio de Janeiro of 1992, public hearings concerning the management of wildlife disease in Wood Buffalo National Park, and the U.S.-Mexico border controversy over the impact of industrial development on birth defects in Cameron County Texas.

Peterson is very perceptive in the first and third sections of her book. Her discussion of the historical genesis and development of the idea of sustainable development would be useful for anyone wishing to know more about this concept. The case studies provide interesting accounts of the various controversies. These accounts would be a valuable resource for anyone doing research into these particular controversies, or research into the kinds of

Metaphysics, Epistemology, and Technology, Volume, 19, pages 401–402.

ISBN: 0-7623-0681-5

practical difficulties that people participating in organized discussions of complex community development issues must face generally.

I have less to recommend about the discussion of the theories of Burke and Luhmann. My reticence about this section of Peterson's book has nothing to do with her synopsis of the work of these two social theorists which, as with the other two sections, is well written and perceptive. Burke and Luhmann take the idea of communication as a 'constitutive force' seriously, and "understand how people mutually create the multiple social realities within which they live" (p. 35). Any appreciation one will have for the middle section of Peterson's book will depend on the sympathy one has for this idea. I find that this approach to communication, which places such emphasis on content, rather narrow. Compared to a theory like that of Marshall McLuhan, the theories of Burke and Luhmann appear too open to naive hopes that the technological biases set up by the media of communication can be easily overcome through the development of better "tools for exploring the expansive social dramas formed around terms such as *sustainable development*" (p. 41). The result is that, instead of providing a critical "analytical frame for discovering aspects of our social constructs which need revising" (p. 45), the discussion of the theories of Burke and Luhmann seems more like a defense of the necessity for the services of experts in the discourse of environmental negotiation, Peterson's own field of specialization.

JAMES GERRIE
University of Guelph, Canada

THE ARMORING OF UNIDISCIPLINARITY

Eric Ramsey

The Long Path to Nearness; A Contribution to a Corporeal Philosophy of Communication and the Groundwork for an Ethics of Relief.
Atlantic Highlands, NJ: Humanities Press, 1998. Pp. xiv + 145.

Ramsey has set about to develop a new materialist corporeal philosophy of communication directed toward a participatory democracy and liberation politics. He attempts to incorporate the body into a philosophy of communication and proposes the foundation of an ethics of relief understood in both senses of the word: relieving and three-dimensionality. How does Ramsey attempt to achieve this ambitious goal? With the help of Ernst Bloch, Wilhelm Reich, Martin Heidegger, Maurice Merleau-Ponty, Calvin O. Schrag, and Iris Young.

In Chapter 2, 'Our Impatience for Liberty: Approaching Communication from the Point of View of Possibility', Ramsey addresses the problem of communication focusing on possibility and its actualization, based exclusively on Ernst Bloch's writings. He distinguishes both active and passive moments of possibility as well as potentiality and capability, concluding that "potentiality without capacity is stasis and capacity without potentiality is impotent, and possibility without both is not" (p. 22). His conclusion makes sense. Moreover, it is common sense. One could certainly ask why exactly does he claim there is an impatience for, rather than fear of, liberty? An obligatory reference here would obviously have been Erich Fromm's well known classic Escape from Freedom (1941), among others.

In Chapter 3, 'Corporeality, Possibility and Character Structure', Ramsey is concerned with the active moment of possibility from the point of view of

Metaphysics, Epistemology, and Technology, Volume, 19, pages 403–405.
2000 by Elsevier Science Inc.
ISBN: 0–7623–0681–5

Wilhelm Reich's concept of 'armoring'. This concept describes "the bodily rigidity of persons who embody the lessons learned from practices engaged against them and the prohibitions of actualizing many possibilities of their worlds" (p. 25). Reich's concept of 'armoring' is certainly still fruitful and worth reconsideration, and it is Ramsey's merit to bring it to attention for contemporary theoretical efforts in dealing with more corporeal approaches. This corporeal memory which, according to Reich, "expresses itself in character attitudes, as well as muscular attitudes," restricts and heavily conditions what we see ourselves capable of achieving, as against apparently pure and unlimited volition. It is unlikely, however, that anyone would seriously believe nowadays that agency is unlimited and freedom unconstrained. Conditions that have a crucial role in restricting agency were those treated by Foucault regarding the construction of subjectivity. These other conditions should have been considered as well in this chapter if indeed character structure was the issue at stake.

Chapter 4, 'The Appearance of Communication', is strangely based on Heidegger's concept of Dasein in *Being and Time*. It is strange because communicative acts are not precisely Heidegger's most interesting contribution to philosophy. Ramsey questions Heidegger's notion of discourse for not being 'nuanced' enough, and tries to add further distinctions, such as what he terms 'unreflected pre-thematic communication' (p. 45), which in part incorporates Reich's idea of armoring. This 'unreflected pre-thematic communication', as Ramsey calls it, is opposed to 'thematic communication' directed to the other by means of communicative proximity. Again, it is difficult not to resort to the results of semiotics, linguistics, discourse analysis, pragmatics, and particularly speech act theory (Austin and Searle) for dealing with this subject, since what is at stake here is agency through language. Moreover, Ramsey speaks of face-to-face interactions but there is no mention of Goffman or any of his followers either.

Ramsey states that "the other who is before me always has a dignity independent of the historical force of Anyone, and this is what demands of me a responsibility to the other outside of the reified ways of doing and being in the world" (p. 52). As in the previous chapters, one is left with the sense that at least a mention of how Levinas deals with the problem of responsibility and alterity as a critical development of Heidegger is necessary.

The author attempts a phenomenological justification of character structure in Chapter 5 by means of Merleau-Ponty's concept of 'habit body' and especially by Iris Young's concept of 'feminine motility' as social values instilled within the body. He ends by stating that interpretive agency is linked up with the active moment of possibility with respect to communication, while the concept of embodiment brings together concerns of the passive moment of possibility (p. 68). This is as far as the author has taken us into a corporeal philosophy.

In addition to his previous dichotomies (actuality/possibility, passivity/activity, materiality/subjectivity, potentiality/capability, thematic/unreflected pre-thematic), Ramsey provides yet another dichotomy in Chapter 6: calculative vs. meditative thinking relative to Heidegger's 'challenging-forth' and 'bringing-forth', respectively. As an example of the former, he uses scientific thought; as illustrative of the latter, he appeals to the aesthetic, particularly poetry, through which, for Heidegger, the world can 'come-forth'. Beside the oversimplification of the differences between aesthetic and scientific processes, it was in this chapter where Ramsey was supposed to establish the groundwork for an ethics of relief – but I must have missed it along the way.

In the very last chapter, Ramsey attempts to make a contribution to a corporeal philosophy of communication. He sees people as preceded by a 'flat world' of habits and fixed expectations he calls 'a-whereness' (a confusing term because it does not imply what it phonetically calls for, i.e. 'awareness') that determines act intentions and communicative intents. The main contribution I could find in this chapter was his distinction between intentionality as what keeps us connected to the world, practical intention, which is motivational, and intent, which is communicative (p. 99).

All in all, Ramsey's volume is densely populated by question begging, insufficiently argued and lacking very basic and more contemporary references to the problems addressed. The text seems to be less about corporeal philosophy and communication than about agency itself. As such, Ramsey's book does not fulfill its ambitious project of grounding an ethics and constructing both a corporeal philosophy and a new approach to communication. Its insights into the corporeal and phenomenological are too thin. For this project perhaps a less unidisciplinary philosophical and more transdisciplinary approach would have helped, if the author really intended to go beyond philosophers' problems exclusively and into philosophical contributions to other professionals working along the lines of communication and corporeality from various perspectives and fields of inquiry.

KATYA MANDOKI
Universidad Autonoma Metropolitana, Mexico

REAL-WORLD ETHICS FOR ENGINEERS

Caroline Whitbeck

Ethics in Engineering Practice and Research. Cambridge, MA: Cambridge University Press, 1998. Pp. xx + 330.

This fine textbook in engineering ethics will be useful for courses in engineering ethics, and as a supplementary resource for engineering and science instructors who wish to infuse attention to ethics into their courses. It will also interest students of philosophical ethics who wish to consider a refreshing approach to case-based, 'applied' ethics.

Its most prominent virtues are described on the frontispiece, which compares ethical problems to design problems, and advertises "a real-world, problem-centered approach to engineering ethics, using . . . open-ended scenarios and case studies to develop skill in recognizing and addressing ethical issues." This approach is strengthened by a link to the web-based resources of the Ethics Center for Engineering and Science (http://ethics.cwru.edu), where even more such cases, professional codes and standards, and supplementary essays by the author (including one on applied ethics and the use of cases) may be found. The cases and resources alone are worth the price of the book, and might be enough to justify its use as a text apart from its other strengths. Almost all are either direct reports of actual events, or are closely based on actual events, professional standards, and the like.

After a fifty-two-page introduction to ethical concepts, and a chapter on 'Ethics as Design', Whitbeck covers topics from professional responsibility in general, and the specific professional responsibilities of engineers, to workplace rights and responsibilities, research integrity, the environment, use

Metaphysics, Epistemology, and Technology, Volume, 19, pages 407–409.

ISBN: 0-7623-0681-5

of experimental subjects, and questions about credit and intellectual property. An epilogue, 'Making a Life in Engineering and Science', gives thumbnail biographies of two engineers (Miguel Barrientos and Jim Melcher) whom the author presents as positive models.

Most interesting to me, as a philosopher and teacher of ethics, was the chapter on 'Ethics as Design'. This analogy seems a brilliant way of introducing engineering students to the study of ethics. In Whitbeck's hands it also becomes a contribution to theoretical ethics. To see ethics as design means recognizing that "People confronted with ethical problems must do more than simply make judgments; they must figure out what to do. (This is the reason for calling people 'agents')." Yet in Whitbeck's judgment (she cites Stuart Hampshire to the same effect), ethics courses have largely neglected ethical practice; at best they have helped students to be armchair moral critics (p. 53). Even the use of practical cases has usually been designed to elicit particular moral judgments, rather than to give practice in "figuring out what to do."

But can a class provide anything more than armchair analysis, when values are plural and ethical experiments often impossible or immoral? Whitbeck points the way to what can be done. She provides longer real cases (e.g. the Challenger space shuttle disaster), where students can learn from a clearly described real world situation and a known outcome (like Harvard Business School cases). By contrast, the concluding question of her open-ended scenarios, "What would you do?," leaves readers asking, "I'm not sure; can you tell me more?" I found this a little disconcerting sometimes; but that seems to be the point. Figuring out what to do is a design problem. A class can indeed give engineering students a better orientation to such 'ethical design problems' in their profession, and help them practice some elements of solution-finding.

Studying the ethics/design analogy itself proves helpful. As Whitbeck says, there is usually no one best solution to a design problem, but many better and worse ones, and some that won't work at all. All solutions, however, must achieve a desired end, meet specifications, be reasonably safe, and consistent with background constraints (pp. 56–60). She derives other lessons for ethics from the design/ethics analogy: (1) "Consider unknowns and uncertainties in the situation." The problem definition will likely change as you start trying to find a solution; and the solution may need to be a flexible one, because of remaining uncertainties (p. 62). (2) Finding a solution is different from defining a problem, and may require more information (p. 63). (Contrast the famous Heinz case whose use by Kohlberg is criticized by Gilligan; a child who tries getting more information before announcing a solution is ranked at a lower developmental stage, because the researcher assumes that once the problem is defined, nothing but an ethical judgment is required.) (3) Time pressure may require pursuit of several solutions simultaneously, so that when one fails, another may still be completed (pp. 65–66). (4) Ethical problems are dynamic; hence both they and their solutions will change over the course of trying to deal with them (pp. 66–68). Space does not permit me further comment on these interesting ideas here. But readers may wish to read Whitbeck's additional essay on the subject, available at the above-mentioned web page.

The text contains more precise definitions of ethical terms than seems necessary. Do we really need technical definitions of negligence, stupidity, incompetence, and carelessness (p. 90)? Would not a definition of the morally relevant features of such traits be sufficient? Thus I found the definition-rich introductory section tedious, and very occasionally I felt that theoretical issues were being settled by these definitions rather than by argument (e.g. in the discussion of preferences and values, pp. 3ff.).

I was also disappointed by the lack of attention to engineers' responsibility for the global impact of technology on society. Whitbeck correctly observes that engineers cannot be held accountable for all uses of their inventions (e.g. for all the social effects of new medical technologies), but only for reasonably foreseeable uses (pp. 122ff.). Nevertheless, through professional associations, engineering schools, publications, participation on public policy advisory boards, and in many other ways, engineers have a unique opportunity to shape the impact of technology on society; and as technologists they have a special professional responsibility to do so. The one chapter that moves furthest in this direction, on responsibility for the environment, opens some important issues, and connects realistically to a variety of corporate attitudes. But it is short, gives relatively few cases, provides only a cursory analysis of the Exxon Valdez case, and in general does not lead students as far as it might.

Perhaps a future edition will rectify these shortcomings. Whitbeck herself seems drawn to do so. The two 'lives' with which she closes, for example, are precisely examples of engineers who have gone 'outside the box' to take personal responsibility for the effect of engineering science on society, in the areas of military technology and of appropriate technology for third world development. It is the commitment, vision, and integrity which they showed in doing this for which she most seems to admire them.

DAVID CLOWNEY
Rowan University

REMEMBERING THE NAMELESS

Edith Wyschogrod

An Ethics of Remembering: History, Heterology, and the Nameless Others. Chicago: University of Chicago Press, 1998. Pp. xxii + 280.

How history is to be represented after the Holocaust is a central preoccupation of recent poststructuralist thought. Many of the debates have focused on the difficulty of testifying on behalf of those whose memory lies in ruins – shattered by traumatic circumstances that have become all too familiar in this century. How does the historian do justice to the voiceless memory of the murdered victims of the camps, or to the fragmented psyches of the survivors? For some philosophers, however, the problem runs even deeper than this question. For these thinkers – among whom Edith Wyschogrod is regarded as one of the more respected voices – the traumas witnessed in this century have effected an unparalleled transformation of western consciousness.

The development and proliferation in this century of mass-murder technologies have derailed, or at the very least, brought conventional representations of subjectivity and history into question. Maurice Blanchot has aptly described our present epoch as marked by *the disaster*, a term he employs in its etymological sense (*dés-astre*); that is, we find ourselves without a guiding beacon or star with which to interpret our experiences. Whereas previous generations could to a certain extent naively comfort themselves during periods of human-made disasters and natural catastrophes with the help of elaborate theodicies, these traditional forms of rationalization are hardly tenable given the continued presence of hatred and violence in our supposedly rational century. This skepticism presents an invaluable opportunity to reflect upon a conception of ethics that does not make use of some *Hinterland* or idealized conception of 'God' that will magically balance the scales of justice

Metaphysics, Epistemology, and Technology, Volume, 19, pages 411–413.

ISBN: 0–7623–0681–5

at the end of history. At the same time the ontological disaster presents us, in Wyschogrod's estimate, with new hazards and temptations. Wyschogrod assumes a unique position in post-Holocaust literature because she is one of only a small number of thinkers who insist on working within the traditional framework and vocabulary of Western philosophy in order to rethink the central concepts of subjectivity and history.

In this challenging book, Wyschogrod carries out an extended conversation with an imaginary interlocutor whom she dubs the 'heterological historian'. She summarizes and amends the ideas of continental authors like Heidegger and Derrida, and even engages with issues and concerns that are primarily associated with Anglo-American writers like Dennett and Penrose, in the hope of helping the heterological historian to orient him or herself within the tenebrous region of the present 'cataclysm' – Wyschogrod's name for the disaster of being.

While Wyschogrod is influenced by the ideas of several important European philosophers, her own thought bears the unmistakable signature of her mentor and teacher, the Jewish philosopher, Emmanuel Levinas. Like Levinas, Wyschogrod's present and past work (1985, 1990) strives to articulate a non-moralizing ethics, a vision of ethicality that is not rooted in the covert aggression that Nietzsche aptly calls *ressentiment*. While Levinas is certainly not her only point of reference, his thought serves as her primary check against some of the more questionable features of recent European and analytic philosophy. By questioning the meaning of history, some writers – such as the French sociologist, Jean Baudrillard – have gone so far as to reject the very idea of history. While there is some merit to the Nietzschean unmasking of the different ways that theodicy is reintroduced into secular models of historical representations, unless these suspicions are themselves *limited* or conditioned by ethical concern for others, the critique of the philosophy of history easily succumbs to mindless nihilism.

Wyschogrod reserves her strongest criticisms for the tendency, especially conspicuous in the area of cultural studies, to either aestheticize the 'cataclysm'; or to uncritically celebrate the vertigo and frenetic pace of the postmodern experience. Similarly, she raises a number of important caveats regarding certain trends in analytic thought. For example, she cogently argues that far from breaking with traditional conceptions of the mind, the logic that subtends the idea of artificial intelligence is anticipated by texts that date back to antiquity and the Renaissance (like the anonymous *Ad Herrenium*). She invites her readers to reexamine these earlier texts in order to better appreciate the limitations of contemporary thinking around the nature of memory and consciousness.

Some readers might be put off by Wyschogrod's style. It takes time to warm up to her sometimes overwrought prose. The patient reader, however, will be rewarded by the author's impressive grasp of the Western philosophical tradition, as well as her unswerving devotion to the principle of ethical reparation. To make the task easier for the heterological historian, Wyschogrod provides useful summaries at the end of each chapter.

JOHN CARUANA
York University, Canada

REFERENCES

Wyschogrod, E. (1985). *Spirit in Ashes: Hegel, Heidegger, and Man-made Mass Death*. New Haven, CT: Yale University Press.

Wyschogrod, E. (1990). *Saints and Postmodernism: Revisioning Moral Philosophy*. Chicago: University of Chicago Press.

NOTES ON CONTRIBUTORS

Agustin Araya is an Associate Professor in the Department of Mathematics and Computer Science at the California State University, San Jose. His work focuses on the ontological analysis of computing and communications technology.

Jean-Yves Beziau has a Ph.D. in both mathematical logic and philosophical logic. He has been a Visiting Fellow of the Department of Logic of the University of Wroclaw (Poland), and a Fulbright Research Scholar at the Department of Mathematics of UCLA. He is presently Visiting Researcher of the National Laboratory for Scientific Computing (Brazil).

Mary Bloodsworth is an Assistant Professor of Philosophy and Women's Studies at Washington State University. She is interested in contemporary continental philosophy, feminist philosophy, ancient philosophy, and ethics. She has recently published in *Ethics*, *Contemporary Sociology*, and *Gender and Society*.

Albert Borgmann has taught philosophy at the University of Montana since, 1970. His special area is the philosophy of society and culture with emphasis on technology. Among his publications are *Technology and the Character of Contemporary Life* (1984), *Crossing the Postmodern Divide* (1992), and *Holding On to Reality* (1999).

Philip Brey is an Assistant Professor in the Department of Philosophy at the University of Twente, The Netherlands. His research focuses on the social, cultural, and experiential role of technical artifacts, and on philosophical aspects of information and communication technology.

Rafael Capurro is a Professor of Information Management and Information Ethics at Fachhochschule Stuttgart, University of Applied Sciences, and a lecturer at the Institute of Philosophy, Stuttgart University. He has numerous publications dealing with foundations of information science, information ethics, hermeneutics, and philosophy of media.

John Caruana is presently completing a dissertation in the Social and Political Thought Program at York University, Toronto, Canada. He is also an instructor in the philosophy department at Ryerson Polytechnic University, Toronto.

Drew Christie is Associate Professor of Philosophy at the University of New Hampshire, and part of the Natural Resources Ph.D. Program. He teaches environmental philosophy, logic, philosophy of law, and social and political philosophy. His research interests also encompass American pragmatism (old and new), and the philosophy of education. Recent works include 'Dewey and the Splintered Vision' and an encyclopedia entry on Richard Rorty.

David Clowney is Associate Professor of Philosophy at Rowan University, where he has taught since, 1988. He is a graduate of Temple University (Ph.D., Philosophy, 1988), Westminster Seminary (M.Div., 1971), Wayne State University (M.A., Philosophy, 1967), and Calvin College (B.A., 1964). His areas of specialization are theoretical and applied ethics, aesthetics, and the philosophy of religion.

John Collier has been Research Associate in the Complexly Organised Adaptive Systems Group for the past four years. Previously, he was a Senior Lecturer at the University of Melbourne. He has taught Philosophy at a number of universities in the United States and Canada. He is best known for his work on information theory and in the philosophy of biology.

José Manuel de Cózar is an Associate Professor at the University of La Laguna, Spain. His research interests center on applied philosophy of science and technology. He has published several papers on issues such as evolutionary epistemology, pragmatics, and the role of technology in the quality of life. He is currently editing a book which examines the relationship between technological development, civilization and barbarism in the 20th century.

Stephen Crowley is a Ph.D. candidate in the Philosophy Department at Indiana University with research and teaching interests in the history and philosophy of science, epistemology, skepticism, and ancient philosophy.

Stuart Dalton is an Assistant Professor of Philosophy at the University of Hartford. His interests include contemporary continental philosophy, philosophy of technology, and aesthetics. He has recently published on Nietzsche, Kant, and Freud.

Cornelis de Waal is Visiting Assistant Editor for *The Writings of Charles S. Peirce: a chronological edition* (1982–), at the Peirce Edition Project, Indiana University, Indianapolis.

James Gerrie is with the Department of Philosophy at the University of Guelph, Canada. Gerrie is interested in all aspects of the field of philosophy of technology, but especially with questions about the ethical criticism of technology.

Bill Hook is a philosophy graduate student at the University of Montana. He has worked as a computer programmer and systems manager for eleven years. His research interests are in the philosophy of technology, with special attention to information technology.

Richard Liebendorfer is interested in Wittgenstein and the philosophy of language and mind.

Andrew Light is Assistant Professor of Philosophy and Environmental Studies at the State University of New York, Binghamton, and Resident Fellow at the International Center for Advanced Studies at New York University. Light is the author of over thirty articles and book chapters on environmental ethics and philosophy of technology, and has edited or co-edited six books, most recently *Technology and the Good Life* (forthcoming). He is also the editor of *Philosophy and Geography.*

Katya Mandoki is tenured professor of design in Tlalpan, Mexico. She has taught courses on aesthetics, theory of design, philosophy of art, semiotics, and visual and corporeal rhetorics. Mandoki has presented numerous papers in national and international conferences related to art and communication studies, and published in academic journals and cultural publications.

Craig Murray is a lecturer in psychology at Liverpool Hope University College. He has published work on the phenomenology of embodiment, psychology and technology, cyborg discourse, and cyber-ethics and methodology. His most recent work is 'The Experience of Body Boundaries by Siamese Twins', in *The British Journal of Psychology* (forthcoming).

Michael Neumann is an Associate Professor of Philosophy at Trent University. He publishes mainly in the fields of ethics and political philosophy.

Yannis Niadas is a civil engineer and hydrologist currently doing research and working toward his Ph.D. in the Program on Philosophy and History of Science and Technology, run jointly by the Department of Philosophy and History of Science of the University of Athens, and the Department of Engineering Science of the National Technical University of Athens. His research interests include the epistemology of engineering and technology, philosophy of engineering design, and technology-nature interactions.

Roderick Nicholls is an Associate Professor of Philosophy at the University College of Cape Breton in Nova Scotia, Canada.

Theodore John Rivers holds a doctorate in medieval history, and is interested in the history and philosophy of technology. His works include *Law and the*

Limits of Freedom: A Philosophical Exposition of Their Relationship (1984) and *Contra Technologiam: The Crisis of Value in the Technological Age* (1993).

David Roberts is working on a Ph.D. in philosophy at the University of Alberta. He received his M.A. in philosophy from the University of Montana in, 1998 with a thesis on moral philosophy and cognition. Roberts has been Assistant Editor of *Philosophy and Geography.*

Brian Rosmaita is an Assistant Professor of Philosophy at Kent State University. His interests include philosophy of mind, philosophy of logic, the history and philosophy of science, and the theoretical and practical aspects of teaching with technology.

Raphael Sassower is a Professor of Philosophy at the University of Colorado, Colorado Springs. He is the author of *A Sanctuary of Their Own: Intellectual Refugees in the Academy* and *The Golden Avant-Garde: Idolatry, Commercialism and Art (Cultural Frames, Framing Culture)*, among other books.

Jessica Sewell is a doctoral candidate in Architecture at the University of California, Berkeley, where she is writing a dissertation entitled 'Gendering the Spaces of Modernity: Women and Public Space in San Francisco, 1890–1917'.

Sergio Sismondo is the author of *Science without Myth: On Constructions, Reality, and Social Knowledge* (Albany, NY: SUNY Press, 1996), and co-author with Boris Castel of *The Art of Science* (Broadview, forthcoming). He is currently working on issues concerning scientific objectivity and the objects of scientific knowledge. He teaches at Queen's University, in Kingston, Canada.

Karina Stokes is the Senior Editor for the University of Texas (Houston) Health Science Center, Department of Surgery, Organ Transplant Division. Her work encompasses cognitive science, linguistics, rhetoric, and artificial intelligence. She is the author of *A Dialogic Model for Natural Language Cognition* (1991), and has written for *International Journal of Applied Philosophy and Hormone Research.*

Ton Van der Valk is working on a Ph.D. in philosophy at the Free University of Amsterdam. He has previously studied Philosophy of Technology at Twente University of Technology, The Netherlands. His recent publications include 'The Mutual Embarrassment of Science and Society and the Shortcomings of the Precautionary Principle' and 'Cloning as a Test-case of Autonomous Technology'. Van der Valk is also interested in the change of the educational system in the rising tide of information and communication technologies.

Peter-Paul Verbeek is an Assistant Professor in the Department of Philosophy at Twente University, The Netherlands. His research concerns technical mediation and the relation between people and artifacts in the technological culture. He teaches philosophy of technology and philosophy of communication.

Karen Wiley is an Associate Professor of Liberal Arts and International Studies at the Colorado School of Mines. Her research involves the politics of science and technology.

INDEX

The index is to all conceptually significant personal names, real and fictitious, including (when appropriate) their named influences and schools of thought. In some cases expanded identifiers are supplied beyond what is available in the text itself. The index includes authors and reviewers (suitably noted) in this volume, and gives the pages for their contributions; authors reviewed (again, with suitable notation) are indexed as well. The abbreviations "n" and "nn" indicate note and notes, respectively. Although names in notes are indexed, names in reference lists are not; the Preface and Notes on Contributors are not indexed either.

Printed and bound by CPI Group (UK) Ltd, Croydon, CR0 4YY
16/09/2026
14963401-0006